Raymond Chiong and Sandeep Dhakal (Eds.)

Natural Intelligence for Scheduling, Planning and Packing Problems

Studies in Computational Intelligence, Volume 250

Editor-in-Chief
Prof. Janusz Kacprzyk
Systems Research Institute
Polish Academy of Sciences
ul. Newelska 6
01-447 Warsaw
Poland
E-mail: kacprzyk@ibspan.waw.pl

Further volumes of this series can be found on our homepage: springer.com

Vol. 230. Uday K. Chakraborty (Ed.)
Computational Intelligence in Flow Shop and Job Shop Scheduling, 2009
ISBN 978-3-642-02835-9

Vol. 231. Mislav Grgic, Kresimir Delac, and Mohammed Ghanbari (Eds.)
Recent Advances in Multimedia Signal Processing and Communications, 2009
ISBN 978-3-642-02899-1

Vol. 232. Feng-Hsing Wang, Jeng-Shyang Pan, and Lakhmi C. Jain
Innovations in Digital Watermarking Techniques, 2009
ISBN 978-3-642-03186-1

Vol. 233. Takayuki Ito, Minjie Zhang, Valentin Robu, Shaheen Fatima, and Tokuro Matsuo (Eds.)
Advances in Agent-Based Complex Automated Negotiations, 2009
ISBN 978-3-642-03189-2

Vol. 234. Aruna Chakraborty and Amit Konar
Emotional Intelligence, 2009
ISBN 978-3-540-68606-4

Vol. 235. Reiner Onken and Axel Schulte
System-Ergonomic Design of Cognitive Automation, 2009
ISBN 978-3-642-03134-2

Vol. 236. Natalio Krasnogor, Belén Melián-Batista, José A. Moreno-Pérez, J. Marcos Moreno-Vega, and David Pelta (Eds.)
Nature Inspired Cooperative Strategies for Optimization (NICSO 2008), 2009
ISBN 978-3-642-03210-3

Vol. 237. George A. Papadopoulos and Costin Badica (Eds.)
Intelligent Distributed Computing III, 2009
ISBN 978-3-642-03213-4

Vol. 238. Li Niu, Jie Lu, and Guangquan Zhang
Cognition-Driven Decision Support for Business Intelligence, 2009
ISBN 978-3-642-03207-3

Vol. 239. Zong Woo Geem (Ed.)
Harmony Search Algorithms for Structural Design Optimization, 2009
ISBN 978-3-642-03449-7

Vol. 240. Dimitri Plemenos and Georgios Miaoulis (Eds.)
Intelligent Computer Graphics 2009, 2009
ISBN 978-3-642-03451-0

Vol. 241. János Fodor and Janusz Kacprzyk (Eds.)
Aspects of Soft Computing, Intelligent Robotics and Control, 2009
ISBN 978-3-642-03632-3

Vol. 242. Carlos A. Coello Coello, Satchidananda Dehuri, and Susmita Ghosh (Eds.)
Swarm Intelligence for Multi-objective Problems in Data Mining, 2009
ISBN 978-3-642-03624-8

Vol. 243. Imre J. Rudas, János Fodor, and Janusz Kacprzyk (Eds.)
Towards Intelligent Engineering and Information Technology, 2009
ISBN 978-3-642-03736-8

Vol. 244. Ngoc Thanh Nguyen, Rados law Piotr Katarzyniak, and Adam Janiak (Eds.)
New Challenges in Computational Collective Intelligence, 2009
ISBN 978-3-642-03957-7

Vol. 245. Oleg Okun and Giorgio Valentini (Eds.)
Applications of Supervised and Unsupervised Ensemble Methods, 2009
ISBN 978-3-642-03998-0

Vol. 246. Thanasis Daradoumis, Santi Caballé, Joan Manuel Marquès, and Fatos Xhafa (Eds.)
Intelligent Collaborative e-Learning Systems and Applications, 2009
ISBN 978-3-642-04000-9

Vol. 247. Monica Bianchini, Marco Maggini, Franco Scarselli, and Lakhmi C. Jain (Eds.)
Innovations in Neural Information Paradigms and Applications, 2009
ISBN 978-3-642-04002-3

Vol. 248. Chee Peng Lim, Lakhmi C. Jain, and Satchidananda Dehuri (Eds.)
Innovations in Swarm Intelligence, 2009
ISBN 978-3-642-04224-9

Vol. 249. Wesam Ashour Barbakh, Ying Wu, and Colin Fyfe
Non-Standard Parameter Adaptation for Exploratory Data Analysis, 2009
ISBN 978-3-642-04004-7

Vol. 250. Raymond Chiong and Sandeep Dhakal (Eds.)
Natural Intelligence for Scheduling, Planning and Packing Problems, 2009
ISBN 978-3-642-04038-2

Raymond Chiong and Sandeep Dhakal (Eds.)

Natural Intelligence for Scheduling, Planning and Packing Problems

Springer

Raymond Chiong
School of Computing & Design
Swinburne University of Technology (Sarawak Campus)
Jalan Simpang Tiga
93350 Kuching
Sarawak, Malaysia
E-mail: rchiong@swinburne.edu.my

Sandeep Dhakal
School of Computing & Design
Swinburne University of Technology (Sarawak Campus)
Jalan Simpang Tiga
93350 Kuching
Sarawak, Malaysia
E-mail: sandeep.dhakal@gmail.com

ISBN 978-3-642-04038-2 e-ISBN 978-3-642-04039-9

DOI 10.1007/978-3-642-04039-9

Studies in Computational Intelligence ISSN 1860-949X

Library of Congress Control Number: 2009934305

Typeset & Cover Design: Scientific Publishing Services Pvt. Ltd., Chennai, India.

Printed in acid-free paper

9 8 7 6 5 4 3 2 1

springer.com

Foreword

I was delighted to be asked to write a few words to set the scene for this volume and I would like to take this opportunity to thank the editors for their kind invitation.

The scope of this book covers many aspects of research into computational search methodologies to underpin intelligent system design. As the title suggests, it has a focus on techniques drawn from nature and it has a broad span of application areas. When reading through this collection of papers, I was particularly struck by the high level of diversity of the techniques, approaches, methodologies and applications. We can see discussions of evolutionary approaches, particle swarm methods, fuzzy reasoning and neural networks. We can see applications across supply chains, machine scheduling, transport and the electricity industry. However, most of all, we can see how inter-disciplinarity cuts across traditional disciplinary borders to facilitate innovative scientific development in methodologies to underpin intelligent decision support search engines.

I am convinced that, in the field of computational search and computational intelligence, inter-disciplinarity will continue to play a major role in leading edge research activity over the next few years. The level of this inter-disciplinarity is far reaching. The fields of Operational Research and Artificial Intelligence overlap to a significant extent, especially within the context of computational intelligence. Moreover, this research field draws upon expertise across mathematics, computer science, business and engineering. In addition, there are important application areas which span a wide variety of other disciplines such as engineering, management, bioinformatics, computational chemistry and many others. All of these communities have significant expertise, from a broad spectrum of different perspectives and all have the potential to make a major impact upon the field of computational intelligence. In terms of scientific advances, an inter-disciplinary approach which draws upon knowledge and innovation from across disciplinary boundaries is far more likely to be effective than if separate research communities work in isolation. I think that this book is indicative of the broad appeal of this research field and its wide span demonstrates how different disciplinary perspectives can address important scientific challenges.

However, the impact of this field is felt not just across different disciplines. It can also be seen across different industries and service sectors. In my view, another major theme and research challenge for the scientific community in computational intelligence and search is in modelling and addressing the complexity of real world problems. Over the years, there has been a significant gap between scientific research and industrial practice in this field. In order to close this gap it is important that there is a close engagement between the research community and the practitioner community. It is important that real world issues feed into key research challenges that play a role in setting the international research agenda. In this book, we see particularly compelling examples such as in freight routing and in electrical load forecasting.

In summary, I think that this book represents a range of research methods and applications which are indicative of some of the research themes which are being addressed by the international research community and which are having an impact on commerce and the service sector. I have enjoyed reading it and I hope that you find it as interesting as I have.

July 2009

Professor Edmund Burke
School of Computer Science
University of Nottingham, UK
http://www.cs.nott.ac.uk/~ekb/

Preface

Scheduling, planning and packing are ubiquitous problems that can be found in a wide range of real-world settings. These problems transpire in a large variety of forms where a set of actions or operations to be performed are often constrained by various factors such as the allocation/assignment of resources and time as well as other optimisation criteria. Due to their enormous socio-economic impact, for many years, a lot of efforts have been devoted to automating the processes of scheduling, planning and packing using different kinds of methods. However, poor scaling and the lack of flexibility of many of the conventional methods coupled with the fact that many real-world problems of these sorts nowadays tend to be of large scale, dynamic and full of complex dependencies have made unconventional ways of tackling them necessary.

Natural intelligence includes all the systems of control present in nature. This control is demonstrated via neural-control in humans and animals, optimisation through evolution, distributed intelligence in colony species like ants, bird swarms, fish school, etc. These and other mechanisms are utilised by nature for various purposes similar to many of the difficult problems we face on a daily basis. Since these mechanisms have stood the test of time for millions of years, looking at them for inspiration to solve our own problems is only natural.

Given the variety of scheduling, planning and packing problems, and the numerous natural intelligence based approaches for solving these problems, this book provides merely a glimpse of what is available in the fields. It comprises 12 chapters which have assembled not only some of the most intriguing applications in scheduling, planning and packing, but also new perspectives, tutorials and additions to the methodology of natural computing. While some approaches, issues and results are similar in certain manner for the various problems presented, others differ considerably. It is our hope that this volume will be a useful resource for academics, researchers and practitioners involved in any aspect of the problem domains.

The book begins with the first chapter by *Ibrahimov et al.* that presents a discussion of some optimisation issues from a business perspective in the context of the supply chain operations. Two real-world business optimisation issues, one representing a local perspective and another representing a global perspective, are used as examples to emphasise the importance of global optimisation. The aim, here, is not a comparative study of these two optimisation issues; rather, it is to illustrate the concepts associated with both local and global optimisation. The authors argue that global optimisation from a business perspective can be distinguished from the traditional global optimisation in terms of its search space. This abstraction along with a proper investigation of the inter-dependence of the components involved in global optimisation, as presented in this chapter, will no doubt be useful for studying global optimisation problems.

In the second chapter, *Weise et al.* present the freight transportation planning component of the *in.west* project. The problem presented in this chapter is unlike other planning problems seen in the literature because of its size as well as the number of constraints involved. With the aim of minimising the total distance travelled as well as the overall cost, and maximising resource-utilisation, an evolutionary algorithm complete with various custom, intelligent search operations was applied to the problem. The algorithm was tested on real-world data supplied by the project's partner organisations and significant performance improvements were observed. Additionally, different parameter settings for the evolutionary algorithm were studied and their significance was statistically verified. This chapter will, no doubt, be a valuable read for anybody attempting to attack problems of similar nature, scale and complexity.

In the next chapter, *Hasan et al.* use a Genetic Algorithm (GA) to solve Job Shop Scheduling Problems (JSSPs) with the objective of minimising the makespan while satisfying a number of hard constraints. The authors have created a new approach by hybridising GA with several priority rules such as partial reordering, gap-reduction, restricted swapping, etc. Experimental studies on 40 benchmark problems from the literature show that this new hybrid GA is able to achieve optimum results for 27 of those problems and also outperforms other approaches found in the literature for the remaining problems. Notwithstanding these results, the authors also point out several areas where their approach could be further improved to better deal with real-world problems. Nevertheless, the GA proposed in this chapter is a significant contribution to the research into solving JSSPs.

With the advent of new technologies in the manufacturing systems, limited capacity storage between machines has become profitable. Therefore, in the following chapter, *Eddaly et al.* present an application of the Estimation of Distribution Algorithm (EDA) to a Flowshop Scheduling Problem (FSP) under limited buffer with respect to makespan criterion. After building the probabilistic model taking into consideration characteristics of the FSP such as the order of jobs and similar blocks of jobs in the sequence, the authors employ a skewed variable neighbourhood search algorithm to improve the solution-quality. The experiments demonstrate that the approach presented in this chapter is able to obtain better results in a short time-span than genetic algorithms, particle swam algorithms and several others in the literature.

The fifth chapter, by *Lenaerts and Defaweux*, proves the suitability of the Evolutionary Transition Algorithm (ETA) for solving hierarchically decomposable problems defined by hierarchical if-and-only-if (H-IFF). The authors discuss the behaviour of the ETA on hierarchical problems of varying sizes and modular complexities in depth and compare the results to those obtained using the

Symbiogenetic Model (SEAM). The chapter's major conclusions are that parameter tuning does not have a significant impact on the algorithm's success-ratio, the time required to develop a solution varies in proportion to the genotype size or the domain size of the variables, and the problem's modularity is not an issue for the ETA. The dynamics of the ETA and SEAM on shuffled H-IFF problem instances have also been compared in this chapter with the observation that ETA requires fewer evaluations to reach its goal and that both processes' dynamics differ a lot. This leads to the conclusion that the ETA is more general as it does not require any adaptation to address higher modularity. These findings should pave the way for the use of ETA on various other constraint-optimisation problems encountered in the real world.

Palit et al., in the next chapter, present a novel hybrid system based on the neural implementation of fuzzy logic model of the non-linear and chaotic electrical load time series. This approach has been used for short-term and mid-term electrical load forecasting. The experimental results prove that the neural-fuzzy network trained with improved back-propagation algorithm provides sufficiently accurate modelling and forecasting. The authors also observed that the application of fuzzy clustering and least square error estimation technique on electrical load data to select the network's initial parameters leads to a significant improvement in forecasting. The chapter ends with the conclusion that the neuro-fuzzy approach is preferable to both neural networks and fuzzy logic technologies applied individually.

It has been observed that the Particle Swam Optimisation (PSO) algorithm struggles in problem spaces containing multiple optima. The following chapter, by *Tim Hendtlass*, applies the Waves of Swarm Particles (WoSP), a variant of the PSO, to optimise quantised problem spaces. This algorithm combines the search abilities of both particle swarm and evolution to sequentially explore the multiple optima in the search space. The resulting algorithm has been applied to a highly constrained problem with many constraint-violating solutions but few constraint-satisfying solutions and is able to obtain considerably good results. Even though the WoSP may be outperformed by certain highly optimised problem-specific solutions, it is very promising for problem areas which lack any specialist customised software.

In the next chapter, *Andrew Koh* applies a co-evolutionary particle swarm algorithm for BiLevel Variational Inequalities (BLVI) to study the equilibrium decision variables offered by participants in a deregulated highway transportation sector. Assuming that the participants are similar to players in a Cournot-Nash game in choosing their decision variables and their actions are constrained by a variational inequality describing the equilibrium routing pattern of the highway users, the Nash equilibrium is used to model the behaviour of players in the resulting BLVI. A heuristic co-evolutionary particle swarm algorithm is then used to evolve strategies for each player that are robust against the other players' strategies. This approach was tested on two examples from the literature and the solutions mentioned in the literature were easily obtained. Thus, the results illustrate the potential application of this method for the class of intrinsically non-smooth optimisation problems.

Bülent Çatay, in the next chapter, applies Ant Colony Optimisation (ACO) to the Vehicle Routing Problem with Pickup and Delivery (VRPPD). The ACO approach proposes a new visibility function which attempts to capture the 'delivery' and 'pickup' nature of the problem. This approach is then used to address two types of VRPPD: Mixed VRP with Backhauls (MVRPB) and VRP with Simultaneous Pickups and Deliveries (VRPSPD) which have a growing practical relevance in the reverse logistics literature. Promising results were obtained compared to the results available in the literature and in some cases, the solutions were better than the best solutions mentioned in the literature.

The next chapter by *Salto et al.* is a tutorial chapter that provides an overview of Evolutionary Algorithms (EA) and Ant Colony Optimisation (ACO) when applied to the two-dimensional strip packing problem. Two hybrids of these two approaches are also studied in this chapter, those with Modified First-Fit (MFF) and Simulated Annealing (SA). The performances of EA and ACO along with the above-mentioned hybrids are studied by employing a Genetic Algorithm (GA) and an Ant Colony System (ACS). The studies reveal that the ACS with MFF is able to outperform the other combinations for the ACS. Similarly, it was observed that the GA+SA combination produces better results than other hybrids for the GA as well as the ACS+MFF combination in terms of both efficiency and solution-quality.

In the penultimate chapter, *Potter et al.* address the problems related to diagnosis (Multiple Fault Diagnosis), configuration (Network Configuration), planning (Forest Planning) and pathfinding (Snake in the Box Problem) by implementing four nature-inspired algorithms: Genetic Algorithm (GA), Discrete Particle Swarm Optimisation (DPSO), Extremal Optimisation (EO) and Raindrop Optimisation (RO). The chapter presents technique setup, issues with adapting the various techniques to the selected problems, and the emerging results produced by the techniques. The experimental results reveal that all these algorithms have their own areas of speciality and tend to perform strongly in these areas compared to other problem areas whose characteristics are not particularly suited to the algorithms. The major contribution of this chapter is that DPSO, EO and RO have not been applied to the majority of the exemplar problems presented in this chapter, making this the first time most of these results have appeared in the literature.

The final chapter, by *Maduriera et al.*, addresses the resolution of dynamic scheduling by means of meta-heuristics and multi-agent systems. Several decision-making entities, called Resource Agents, which are distributed within the manufacturing system, interact with each other to obtain optimal or near-optimal global solutions. These agents manage their internal behaviour and relationships with other agents via cooperative negotiations in accordance with business policies defined by the user manager. The main characteristic of this approach is that the machines first find local solutions followed by interaction with other machines to produce global schedules. Similarly, the authors present a population/solution regeneration mechanism to adapt the solutions to disturbances in the system. This chapter is a valuable contribution to the resolution of more realistic scheduling problems, especially the extended job shop scheduling problems.

As can be seen from these chapters, scheduling, planning and packing in their wide variety of forms are critical problems in today's world. Natural intelligence provides various approaches to automatically find optimal or quasi-optimal solutions for many of these problems within reasonable amount of time. This special volume highlights many important issues and provides imperative new insights into the use of natural intelligence for solving practical problems. In closing, we would like to thank all the authors for their excellent contributions. We also wish to acknowledge the help of the editorial review board as well as some authors who also served as reviewers, without whose support this book project could not have been satisfactorily completed. Special thanks go to all those who provided constructive and comprehensive review comments, as well as those who willingly helped in last-minute urgent reviews. We would also like to offer our gratitude to Prof Edmund Burke for writing the foreword. A further special note of thanks goes to Dr Thomas Ditzinger (Engineering Senior Editor, Springer-Verlag) and Ms Heather King (Engineering Editorial, Springer-Verlag) for their editorial assistance and professional support. Finally, we hope that the readers would enjoy reading this book as much as we have enjoyed putting it together.

May 2009

Raymond Chiong
Sandeep Dhakal

Editorial Review Board

Contents

Global Optimization in Supply Chain Operations

Maksud Ibrahimov, Arvind Mohais, and Zbigniew Michalewicz

Abstract. This chapter discusses some optimization issues from a business perspective in the context of the supply chain operations. We note that the term "global optimization" may have different meanings in different communities and we look at it from the business and classical optimization points of view. We present two real-world optimization problems which differ in scope and use them for our discussion on global optimization issues. The differences between these two problems, experimental results, the main challenges they present and the algorithms used are discussed. Here, we claim neither uniqueness nor superiority of the algorithms used, rather the main goal of this chapter is to emphasize the importance of the global optimization concept.

1 Introduction

Many aspects of commercial and industrial processes are subject to optimization, and this fact has been the source of many applications of modern heuristic methods. For example, the sequencing of jobs can be optimized in the classical job shop scheduling problem [2, 14, 4], there are many algorithms that address different kinds of planning and cutting problems [8, 7], many methods were developed to solve different types of transportation issues [11], as well as allocation and distribution problems [3, 17]. These are all examples of problems in which a globally optimal solution is sought for a specific and fairly narrow and focused problem. In real world business situations, however, it is often the case that the potential for optimizing operations and profits can be increased by considering broader subsets of the business, and

Maksud Ibrahimov and Zbigniew Michalewicz
School of Computer Science, University of Adelaide, Adelaide, SA 5005, Australia
e-mail: maksud.ibrahimov@adelaide.edu.au,zbyszek@cs.adelaide.edu.au

Arvind Mohais
SolveIT Software, Level 12, 90 King William Street, Adelaide, SA 5000, Australia
e-mail: am@solveitsoftware.com

R. Chiong, S. Dhakal (Eds.): Nat. Intel. for Sched., Plan. and Pack. Prob., SCI 250, pp. 1–28.
springerlink.com

by looking at how major processes affect each other, in addition to the traditional approach of seeking independent opportunities for improvement. For example, in the case of job shop scheduling, one might also look at the logistics of the provision of raw materials for the shop, as well as the details of post-production inventory shipping, marketing and retailing. From a *business* perspective, this is *global optimization.*

Lawrence Davis, one of the experts in supply chain optimization , describes this situation as follows [9].

> "...companies should be able to use solutions to optimize across their sourcing and procurement, production and distribution processes all at the same time. Instead, supply chain solutions tend to break those functions out into separate modules, each of which runs separately from the others. The answer you get when you first do production and then do distribution or any of these functions independently is not as good as the answer you get when you do them together."

Consider an example in which a company produces and distributes certain types of goods. Suppose that the company's factory is very efficient and, in a given month, produces an exceptionally large quantity of finished goods. From a local point of view this component of the supply-chain operation (i.e. the factory) performed excellently. However, if we look at the broader picture of global production, if the excess goods are not readily distributed and consumed, the company will need to use storage space which will cost extra money, thus making this locally good production run sub-optimal from a global point of view.

It is possible to take the idea of global optimization from a business perspective to increasingly greater heights by taking into consideration more and more factors that affect a business' profitability. However, practical limits are often encountered when the scope of the optimization problem begins to encroach on boundaries of another business entity's operations. For example, it may be that the provision of raw materials to a job shop is a matter handled by a different business entity (i.e. the component) who may not be willing to divulge minute details of their operations.

In this chapter, we will look at characteristics of global optimization from a business perspective, and use two real-world business examples that we have solved to illustrate the distinction between a relatively simple situation that focuses on one aspect of a business' operations, and another, in which several components are considered simultaneously. The business domains of each example will be presented in depth, along with details of the evolutionary algorithms that were used to solve the problems. The data for each example was collected from actual businesses, but has been altered systematically to protect the privacy of the entities involved. These examples will allow the reader to appreciate the use of nature-inspired techniques in solving the problems. Our broader goal, however, remains to highlight the differences between the two types of problems, and how those differences might influence the solutions to the problems using evolutionary algorithms.

When we make the distinction between local and global optimization from a business perspective, the idea of global optimization in the classical sense still holds in both cases. If we try to optimize over the entire supply chain, then in terms of the objective function being considered, we still seek a classical global optimum. On the other hand, if we restrict our attention to optimizing the operations of a single factory, then this is local optimization with respect to the broader picture of the supply chain; but nevertheless we also seek a classical global optimum for the objective function associated with the optimization of the factory. The distinction between local and global optimization from a business perspective lies in the differences of their respective search spaces.

The general problem of global optimization, in the *classical* sense, is to find $\mathbf{x}$ so as to

$$optimize f(x), x = (x_1, \ldots, x_n) \in \mathcal{F} \subseteq \mathcal{S}, \quad (1)$$

where $\mathcal{S}$ defines the search space and $\mathcal{F}$ defines *feasible region* which is bounded by the set of m constraints: $g_j(\mathbf{x}), \forall j = 1, \ldots, m$.

That is, one is seeking to find an element in the search space $\mathcal{S}$ that optimizes the objective function.

Usually, the search space $\mathcal{S}$ is quite simple: for numerical optimization it is $\mathcal{S} \subseteq \mathbb{R}^n$, and for combinatorial optimization problems, $\mathcal{S}$ is discrete in nature, like a set of permutations, or a set of edges of a graph. An example of a combinatorial optimization problem that one might encounter in a commercial setting is finding a permutation of n jobs to be executed on a machine that would minimize the amount of time spent re-configuring the machine between jobs (or from another perspective, maximizes the amount of time the machine is in use).

In silo[1]-focused optimization, the search space $\mathcal{S}$ is typically built from a single type of variable. For example, as above, $\mathcal{S}$ may consist of all permutations of the set $\{1, 2, \ldots, n\}$. Business global optimization problems, on the other hand, typically have a more complex search space that consists of compositions of simpler search spaces. We will refer to these types as *single component* and *multiple component* search spaces respectively.

Some typical characteristics of a problem that suggest that there might be an opportunity for global optimization from a business perspective are the following:

- *Dependence or partial dependence among components.* Problem specific constraints and business rules tie single components together creating dependencies among them. Note that constraints here can be *intracomponent constraints*, which apply to a single component itself and *intercomponent constraints*, that actually make these dependencies. Intracomponent constraints are the classic type of constraints. Intercomponent constraints

[1] In supply chain business terminology, the word "silo" is used to denote a single component of the business.

have a slightly different purpose in global optimization compared to the classic one.

- *Multiple objectives.* Having many single components in the problem, each component having its own objective or even several objectives, yields a multiple objective optimization problem. From a business perspective, global optimization problems are multi-objective problems, but they are distinguished by the fact that the objectives are related in a hierarchical manner with some objectives related to single silos, and other higher-level objectives being compositions of lower-level objectives.
- *Need to understand complex what-if scenarios.* Businesses and their competitors are rarely static. Instead, they are dynamic and the business environments are constantly changing. Thus, companies want to make some decisions and changes to their operations. There is a large need for businesses to foresee how a particular change or decision will impact their priorities, profit and other objectives. Complex multi-component systems with a large network of dependencies are a lot harder to understand and predict than a single component system.

Examples in this chapter concentrate mostly on the first point (i.e., Dependence or partial dependence among components). Although one of the problems has multiple objectives, we do not treat our problems as such, as the main purpose of this chapter is to underline the importance of the concept of global optimization from a business perspective. The third point (i.e., what-if scenarios) is not considered in this chapter.

It is important to understand that the term "global optimization", as used here, is not a method to solve supply chain management problems but rather a concept of viewing multi-silo supply chain problems from a broader perspective.

In recent years, there has been an increased interest in solving supply chain management problems using evolutionary algorithms. David Naso et al. [13] look at the problem of coordination of just-in-time production and transportation in a network of partially independent facilities of ready-mixed concrete. They optimize the network of independent and distributed production centers serving a number of customers distributed across a certain geographical area. This problem, which has high complexity and strict time delivery constraints, is approached with a meta-heuristic based on a hybrid genetic algorithm with combined constructive heuristics.

Altiparmak et al. [1] propose algorithms using mixed-integer, non-linear programming model for multi-objective optimization of a supply chain network based on the real world problem of a company that produces plastic products. They compare three approaches to find the set of Pareto-optimal solutions and discuss the pros and cons of each of them.

Zhou et al. [16] present a novel genetic algorithm to solve bi-criteria, multiple warehouse allocation problem. The proposed method finds the Pareto-front of a wide range of non-dominated solutions without the arbitrary determination of weighting coefficients.

In Vergara et al. [15], an evolutionary algorithm was developed for dealing with the coordination of supply chain flows between different members of the chain. The authors of that paper recognized the importance of an overarching algorithm that optimizes the whole system. Their work looked at the flow of materials in terms of supply and demand, but did not consider the intricacies of production within each business silo.

Even much earlier, researchers had been working on this type of problem as can be seen in the following references. Lee and Choi [5] apply genetic algorithms to solve a single machine scheduling problem with distinct dates and attempt to minimize all penalties. This method produces near optimal solutions, which they prove by comparison with an exact algorithm. Lee et al. [6] address the problem of inventory management of a refinery that imports several types of crude oil and proposes a mixed-integer linear programming model. Martin et al. [10] create a linear programming model to optimize flat glass production.

The rest of the chapter is organized as follows: the next two sections introduce two real-world optimization problems, their difficulties, algorithms that solve them, and the experiments conducted. Section 4 discusses the differences and similarities between these two problems. Section 5 concludes the chapter.

2 Optimization of the Real-World Glass Company Operations

In this section, we present a real world example of the glass item production optimization problem, its proposed solution, and the experimental results.

2.1 Problem Statement

Here, we consider a real world example of an optimization problem that focuses on a single silo of a business' operation. Assume that 'Max Glass' is a company that produces glass products. It uses machines called *powerpacks* to produce molten glass from raw materials, and other *moulding machines* to make moulds and uses the molten glass to make finished products. This process is illustrated in Figure 1.

The glass-producing factory uses several types of glasses to make a variety of finished products. Any powerpack may be used to prepare any type of molten glass. However, when changing between glass types on the same powerpack, a cleaning operation is needed. The extent of this cleaning procedure varies. Sometimes melting of new materials can start without any cleaning, but sometimes a careful and thorough procedure is required when switching between materials. This operation consumes time and money, and should be avoided as far as possible.

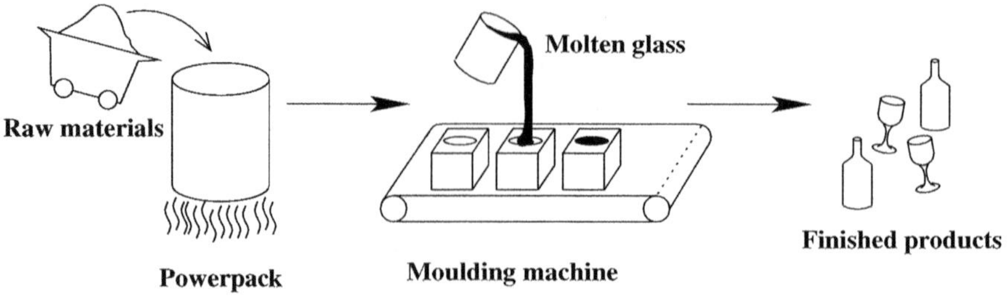

Fig. 1 The process of creating glass products

The moulding machines, also called *production lines*, are not all the same. Some products can be produced only on one machine, whereas others can be produced on more than one. In the latter case, however, production rates may differ due to different efficiency rates. Production lines use special implements called *tools* to make moulds, and naturally different products require different tools. As before, time and money are consumed when there is a need to change tools.

Other business rules and constraints that characterize the problem are the following:

- Powerpacks and production lines can only operate at certain times of the day and at certain days of the week.
- Powerpacks operate by *batch loads*. This means that they produce a specific tonnage of molten glass in a batch, and that smaller amounts are not permitted. So, for example, assuming a batch size of 8 tonnes, if we need to have just 5 tonnes of material for a particular job, we would be forced to produce 8 tonnes and there would be 3 tonnes of wastage.
- No production run is perfect and some products can be defective. To take this issue into account, a small percentage of surplus is usually planned.
- Certain different products cannot be produced around the same time because they are too similar in appearance, and since all products end up on a common polishing line, different products that look very similar cannot be easily distinguished.

2.2 Customers' Orders

For a given period of time, say one week, there are a number of customer orders that need to be produced. Suppose that for such a period of time, we have n orders placed by n customers $C_1, C_2, \ldots, C_n$. Each customer's order consists of a set of ordered items, with associated quantities for each. Assuming that the m possible products that can be produced are $I_1, I_2, \ldots, I_m$, Table 1 shows what the list of orders might look like. In Table 1, each cell is called an *order item* (OI).

Table 1 Demand table showing how many of each item are needed by each customer during a given period

	I_1	I_2	...	I_m
C_1	5500			15500
C_2	20000			
$\vdots$	$\vdots$	$\vdots$		$\vdots$
C_n		40000		32500

Given the above description of the application domain, the problem can now be stated as follows:

> Given a set of order items $\{OI_1, OI_2, \ldots, OI_k\}$ to be completed during a specific time frame, and a set of production lines $\{P_1, P_2, \ldots, P_l\}$, allocate each order item to a particular production line in such a way as to have a feasible schedule in which as many order items are completed as possible, and the production costs are minimized.

2.3 Solution

Figure 2 shows a screen of a system that was developed to solve the sequencing problem described above. It shows three production lines called Machine #1,

Fig. 2 The application's main screen with production lines (A), powerpacks (B), ordered items (D). Here, grayed areas represent hours when a particular machine cannot work (C). An ordered item's selected properties are shown in the bottom box (E). (F) represents the load of a particular powerpack at particular time.

Machine #2 and Machine #3, and two power packs called Supplier #1 and Supplier #2 for melting glass. In this figure, each order item is represented by a rectangle and each glass type is represented by a different color. Some orders have been scheduled for production and some are still unassigned. In the middle section of the figure is a table that shows the details of all the orders, and the bottom part of the figure shows further details of a single selected order, such as the production lines where can it be produced, the rate of production, the tools required and the quantity of finished products needed.

2.4 Representation

The following version of evolutionary algorithm was used for this problem.

Solution Representation
We represent our solution indirectly in the following way

$$X = (x_1, \ldots, x_v) \quad (2)$$

where each x_i is a pair $(OI_k, P_l), OI_k \in OI, P_l \in P$. The pair (OI_k, P_l) means that order item OI_k will be executed on machine P_l.

Main Algorithm Loop
After the population of individuals has been initialized, the evolutionary algorithm enters the main loop where the population is *evolved* until a specific criterion is satisfied. Each individual is assigned a fitness value by an evaluation function which is designed so that its value is indicative of how well the individual solves the problem. The evaluation functions, in our work, makes use of penalty functions which essentially assign a value to an individual in proportion to how badly the individual violates some constraint.

During the main loop, individuals with higher fitness are probabilistically given more chances to *reproduce*, that is, to make modified copies of themselves to replace them if modified copies have better fitness values. The changes are made to an individual by means of *operators*, which are functions that take one or more individuals and combine, alter and manipulate their structures in a partially random way. After operators have been applied, some of the resulting new individuals are passed through *repair* functions, that attempt to randomly, but intelligently alter their structure so as to reduce the extent of their constraint violations.

Decoders
From the description of the solution structure, the order in which these jobs will be executed is not clear. This decision is the job of the *decoders*. It works in the following way. The ordering of tuple x is taken as a priority ordering. Starting with a blank state, i.e. with all machines marked as completely unoccupied, x_1 is given first priority to find a place on the machine.

Suppose that $x_1 = (OI_{k_1}, P_{l_1})$, then the decoder looks at the timetables of machine P_{l_1} and finds the first suitable time slot in which to place the ordered item OI_{k_1}. If there is no possible time slot in machine P_{l_1}, then other machines are considered. Next the decoder processes x_2, and so on until the last element.

Operators

- *Swap order item operator:* When this operator is applied to an individual, it randomly picks two elements, for example $x_a = (x_a.OI, x_a.P)$ and $x_b = (x_b.OI, x_b.P)$, and swaps the order items in each pair. So, after the operator is applied, these elements change to $x_a = (x_b.OI, x_a.P)$ and $x_b = (x_a.OI, x_b.P)$. Figure 3 illustrates this.

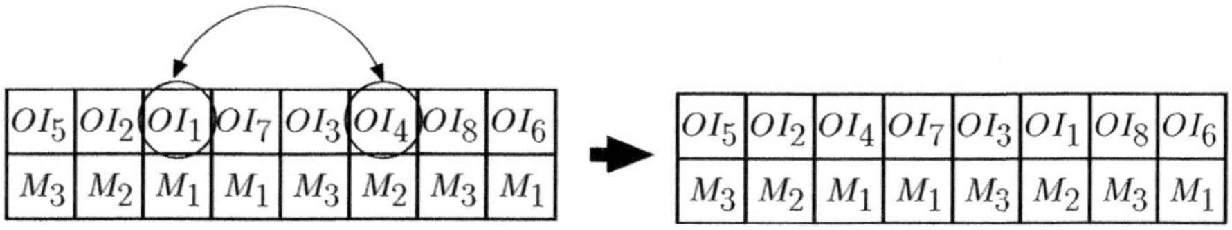

Fig. 3 Swap order item operator. Vector x before and after the operator.

- *Swap machines operator:* This operator is very similar to the previous one with the only difference being that machines are swapped instead. Hence if the operator randomly chooses element $x_a = (x_a.OI, x_a.P)$ and $x_b = (x_b.OI, x_b.P)$, then after swapping, they become $x_a = (x_a.OI, x_b.P)$ and $x_b = (x_b.OI, x_a.P)$.
- *Swap pairs operator:* This is another variation of the two previous swap operators. In this case, both the order item and the machine are swapped, so basically, we change the priorities of the selected pairs. Hence, if the operator picks elements $x_a = (x_a.OI, x_a.P)$ and $x_b = (x_b.OI, x_b.P)$, after execution of the operator, they become $x_a = (x_b.OI, x_b.P)$ and $x_b = (x_a.OI, x_a.P)$
- *Move pair left operator:* This operator picks a random element x_i and moves it a random number of places to the left, thus randomly increasing its priority. Figure 4 illustrates this.
- *Mutate machine operator:* This operator randomly chooses a pair $x_i = (x_i.OI, x_i.P)$ and randomly changes its machine to another machine which is capable of producing its order item $x_i.OI$.
- *Prioritize late order operator:* This operator searches for orders that are scheduled to be completed later than their due date according to current decoded individual. It randomly increases the priority for each such order by shifting them left.

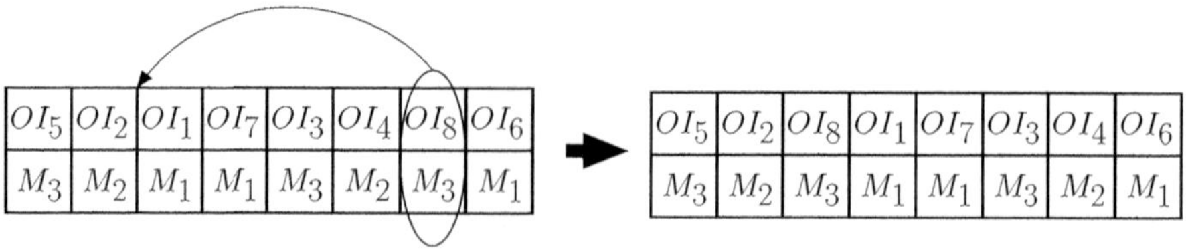

Fig. 4 Move order to the left operator

- *Move order to a less occupied machine operator:* This operator is used to decrease the load of one machine by moving orders to others. It randomly chooses a pair $x_i = (x_i.OI, x_i.P)$, and if there is a machine capable of producing $x_i.OI$, which is less loaded than $x_i.P$, it assigns the order to the new machine.
- *Group orders with the same tool operator:* A *group* is a sequence of consecutive pairs that share a common value. There are two types of groups: tool groups and material groups. This operator randomly chooses a pair and then finds a tool group that this element belongs to by searching to the left and to the right. Then it tries to find another tool group that has the same tool, and places the two groups together. Figure 5 illustrates this.

Fig. 5 Group orders operator

- *Group orders with the same material operator:* This operator is similar to previous one, except that it uses glass type as the grouping criterion.

Penalties

The evolutionary algorithm uses the following penalties:

- *Late penalty:* This penalty is used to avoid late orders. The value returned by this penalty is the sum of the number of hours all the orders are late by.
- *Early production penalty:* Orders that are completed excessively early are also not preferable because of the need to store them somewhere until the delivery date. This penalty is not as severe as *late penalty*, but is nonetheless needed.
- *Material change penalty:* As discussed earlier some cleaning is needed when changing materials in the powerpacks. Therefore, it is preferable to have as few material changes as possible. This penalty penalizes individuals in proportion to the number of material changes in their phenotype.

- *Tool changes penalty:* This is similar to the material change penalty, except that individuals are penalized in proportion to the number of tool changes in their phenotype.
- *Surplus material penalty:* As previously mentioned, powerpacks produce molten glass in batches. If all of the material in a batch is not used, the leftover would be wasted which is not desirable. The penalty applied to individuals is in proportion to the total wastage that occurs in their phenotype.

Evaluation Function
The evaluation function of an individual X is calculated in the following way:

$$F(X) = -\sum_{i} (R_i \times Penalty_i(X)) \tag{3}$$

where R_i is a $Penalty_i(X)$ coefficient. Our evolutionary algorithm will try to maximize this function $F(X)$.

2.5 Experiments and Results

The evolutionary algorithm as described above was implemented in Java. We used a Core 2 Duo T9300 machine with 2 GB of RAM running Windows Vista to test our solution. We used a dataset with 3 production lines, 2 power packs and 327 ordered items. The optimization runs took 5-6 minutes on average. The following parameters were applied:

- Population size = 2000
- Mutate machine operator probability = 0.5
- Swap probability = 0.5
- Move probability = 0.5
- Metal changes penalty coefficient = 100.0
- Tool changes penalty coefficient = 500.0
- Surplus metal penalty coefficient = 0.1
- Tournament selection of size 2
- A steady state population was used

Figure 6 shows the fitness over generations in a sample run and the best run. The delay penalty function over generations is shown in Figure 7.

We made 90 runs of our algorithm on the same dataset. This experiment shows stable results with median of -4.31E+08 and standard deviation of 699205.8988.

The software developed is currently in use at the company as an optimization tool and decision support system. The modeling of different situations and what-if scenarios becomes a rather easy task which can lead to better production-strategy improvements.

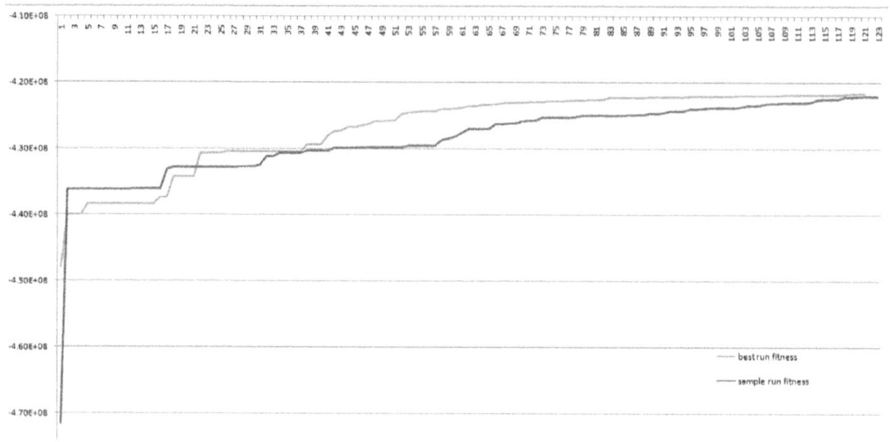

Fig. 6 Sample run. Evaluation function over generations.

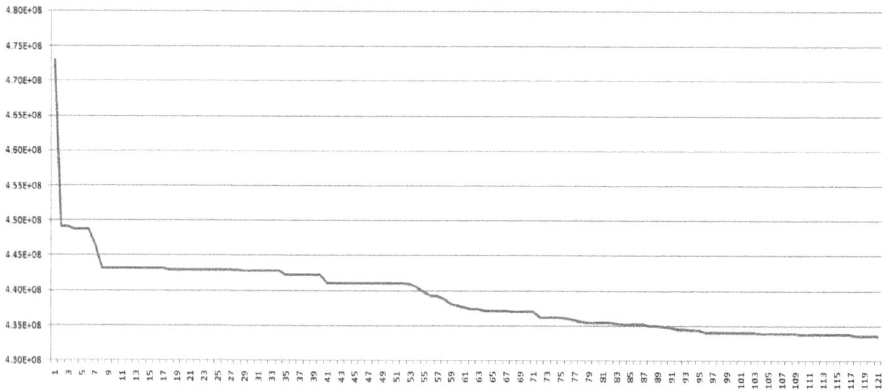

Fig. 7 Sample run. Penalty over generations.

3 Optimization of Operation of a Real-World Steel Company

In this section, we present a real world example of the steel sheet production optimization problem, then the proposed solution, and finally the experiments conducted on this problem.

3.1 Problem Statement

Suppose that a company called Global Steel has several plants across Australia and specializes in producing sheet steel. Sheet steel products are basically thin sheets of steel that have been coiled into rolls. There are many different types of products that the company can manufacture, that differ in the chemical composition, density and width. Furthermore, the rolls can be configured in different diameters and cores (see Figure 8). When a plant manufactures a roll of steel, the width of the roll cannot be changed. However, customers typically place orders for smaller width, and so, large rolls must be cut into pieces along their width to satisfy these orders. This requires orders of similar configuration to be grouped together before cutting the sheets (see Figure 9).

The constraints and business rules of this problem are listed below; many of which are similar to the previous problem.

- Each plant's daily work hours are limited.
- Each plant can produce a subset of all possible goods. Some products can be produced at multiple plants, but the cost and rate of production is different at different plants.
- Each plant is equipped with a particular number of 'knives', which are used to cut large rolls into smaller ones. Each plant can only produce sheets of fixed width and has special knives to cut to the width needed.
- Each plant has a fixed daily operating cost which is the cost of running the plant, and does not depend on the actual products produced.
- After production, products are delivered to customers, and this incurs a transportation cost. There are defined transportation costs between each customer and plants. Some plants cannot ship to some customer due to specific business rules.
- Due to shipping constraints, export orders must be produced and shipped from a single plant; whereas for others, part of the order can be produced at one plant and other parts at other plants.
- Orders are produced in batches by the plants, which will be explained in detail later in this section.

In order to understand the problem better, let's look at a simplified example. Suppose we have 5 plants: P_1, P_2, P_3, P_4, P_5 and 7 customers C_1, C_2, …, C_7. Let's say that customer C_2 has an order for 5 tonnes of product I_1, and 8 tonnes of I_3. Customer C_6 wants 15 tonnes of I_1, 3 tonnes of I_2, and 10 tonnes of I_3. I_1 can be produced only at P_2, I_2 can be produced at P_1 and P_4 and I_3 can be produced at P_2, P_3 and P_5 although with different costs of production (see Table 2(a)). Table 2(b) shows the products that can

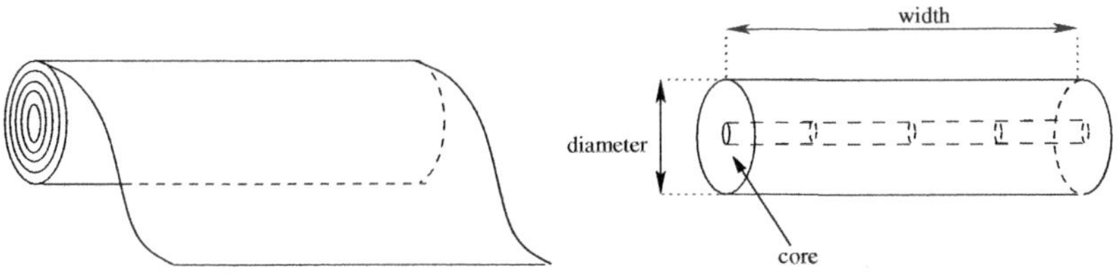

Fig. 8 A roll of sheet steel

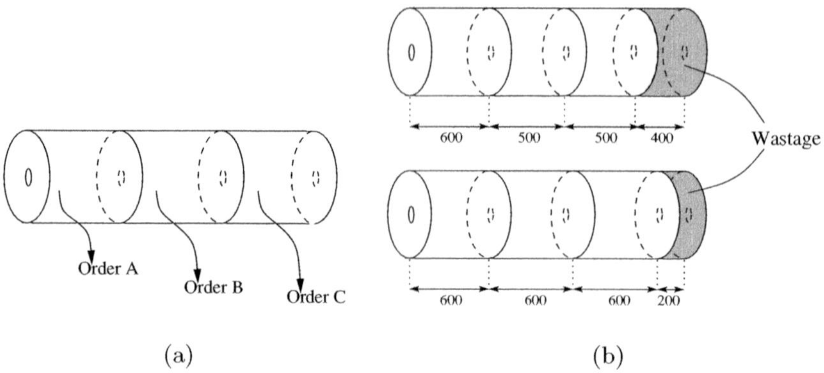

Fig. 9 (a) Batching of orders into a single roll. (b) Wastage resulting from the batching process. 4 rolls of 600mm, and 2 of 500mm needed.

be produced at each plant, as well as the tonnes per hour rate of production for each.

Suppose that we have to complete these orders in a time frame of 100 hours. Let's look at one possible way of distributing this workload amongst the plants. As I_1 can only be produced at plant P_2 we are forced to assign all 100 tonnes of I_1 there. This then implies that the required 30 tonnes of I_2 must be produced at P_4, using up 20 hours of the available time. 280 tonnes of I_3 still need to be produced. P_2 could produce 200 tonnes of it, P_3 could produce 150 tonnes and P_4, because it only has 80 hours of available time left, could produce 120 tonnes. Thus the 280 required tonnes of I_4 can be split amongst these three plants in various ways.

Plants can produce rolls of sheet steel of a certain width only, which is generally much larger than the widths typically required by customers, and therefore, ordered items that have the same type, core, and diameter must be batched into a single roll. That is, a large roll is cut into pieces to satisfy these orders. This is illustrated in Figure 9(a).

Table 2 (a) Demand table: The quantity of each item ordered by each customer. (b) Production capacity table: Tonnes per hour production rates of each plant for different products.

(a)

	Product		
Customer	I_1	I_2	I_3
C_2	50		180
C_6	50	30	100
Total	**20**	**3**	**18**

(b)

	Plant			
Product	P_1	P_2	P_3	P_4
I_1	1.0			
I_2	1.0			1.5
I_3		2.0	1.5	1.5

This process of batching sometimes results in wastage. To illustrate, suppose that customers C_1 and C_2 need rolls of product I_1, of the same diameter and with the same type of core. C_1 needs 2 rolls of width 500 mm, and C_2 needs 4 rolls of width 600 mm. Let's assume that the only plant that can produce I_1, must produce rolls of steel of width 2000 mm. In this situation, batching can be done as shown in Figure 9(b). The plant can produce two rolls of I_1 of width 2000mm. One can be cut into 4 pieces: 3 pieces of 600mm and 1 piece of 200mm. The 3 pieces of 600mm can be used for orders, but the remaining 200mm is wasted. It may be recycled, but a loss is incurred nevertheless. The second roll of 2000mm can be cut into widths of: 600mm, 500mm, 500mm, 400mm, the first 3 of which would go towards satisfying orders, and the last piece ends up as wastage. Some wastage is unavoidable, but clearly it is desirable to have as little wastage as possible in order to maximize production efficiency and hence profit. Intelligent grouping of orders can minimize wastage.

Plants use cutting tools called knives, physically arranged in parallel, to cut large rolls into smaller ones. Different plants are equipped with different numbers of knives, and this presents another constraint to consider when batching orders. A plant equipped with n knives can cut a roll into at most $n+1$ pieces. In the above example (Figure 9(b)) cutting a large roll into four small ones, would have required three cuts and hence three knives.

The goal of the problem is to find a set of assignments of orders to plants that optimizes the overall profit of the company. Profit is determined by the following formula:

$$Profit = S - C_m - C_t \tag{4}$$

where S is the *selling price* of the goods produced, C_m is the *cost of manufacturing* those goods, and C_t is the *cost of transporting* the goods to customers. Selling prices may vary for the same product depending on the market which it is being sold. The cost of manufacturing a particular order item is the sum of the cost per hour of producing that grade of sheet steel, and the fixed hourly cost of running the relevant plant for the time it takes to produce the required tonnage. Transportation costs vary with the destination address of the customer. Export orders naturally incur higher transportation costs. These last two components are the negative components of the profit; the bigger they are, lesser is the profit. The selling price is the positive component of profit; it is known before hand, and does not depend on which plant an order is allocated to. Therefore, the goal of the problem is to minimize the transportation and manufacturing costs.

3.2 Solution

The software created for this problem used an evolutionary algorithm to find a solution. It is shown in Figure 10. The software is fully configurable

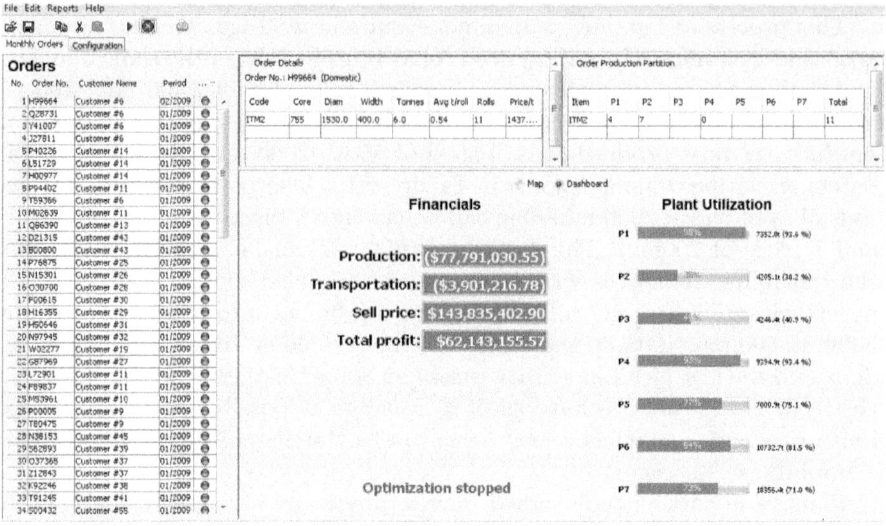

Fig. 10 Steel optimizer's main screen

to cater to various selling prices, transportation and manufacturing costs. Given the projected orders for an upcoming month, it seeks an optimal set of assignments that maximizes the profit of the company. Being a decision support tool, the software also provides financial comparisons with human-proposed allocations. In practice, the software suggests decisions that have better financial bottom lines.

Solution Representation

The representation used for the evolutionary algorithm consists of two parts. The first part is a string of quantities assigned to each capable plant for various order items. Figure 11 illustrates the structure of this representation hierarchically. It can be visualized as a string of s records, one for each order being considered. In turn, each order record can be thought of as consisting of a number of smaller records, one for each item in that order. And finally, each order item record can be thought of as consisting of a number of allocation records, one for each plant that is capable of producing the item under consideration. Such a lowest-level record is called a *plant quantity allocation (PQA)*. More formally, assuming that there are s orders to be satisfied, a candidate solution is represented as $X = (x_1, \ldots, x_s)$, where for $1 \leq i \leq s$, $x_i = (y_1, \ldots, y_{m_i})$, where m_i is the number of items in order i. And further, for $1 \leq j \leq m_i$, $y_j = (z_1, \ldots, z_{k_{ij}})$, where k_{ij} is the number of plants capable of producing item j of order i. Each z_{ij} is a tuple (OI_i^j, p, q); it is a plant quantity allocation in which plant p is allocated q rolls of item j in order i.

The second part of the representation is a string of permutations of the plants in the system. Each permutation represents the preferred order of

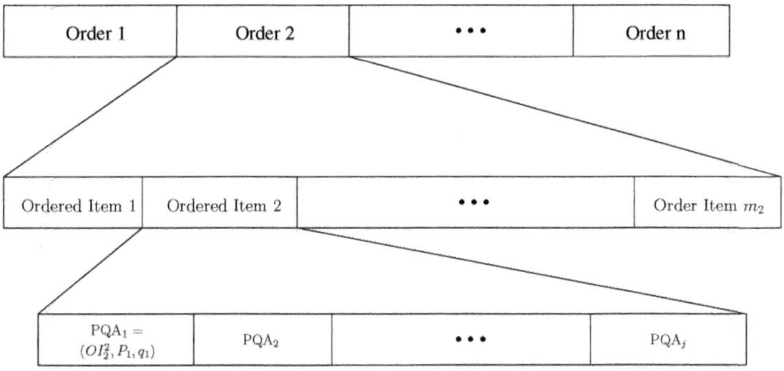

Fig. 11 First part of the representation of an individual

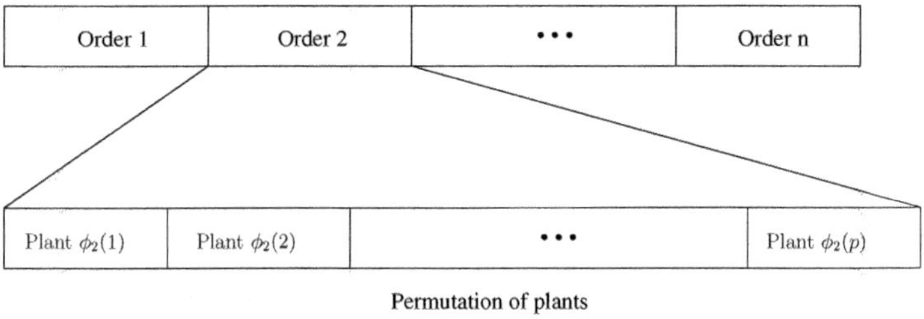

Fig. 12 Second part of the representation of an individual

the plants for use in export orders. This can be formalized as $X_{pref} = \{x_{pref_1}, \ldots, x_{pref_s}\}$, each element x_{pref_i} is a permutation of the set of plants associated with order i. Figure 12 illustrates the second part of the representation.

Population Initialization

At the start of the evolutionary algorithm, a population of randomly initialized individuals is created. First, each order item cell in Figure 11 is initialized by uniformly randomly divvying up the number of rolls of steel required for that order item into the PQA cells lower down in the hierarchy. Second, each order cell in Figure 12 is assigned a randomly generated permutation of the steel plants.

Main Algorithm Loop

After the population of individuals is initialized, the evolutionary algorithm enters the main loop in which the population is *evolved* until a specific criterion is satisfied. Each individual is assigned a fitness value by an evaluation function which is designed so that its value is indicative of how well the individual solves the problem. The evaluation functions in our work make use of penalty functions which essentially assign a value to an individual in proportion to how badly the individual violates some constraint.

During the main loop, individuals of higher fitness are probabilistically given more chances to *reproduce*, that is, to make modified copies of themselves to replace them if the modified copies have better fitness values. The changes made to an individual are accomplished by means of *operators*, which are functions that take one or more individuals and combine, alter and manipulate their structures in a partially random way. After operators are applied, some of the resulting new individuals are passed through *repair* functions, that attempt to randomly, but intelligently alter their structure so as to reduce the extent of their constraint violations.

Penalties

Not every candidate solution that can be constructed as described above would represent a feasible solution. For example, if in a candidate solution, a plant is assigned more work that it could possibly produce in the time frame under consideration, then the solution is infeasible. Many techniques are known to handle different types of constraints [12] and most of them are based on penalty functions. The approach used for this algorithm is also based on penalty functions. Penalties were used to deal with both hard and soft constraints. Penalty functions for hard constraint were given larger weights than those for soft constraints. The penalty functions developed were:

- *Quantity production penalty:* This penalty is used to avoid underproduction or overproduction of an ordered item. If an order item requires N_1 rolls of a given product, and the sum of the PQA quantities for this order item is N_2, then this penalty function returns a value that is proportional to $|N_1 - N_2|$. This is a soft constraint because it is not always possible to produce the exact amount demanded, as plant capabilities are limited and fixed.
- *Plant capacity violation penalty:* If the amount of work allocated to a plant exceeds its capacity for the time period, then this function penalizes the solution in proportion to the over-allocation. This is a hard constraint because a plant cannot produce more than its capacity.
- *Wastage penalty:* Wastage is possible as a result of the batching procedure that occurs during production. This function penalizes in proportion to the number of tonnes of wastage. This is also a soft constraint.
- *Export order single source penalty:* Orders for export customers should be produced at the same plant, and this function penalizes export orders that use multiple plants. This is a hard constraint.

Evaluation Function

Each candidate solution is assigned a fitness value by an *evaluation function* that returns a value that is indicative of how well that candidate solves the problem. For this problem, the fitness function is defined as:

$$f(X) = Profit(X) - \sum_i (R_i \times Penalty_i(X)) \tag{5}$$

X is a candidate solution, and R_i is the coefficient of the penalty function $Penalty_i(X)$. $Profit(X)$ is calculated using three components: manufacturing cost, transportation cost and selling price, as described previously. Our evolutionary algorithm tries to maximize this function $f(X)$, i.e. maximize profit, and minimize constraint violations.

Operators

An evolutionary operator takes one or more candidate solutions and transforms it (or them), into a new candidate solution. This is typically done by using partially random manipulations of the representation. Operators are employed probabilistically. We used several operators to solve this problem. Some of them look like classic mutation and crossover, and some of them are modified versions of these two. The operators used are the following:

- *Quantity crossover:* This is a classic crossover-like operator. It takes two individuals, randomly chooses a crossover point at the boundary of an order item cell, and performs a crossover at this point.
- *Quantity mutation:* This operator is similar to classical mutation. It takes an individual, and probabilistically chooses several PQA cells and changes their quantity values by random amounts within a certain range. The changed amount is chosen randomly in the range $[-\alpha, \alpha]$, where

$$\alpha = \frac{quantityOrdered - quantityAllocated}{|\lambda(OI)|} \tag{6}$$

  where $\lambda(OI)$ is the quantity required for a particular order item.
- *PQA Shuffling:* This operator randomly chooses one or several ordered items and shuffles quantities assigned to plants within each ordered item. It preserves the sum of quantities within one ordered item before and after the shuffling. So, if we randomly chose ordered item OI_k then $G(OI_k) = G'(OI_k)$, where G is the sum of all quantities of ordered item OI_k assigned to all capable plants before applying the shuffling operator and G' is the sum after applying it.
- *Preferred plant operator:* This operator looks at the second part of the representation of an individual, i.e. the permutation of plants, and performs a random transposition on it, i.e. it randomly swaps two elements of the permutation.

Repairers

A *repairer* is an operator that tries to modify a candidate solution in such a way as to eliminate or reduce the number of constraints that it violates. In our evolutionary algorithm, we use a set of problem-specific *probabilistic repairers.* The main difference between these and the classical repairers is that they are not executed each time a solution gets evaluated, but instead are applied probabilistically, which allows for tuning for better performance. The repairers used are:

- *Quantity allocation repairer:* This repairer tries to fix errors in over- and under- allocation of a particular order item amongst capable plants. It works in the following way. We choose an order which has underproduction or overproduction and increase or decrease the quantity produced at each plant. The increment or decrement are not random. The algorithm decides on the changes to be made. This will increase the fitness of individual, but being a random change, some diversity will be maintained.

	P_1	P_2	P_3	P_4	Total
Quantity before repair	6	13	16	9	44
Quantity after repair	5	11	15	8	38

  For instance, suppose that for a particular order item, 36 rolls are needed, but according to an individual about to be repaired, the capable plants have been allocated 44 rolls. The second row shows what might be the result of applying the quantity allocation repairer. Some values increased, some decreased. The new solution is still not perfect, but overproduction has been reduced.
- *Overallocation repairer:* This repairer is different from the quantity allocation repairer. It checks the *total* workload allocated to a plant, to see if it exceeds that plant's capacity for the time period. If it does, the repairer randomly chooses some order item that uses the plant in question, and reduces its quantity allocation by one. It repeats this procedure until the plant can produce its assigned amount in the given time period.
- *Export order source repairer:* This operator checks for export orders in the individual that use multiple plants, and performs repairs according to the preferred plant permutation. For any such order, it tries to assign the whole work to the first capable plant in the permutation list.
- *Final repairer:* Unlike previous repairers, this repairer is run only at the end of evolutionary algorithm to "polish" the solution. It uses non-probabilistic versions of previous algorithms, since at the end of the algorithm, we want to make solutions as good as possible without worrying about diversity of individuals.

Optimization of Batching

One of the challenges of this problem is grouping ordered items into batches. The problem is known in the literature as the *stock cutting problem*. Recall that the problem is the following. We have a set of ordered items all with the same product, core, and diameter, but with different widths. These must be allocated to a plant that can produce rolls of a fixed width which can later be cut into smaller rolls to satisfy the orders. For convenience, let us refer to these large fixed-width rolls as *max-rolls*. We need to group the ordered items into batches in such a way that the number of max-rolls produced by the plant is minimized, thus minimizing the wastage.

For this problem, a greedy approach was taken. The algorithm looks at all order items that are assigned to a particular plant, groups those that are compatible (i.e. that share the same product, diameter and core) and forms batches in the following way. The order item with the largest width is assigned to the first max-roll. Some of the width of this max-roll may be unassigned. The next largest order item is considered next. If there is a max-roll in which it can fit, it will be placed there, otherwise a new max-roll will be created. The same procedure is repeated for all order items. The pseudo-code of this algorithm is shown below:

Algorithm 1. Batching algorithm

```
maxRolls ← ∅
for group ∈ orderItemGroups do
  sortByWidth(group)
  for orderItem ∈ group do
    if canFindAvailableSpaceInMaxRolls(item) then
      placeItemInAvailableMaxRoll(item)
    else
      newMaxRoll ← createNewMaxRoll()
      newMaxRoll.add(orderItem)
      maxRolls.add(newMaxRoll)
    end if
  end for
end for
```

The procedure *canFindAvailableSpaceInMaxRolls(orderItem)* iterates through all available max-rolls and checks if the given *orderItem* can be placed in a max-roll. If it can find such a max-roll, it will return true, if no such max-roll is available, the procedure returns false. The procedure *placeItemInAvailableSpace(item)* places given *item* in the available roll that has been found by *canFindAvailableSpaceInOrders*.

3.3 Experiments and Results

The evolutionary algorithm in this study was implemented in Java, and was tested on real world industrial data. The purpose of the software is to provide decision support to business managers who approve production allocation.

In this subsection we present the results of experimental runs of this algorithm with a view to gaining an understanding of how the algorithm solves the problem.

An experiment was performed on a set of real-world data for a particular month. Due to the stochastic nature of evolutionary algorithms, the experiment was executed multiple times so as to allow for statistically accurate results. The specific parameters used for the experiment are shown below. The penalty coefficients were tuned for the business entity under consideration, and would perhaps not be ideal if this algorithm were to be replicated for another industry or business environment. In such a case, it is likely that re-tuning of the coefficients would be necessary. As such, it would not be prudent to try to interpret the significance of these penalty coefficient values.

- Population size = 250
- Number of trials = 90
- Mutation probability = 0.3
- Overallocation repair probability = 0.1
- Underproduction penalty coefficient = 10 000 000
- Overproduction penalty coefficient = 100 000
- Wastage penalty coefficient = 1
- Plant capacity violation penalty coefficient = 1 000 000 000
- Export order multi source penalty coefficient = 10 000 000 000
- Transportation penalty coefficient = 1
- Tournament selection of size 2
- A steady state population is used

Figure 13 illustrates the evolution of the *profit* variable during the lifetime of a single run of the experiment. From time to time, there is a decrease in the profit value; this corresponds to instances where the algorithm is able to find a solution that dramatically reduces constraint violations, but at the expense of less profit being generated. However, based on the penalty values being generated, the solution is still considered to be superior to what was in hand before. However, the algorithm is eventually able to produce highly profitable solutions.

Figure 14 shows the decrease in the summed penalty function values over the lifetime of a single run of the experiment.

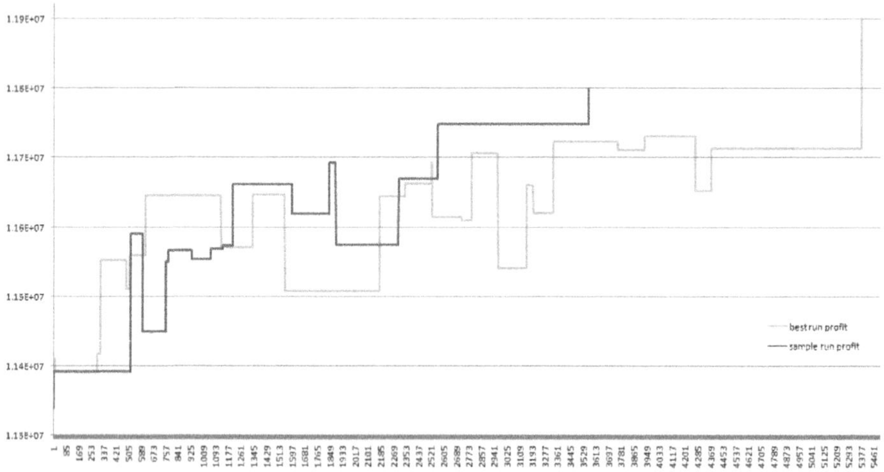

Fig. 13 Sample run. Profit over generations.

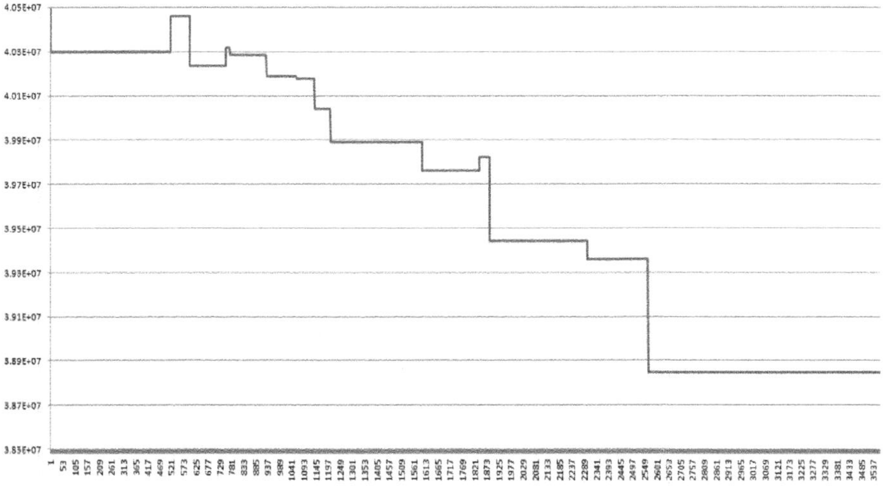

Fig. 14 Sample run. Penalty function over generations.

A 90 run execution shows median 1.18E+07 and 52110.52404 standard deviation of profit.

The assignments generated by our software system produce financial results that are 4% better than those produced manually by the company's experienced team of planners.

4 Discussion

In the previous two sections we described two optimization problems, their challenges, proposed algorithms to solve them, and analyzed their solutions. Though the two problems are somewhat similar in the sense that they both relate to supply chain operations, they have some critical differences from a business perspective. One involves silo-based optimization and the other involves global optimization from a business perspective. Exploring these differences was the main goal of this chapter. It was not our goal to pit the two approaches against each other for the case studies considered, instead we sought to look at these two problems and their solutions by evolutionary algorithm in sufficient detail so that the reader would be able to compare and contrast, at a more-than-theoretical level, the approaches required.

The objective at the glass company was to optimize the operations of a single factory. In this case study, the sequence in which the orders are placed on the machines affects efficiency and hence profitability. A good sequence would not involve much re-configuration and cleaning of the machine between orders. Hence, the goal of the optimizer was to place orders on machines in such a way as to maximize profitability at the factory, by trying to run the machines at maximum capacity so that as many open orders as possible could be satisfied. Besides that, it also aimed to minimize production costs by grouping orders in such a way that raw material changeovers between orders were minimized, and machine tool changes between orders were also minimized.

In the second case study, the objective was to optimize operations over a set of steel mills. In general, for any given order item, there is a choice of one or more factories at which the item can be produced. Therefore, a decision has to be made on how to split the order amongst the factories capable of producing it. The decision as to which one(s) to use, affects the company's profitability. A customer's delivery address may be closer to one factory than it is to another, thus transportation cost is affected. Also, some factories are more efficient than others, so producing the same item at one factory may cost more than the other factories. Thus, one goal of the optimizer, in this case study, was to simply allocate orders to factories in such a way that the cost of production and transportation over all plants was minimized. Another facet of this problem is that before production, each plant has to batch orders, so several ordered items can be produced in one batch. This results in some wastage, and so, another goal of the optimizer was to minimize the wastage over all the plants by batching efficiently.

Comparing these two problems, we can see that in the glass company problem, we are only concerned about optimizing production in one local factory. The scope of the problem does not include any information about availability of raw materials at any given time, nor does the problem deal with the distribution of products to customers, or any logistic issues. (See Figure 15) In the steel company, there are two levels of optimization. First, we try to optimize the production and transportation costs by distributing orders among plants. Second, we also optimize batches at each plant; this optimization is limited to information of orders that have been assigned to that plant and does not take into account any production at other plants, transportation and other costs.

The procedure of packing orders into batches at the steel factory is similar to the first problem at the glass factory; both are unconcerned with optimization at other units, they both optimize in *local* search spaces. After local optimization is complete at every steel plant, information about batched orders is returned to the top level optimization. The top level optimization procedure uses it to help guide optimization at a global level, taking into account the big picture of the whole business system. (See Figure 16) The steel company is an example of **global optimization** from a business perspective, whereas the glass company is an example of **local optimization** from a business perspective.

The experiments performed in this chapter are based on software that was written to solve real-world business problems for actual industrial companies. Values have been transformed to protect private commercial data, but otherwise are representative of actual day-to-day operations at the respective companies. The experiments performed were not for the purpose of demonstrating that one solution approach is better than another, but rather

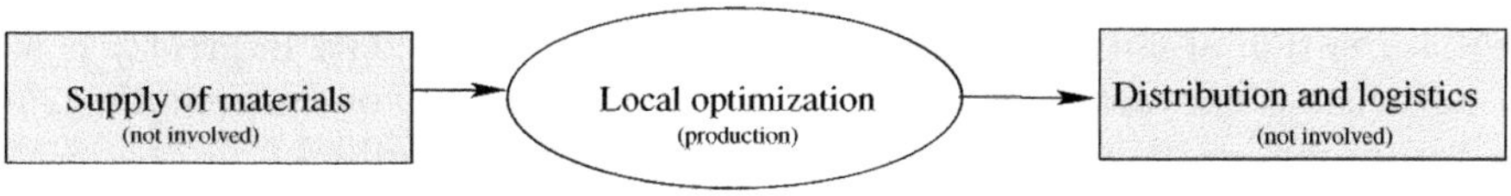

Fig. 15 Local optimization. Production and distribution units are not involved in optimization.

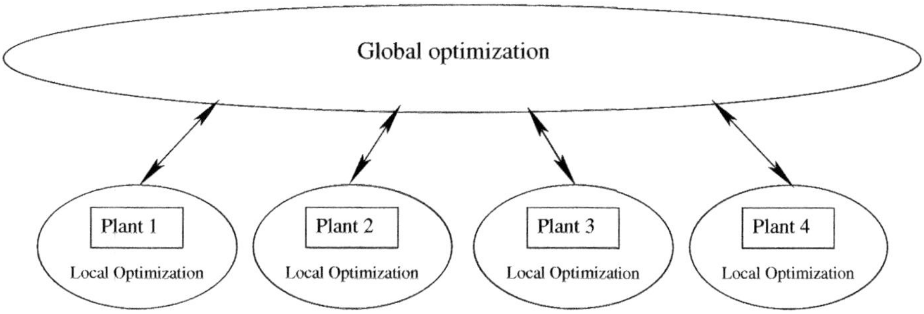

Fig. 16 Global optimization (steel company)

for demonstrating, with appropriate rigor, that our case studies have been solved in stable ways, one using a silo-based optimization approach, and the other using a business-perspective global optimization approach. In the case of optimizing the glass factory, our results indeed show that the changeover penalties are reduced over the course of a run of the algorithm. The fact that the results produced were actually significantly superior to the solution produced by a human being, was actually confirmed by representatives of the company. In the case of optimizing the steel factories, financial results were an integral part of the software's evaluation and show objectively (on numerous data sets) that the results produced by the software were vastly superior to those produced by human planners.

Using these two example problems as a basis, we can try to take a more general point of view, and identify some characteristics of global optimization from a business perspective:

- Two or more levels of optimization. In the steel company problem we have two levels. One is optimization of batches at the plant level, the other one is optimization of the whole system using information from the plants.
- Partial dependence of lower level components on higher level components. In our example, optimization of batches at each plant is only concerned with its own orders, and does not need any external information about freight rates, customers and other plants. It only needs to know about the total tonnages of each product that have been assigned to the plant.
- Limited communication between components. In global optimization problems, different components have limited information exchange due to either business rules or problem specific constraints. In the steel problem,

information about local optimization at each plant is only indirectly communicated between plants by means of the higher level optimizer adjusting total tonnage assignments in response to the fitness values.
- Different search spaces structures. We can think about a global optimization problem as several local optimization problems plus one or more higher level problems that combine the lower level problems. Thus, the biggest difference between local optimization and global optimization is in the search space structure and the sequence of steps needed to evaluate a candidate solution.

Future Improvements
There are several ways in which the global optimization case study can be broadened for future research:

- The steel mill case study could be expanded to consider raw materials' input, as well as secondary processing of steel products that occurs after the rolls of sheet steel leave the factories. It may even be expanded further to include consideration of sales at retail outlets.
- Future work can consider per-factory optimization under the umbrella of global optimization in the business sense. In the case of our sheet steel company, this would mean looking at what happens after the production quantities have been allocated in bulk to the various factories. The exact sequencing of the production could then be optimized locally (i.e. within a single factory) by reducing changeovers between the execution of orders, by reducing the storage of finished products in inventory, by minimizing wastage in the form of small cuts of steel, and by sequencing in such a way that orders are completed before their due dates.
- Finally, instead of only considering per-factory optimization after the higher-level global optimization is complete, it is possible to develop an optimizer that considers fitness values based on both global and per-factory feedback. Such an optimizer could use two-way feedback so that the lower-level optimization affects decisions made at the higher level. This was partially done with the batching procedure, but further work such as sequencing procedures can also be done.

5 Conclusions and Future Work

There exists a large body of research on the application of optimization techniques to the highly-focused aspects of real-world business problems. This research may be suitable for small businesses seeking to optimize their production efficiency and profit. But for bigger businesses, especially ones that are vertically integrated, optimization needs to be looked from a much broader point of view.

Optimization of any of those business' silos would definitely lead to an improvement in their operations in terms of time and money, and separate efforts to optimize multiple silos would also yield benefits. But businesses should be more interested in the optimization of their whole system rather than optimization of single components. A larger opportunity exists for *global optimization from a business perspective.* By modeling the situation in such a way as to have limited communication among the parts of the model representing the various silos, we can achieve a level of optimization that is better than what is attainable by isolating and optimizing individual components.

In this chapter, we considered two real-world business situations. The first looked at an important but isolated aspect of the operation of a glass factory. The second considered a multi-factory sheet steel business, and illustrated how a broader perspective can be taken in such a way as to optimize the operations of all the factories at the same time, as well as some aspects of the operations of each individual factory. These case studies were presented to illustrate some of the concepts associated with "local" and "global" optimization from a business perspective, and not to pit the approaches against each other. In fact, global optimization uses local optimization components as building blocks that optimize each silo.

From an abstract point of view, business-perspective global optimization fits into the framework of traditional global optimization. What distinguishes it as a subclass from traditional business optimization scenarios is that its problems can be formulated as objective functions with more complex-structured search spaces. For example, whereas a traditional problem might have the set of permutation of a finite number of objects as its search space, a business-perspective global optimization problem might have as its search space the vector space of n elements, each of which is taken from the space of permutations of a finite number of objects.

We believe that this abstraction will be useful for studying global optimization problems by allowing the dependence or partial dependence among components to be investigated in a more controlled setting. It would also permit researchers to examine subtle properties of such systems. Our future work will be based on such conceptualizations.

Acknowledgements

We thank the anonymous reviewers for their very insightful comments. This work was partially funded by the ARC Discovery Grant DP0985723 and by grant N516 384734 from the Polish Ministry of Science and Higher Education (MNiSW).

References

1. Altiparmak, F., Gen, M., Lin, L., Paksoy, T.: A genetic algorithm approach for multi-objective optimization of supply chain networks. Computers & Industrial Engineering 51(1), 196–215 (2006); Special Issue on Computational Intelligence and Information Technology: Applications to Industrial Engineering

2. Cheng, R., Gen, M., Tsujimura, Y.: A tutorial survey of job-shop scheduling problems using genetic algorithms—i: representation. Comput. Ind. Eng. 30(4), 983–997 (1996)
3. Coit, D.W., Smith, A.E.: Solving the redundancy allocation problem using a combined neural network/genetic algorithm approach. Computers & Operations Research 23(6), 515–526 (1996)
4. Davis, L.: Job shop scheduling with genetic algorithms. In: Proceedings of the 1st International Conference on Genetic Algorithms, pp. 136–140. Lawrence Erlbaum, Hillsdale (1985)
5. Lee, C.Y., Choi, J.Y.: A genetic algorithm for job sequencing problems with distinct due dates and general early-tardy penalty weights. Computers & Operations Research 22(8), 857–869 (1995)
6. Lee, H., Pinto, J.M., Grossmann, I.E., Park, S.: Mixed-integer linear programming model for refinery short-term scheduling of crude oil unloading with inventory management. Industrial & Engineering Chemistry Research 35(5), 1630–1641 (1996)
7. Levine, J., Ducatelle, F.: Ant colony optimization and local search for bin packing and cutting stock problems. The Journal of the Operational Research Society 55(7), 705–716 (2004)
8. Liang, K.-H., Yao, X., Newton, C., Hoffman, D.: A new evolutionary approach to cutting stock problems with and without contiguity. Computers & Operations Research 29(12), 1641–1659 (2002)
9. Supply & Demand Chain Executive Magazine. Embracing complexity. Toward a 21st century supply chain solution (2008), `http://sdcexec.com/online/printer.jsp?id=9012`
10. Martin, C.H., Dent, D.C., Eckhart, J.C.: Integrated production, distribution, and inventory planning at libbey-owens-ford. Interfaces 23(3), 68–78 (1993)
11. Michalewicz, Z.: Genetic algorithms + data structures = evolution programs, 3rd edn. Springer-Verlag New York, Inc, New York (1996)
12. Michalewicz, Z., Schoenauer, M.: Evolutionary algorithms for constrained parameter optimization problems. Evolutionary Computation 4, 1–32 (1996)
13. Naso, D., Surico, M., Turchiano, B., Kaymak, U.: Genetic algorithms for supply-chain scheduling: A case study in the distribution of ready-mixed concrete. European Journal of Operational Research 177(3), 2069–2099 (2007)
14. Van Laarhoven, P.J.M.: Job shop scheduling by simulated annealing. Operations research 40, 113 (1992)
15. Vergara, F.E., Khouja, M., Michalewicz, Z.: An evolutionary algorithm for optimizing material flow in supply chains. Comput. Ind. Eng. 43(3), 407–421 (2002)
16. Zhou, G., Min, H., Gen, M.: A genetic algorithm approach to the bi-criteria allocation of customers to warehouses. International Journal of Production Economics 86(1), 35–45 (2003)
17. Zielinski, K., Weitkemper, P., Laur, R., Kammeyer, K.-D.: Parameter study for differential evolution using a power allocation problem including interference cancellation. In: IEEE Congress on Evolutionary Computation, 2006. CEC 2006, pp. 1857–1864 (2006)

Solving Real-World Vehicle Routing Problems with Evolutionary Algorithms

Thomas Weise, Alexander Podlich, and Christian Gorldt

Abstract. In this chapter, we present the freight transportation planning component of the *in.west* project. This system uses an Evolutionary Algorithm with intelligent search operations in order to achieve a high utilization of resources and a minimization of the distance travelled by freight carriers in real-world scenarios. We test our planner rigorously with real-world data and obtain substantial improvements when compared to the original freight plans. Additionally, different settings for the Evolutionary Algorithm are studied with further experiments and their utility is verified with statistical tests.

1 Introduction

According to the German Federal Ministry of Economics and Technology [14], the freight traffic volume on German roads will have doubled by 2050 as illustrated in Figure 1. Reasons for this development are the effects of globalization as well as the central location of the country in Europe. With the steadily increasing freight traffic resulting from trade inside the European

Thomas Weise
Distributed Systems Group, University of Kassel, Wilhelmshöher Allee 73, 34121 Kassel, Germany
e-mail: `weise@vs@uni-kassel.de`

Alexander Podlich
Micromata GmbH Kassel, Marie-Calm-Straße 3, 34131 Kassel, Germany
e-mail: `a.podlich@micromata.de`

Christian Gorld
BIBA – Bremer Institut für Produktion und Logistik GmbH, Hochschulring 20, 28359 Bremen, Germany
e-mail: `gor@biba.uni-bremen.de`

R. Chiong, S. Dhakal (Eds.): Nat. Intel. for Sched., Plan. and Pack. Prob., SCI 250, pp. 29–53.
springerlink.com

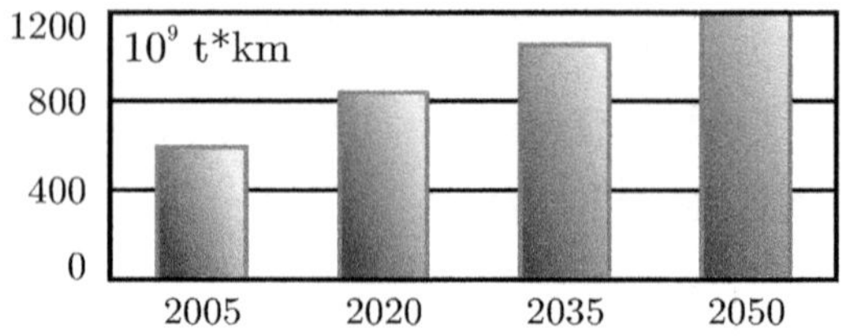

Fig. 1 The freight traffic on German roads in billion tons*kilometer

Union and global import and export [13], transportation and logistics become even more important [7, 45]. Thus, a need for intelligent solutions for the strategic planning of logistics becomes apparent [14]. Such a planning process can be considered as a multi-objective optimization problem which has the goals [49, 54] of increasing the profit of the logistics companies by

1. ensuring on-time collection and delivery of all parcels,
2. utilizing all available means of transportation (rail, trucks) efficiently, i. e., decreasing the total transportation distances by using the capacity of the vehicles to the fullest, while
3. reducing the CO_2 production in order to become more environment-friendly.

Fortunately, the last point is a side-effect of the others. By reducing the total distance covered and by transporting a larger fraction of the freight via (inexpensive) trains, not only the driver's work hours and the costs are decreased, but the CO_2 production also declines.

Efficient freight planning is not a static procedure. Although it involves building an overall plan on how to deliver orders, it should also be able to dynamically react to unforeseen problems such as traffic jams or accidents. This reaction should lead to a local adaptation of the plan and re-routing of all involved freight vehicles whereas parts of the plan concerning geographically distant and uninvolved objects are supposed to stay unchanged.

In the literature, the creation of freight plans is known as the *Vehicle Routing Problem*. In this chapter, we present an approach to Vehicle Routing for real-world scenarios: the freight transportation planning component of the *in.west* system. *in.west*, or "Intelligente Wechselbrücksteuerung" in full, is a joint research project of *DHL*, *Deutsche Post AG*, *Micromata*, *BIBA*, and *OHB Teledata* funded by the German Federal Ministry of Economics and Technology.[1]

In the following section, we discuss different flavors of the Vehicle Routing Problem and the general requirements of the logistics departments which specify the framework for our freight planning component. These specific conditions rendered the related approaches outlined in Section 3 infeasible for our situation. In Section 4, we present an Evolutionary Algorithm for multi-objective, real-world freight planning problems [40]. The problem-specific representation of the solution candidates and the intelligent search operators

[1] See `http://www.inwest.org/` [accessed 2008-10-29].

working on them are introduced, as well as the objective functions derived from the requirements. Our approach has been tested in many different scenarios and the experimental results are summarized in Section 5. The freight transportation planning component described in this chapter is only one part of the holistic *in.west* approach to logistics which will be outlined in Section 6. Finally, we conclude with a discussion of the results and future work in Section 7.

2 Vehicle Routing in Theory and Practice

2.1 Vehicle Routing Problems

The Vehicle Routing Problem (VRP) is one of the most famous combinatorial optimization problems. In simple terms, the goal is to determine a set of routes than can satisfy several geographically scattered customers' demands while minimizing the overall costs [37]. Usually, a fleet of vehicles located in one depot is supposed to fulfill these requests. In this context, the original version of the VRP problem was proposed by Dantzig and Ramser [21] in 1959 who addressed the calculation of a set of optimal routes for a fleet of gasoline delivery trucks.

As described next, a large number of variants of the VRP exist, adding different constraints to the original definition. Within the scope of *in.west*, we first identified all the restrictions of real-world Vehicle Routing Problems that occur in companies like *DHL* and then analyzed available approaches from the literature.

The *Capacitated Vehicle Routing Problem* (CVRP), for example, is similar to the classical VRP with the additional constraint that every vehicle must have the same capacity. A fixed fleet of delivery vehicles must service known customers' demands of a single commodity from a common depot at minimum transit costs [25, 44, 41]. The *Distance Vehicle Routing Problem* (DVRP) is a VRP extended with the additional constraint on the maximum total distance traveled by each vehicle. In addition, *Multiple Depot Vehicle Routing Problems* (MDVRP) have several depots from which customers can be supplied. Therefore, the MDVPR requires the assignment of customers to depots. A fleet of vehicles is based at each depot. Each vehicle then starts at its corresponding depot, services the customers assigned to that depot, and returns.

Typically, the planning period for a classical VRP is a single day. Different from this approach are *Periodic Vehicle Routing Problems* (PVRP), where the planning period is extended to a specific number of days and customers have to be served several times with commodities. In practice, *Vehicle Routing Problems with Backhauls* (VRPB), where customers can return some commodities [41] are very common. Therefore all deliveries for each route must be completed before any pickups are made. Then, it also becomes necessary to take into account that the goods which customers return to the deliverer must fit into the vehicle.

The *Vehicle Routing Problem with Pick-up and Delivering* (VRPPD) is a capacitated VRP where each customer can be supplied with commodities as well as return commodities to the deliverer. Finally the *Vehicle Routing Problem with Time Windows* (VRPTW) is similar to the classical VRP with the additional restriction that time windows (intervals) are defined in which the customers have to be supplied [41]. Figure 2 shows the hierarchy of VRP variants and also the problems which are relevant in the *in.west* case.

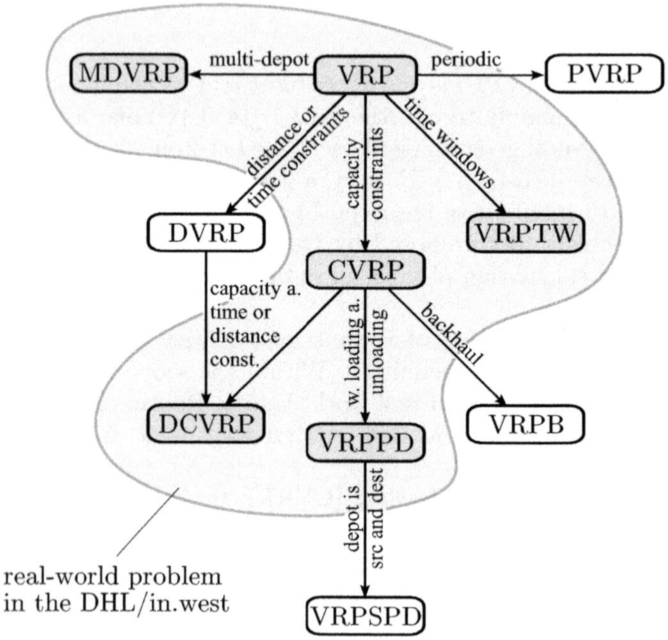

Fig. 2 Different flavors of the VRP and their relation to the *in.west* system

2.2 Model of a Real-World Situation

As it becomes obvious from Figure 2, the situation in logistics companies is relatively complicated and involves many different aspects of Vehicle Routing. The basic unit of freight considered in this work is a swap body b, a standardized container (C 745, EN 284 [15]) with a dimension of roughly $7.5\text{m} \times 2.6\text{m} \times 2.7\text{m}$ and special appliances for easy exchange between transportation vehicles or railway carriages. Logistics companies like *DHL* usually own up to one thousand such containers. We refer to the union of all swap bodies as the set B.

We furthermore define the union of all possible means of transportation as the set F. All trucks $tr \in F$ can carry at most a certain maximum number $\hat{v}(tr)$ of swap bodies at once. Commonly and also in the case of *DHL*, this limit is $\hat{v}(tr) = 2$. The maximum load of trains $z \in F$, on the other hand, is often more variable and usually ranges somewhere between 30 and 60 ($\hat{v}(z) \in [30..60]$). Trains have fixed routes, departure, and arrival times whereas freight trucks can move freely on the map. In many companies, trucks must perform cyclic tours, i.e., return to their point of departure by the end of the day, in order to allow the drivers to return home.

The clients and the depots of the logistics companies together can form more than one thousand locations from which freight may be collected or to which it may be delivered. We will refer to the set of all these locations as L. Each transportation order has a fixed time window $[\check{t_s}, \hat{t_s}]$ in which it must be collected from its source $l_s \in L$. From there, it has to be carried to its destination location $l_d \in L$ where it must arrive within a time window $[\check{t_d}, \hat{t_d}]$. An order furthermore has a volume v which we assume to be an integer multiple of the capacity of a swap body. Hence, a transportation order o can fully be described by the tuple $o = \left\langle l_s, l_d, [\check{t_s}, \hat{t_s}], [\check{t_d}, \hat{t_d}], v \right\rangle$. In our approach, orders which require more than one ($v > 1$) swap body will be split up into multiple orders requiring one swap body ($v = 1$) each.

Logistics companies usually have to service up to a few thousand such orders per day. The *express* unit of the project partner *DHL*, for instance, delivered between 100 and 3000 per day in 2007, depending on the day of the week as well as national holidays etc.

The result of the planning process is a *set* X of tours. Each single tour x is described by a tuple $x = \langle l_s, l_d, f, \check{t}, \hat{t}, \underline{b}, \underline{o} \rangle$. l_s and l_d are the start and destination locations and $\check{t}$ and $\hat{t}$ are the departure and arrival time of the vehicle $f \in F$. On this tour, f carries the set $\underline{b} = \{b_1, b_2, \dots\}$ of swap bodies which, in turn, contain the orders $\underline{o} = \{o_1, o_2, \dots\}$. It is assumed that, for each truck, there is at least one corresponding truck driver and that the same holds for all trains.

Tours are the smallest unit of freight transportation. Usually, multiple tours are combined for a delivery: First, a truck tr may need to drive from the depot in Dortmund to Bochum to pick up an unused swap body sb ($x_1 = \langle \text{Dortmund}, \text{Bochum}, tr, \text{9am}, \text{10am}, \emptyset, \emptyset \rangle$). In a subsequent tour $x_2 = \langle \text{Bochum}, \text{Essen}, tr, \text{10.05am}, \text{11am}, \{sb\}, \emptyset \rangle$, it carries the empty swap body sb to a customer in Essen. There, the order o is loaded into sb and then transported to its destination $o.l_d$ = Hannover ($x_3 = \langle \text{Essen}, \text{Hannover}, tr, \text{11.30am}, \text{4pm}, \{sb\}, \{o\} \rangle$).

Obviously, the set X must be physically sound. It must, for instance, not contain any two intersecting tours x_1, x_2 $\left((x_1.\check{t} < x_2.\hat{t}) \wedge (x_2.\check{t} < x_1.\hat{t}) \right)$ involving the same vehicle ($x_1.f = x_2.f$), swap bodies ($x_1.\underline{b} \cap x_2.\underline{b} \neq \emptyset$), or orders ($x_1.\underline{o} \cap x_2.\underline{o} \neq \emptyset$). Also, it must be ensured that all objects involved in a tour x reside at $x.l_s$ at time $x.\check{t}$. Furthermore, the capacity limits of all involved means of transportation must be respected, i. e., $0 \leq |x.\underline{b}| \leq \hat{v}(x.f)$. If some of the freight is carried by trains, the fixed halting locations of the trains as well as their assigned departure and arrival times must be considered. The same goes for laws restricting the maximum amount of times a truck driver is allowed to drive without breaks and constraints imposed by the company's policies such as the aforementioned cyclic character of truck tours. Only plans for which all these conditions hold can be considered as *correct*.

From the perspective of the planning system's user, runtime constraints are of the same importance: Ideally, the optimization process should not exceed one day. Even the best results become useless if their computation takes longer than the time span from receiving the orders to the day where they actually have to be delivered.

Experience has shown that hiring external carriers for a small fraction of the freight can often reduce the number of required tours to be carried out by the organization's own vehicles and the corresponding total distance to be covered significantly, if the organization's existing capacities are already utilized to their limits. Therefore, a good transportation planning system should also be able to make suggestions on opportunities for such an on-demand outsourcing. An example for this issue is illustrated in Fig. 6(b).

The framework introduced in this section holds for practical scenarios in logistics companies like *DHL* and Deutsche Post. It proposes a hard challenge for research, since it involves multiple intertwined optimization problems and combines several aspects even surpassing the complexity of the most difficult Vehicle Routing Problems known from the literature.

3 Related Work

The approaches discussed in the literature on freight transportation planning can roughly be divided into two basic families: exact and stochastic or metaheuristic methods. The exact approaches are usually only able to solve small instances of VRPs – i. e., those with very limited numbers of orders, customers, or locations – and therefore cannot be applied in most real-world situations. Using them in scenarios with many constraints further complicates the problem [37].

Heuristic methods are reliable and efficient approaches to address VRPs of larger scale. Despite the growing problem dimension, they are still able to provide high quality approximate solutions in a reasonable time. This makes them more attractive than exact methods for practical applications. In over 40 years of research, a large number of heuristics have been proposed for VRPs.

Especially, in recent years the metaheuristic optimization methods have received more and more attention. Well-known members of this family of algorithms which have been applied to Vehicle Routing and freight transportation planning are Tabu Search [26, 3, 6, 10], Simulated Annealing [11, 20], Ant Systems [12, 24], and particularly Evolutionary Algorithms [31, 51, 58, 1, 9].

Ombuki-Berman and Hanshar [35], for example, proposed a Genetic Algorithm (GA) for a Multiple Depot Vehicle Routing Problem. They adopted an indirect and adaptive inter-depot mutation exchange strategy, coupled with capacity and route-length restrictions.

Machado et al. [33] used a basic VRP to compare a standard evolutionary approach with a coevolutionary method. They showed that the inclusion of a heuristic method into evolutionary techniques significantly improves the results. Instead of using additional heuristics, knowledge of the problem domain is incorporated into the search operations in our work.

A cellular and thus decentralized GA for solving the Capacitated Vehicle Routing Problem was presented by Alba and Dorronsoro [1, 2]. This method has a high performance in terms of the quality of the solutions found and the number of function evaluations needed. Decentralization is a good basis for distributing Evolutionary Algorithms, a method for speeding up the evolution which we will consider in our future work.

These methods perform a single-objective optimization enriched with problem-specific constraints. The size of the problems tackled is roughly around a few hundred customers and below 1000 orders. This is the case in most of the test sets available. Examples of such benchmarks are the datasets by Augerat et al. [4], Van Breedam [11], Golden et al. [28], Christofides et al. [18], and Taillard [50] which are publicly available at [43, 23, 36]. Using these (partly artificial) benchmarks in our work was not possible since the framework conditions in *in.west* are very different. Therefore, we could not perform a direct comparison of our system with other approaches mentioned here.

To our knowledge, the problem most similar to the practical situation specified in Section 2.2 is the *Multiple Depot Vehicle Routing Problem with Pickup, Delivery and Intermediary Depots* (MDVRPPDID) defined by Sigurjónsson [48]. This problem, however, does not consider orders and freight containers as different objects. Instead, each container has a source and a target destination and corresponds to one order. Also, all vehicles have the same capacity of one container which is not the case in our system where trucks can usually transport two containers and trains have much higher capacities. The Tabu Search approach developed by Sigurjónsson [48] is similar to our method in that it incorporates domain knowledge in the solution structure and search operations. However, it also allows infeasible intermediate solutions which we rule out in Section 4. It was tested on datasets with up to 16 depots, 40 vehicles, and 100 containers which is more than a magnitude smaller than the problem dimensions the *in.west* system has to deal with.

Confessore et al. [19] define a GA for the Capacitated Vehicle Routing Problem with Time Windows (CVRPTW, see Figure 2) for real-world scenarios with a heterogeneous vehicle fleet with different capacities, multidimensional capacity constraints, order/vehicle, item/vehicle, and item/item compatibility constraints. In *in.west*, the heterogeneity of the vehicles is taken a step further in the sense that trains have totally different characteristics in terms of the degree of freedom regarding the tour times and end points. Furthermore, in *in.west*, orders are not assigned to vehicles but to containers which, in turn, are assigned to trucks and trains.

The general idea of using Evolutionary Algorithms and their hybrids for VRPs has proven to be very efficient [41]. The quality of solutions produced by evolutionary or genetic methods is often higher than that obtained by classic heuristics. Potvin [41] pointed out that Evolutionary Algorithms can also outperform widely used metaheuristcs like Tabu Search on classic problems. He also states that other approaches like Artificial Neural Networks have more or less been abandoned by now in the area of VRPs due to their poor performance on off-the-shelf computer platforms.

In many of the publications listed in this section, it is indicated that metaheuristics work best when a good share of domain knowledge is incorporated. This holds not only for Vehicle Routing, but also in virtually every other application of global optimization [52, 55, 42]. Nevertheless, such knowledge is generally used as an extension, as a method to tweak generic operators and methods. In this work, we have placed problem-specific knowledge at the center of the approach.

4 Evolutionary Approach

4.1 Evolutionary Algorithms

Evolutionary Algorithms (EAs) are a family of nature-inspired optimization algorithms which utilize natural processes such as selection and reproduction in order to refine a set (population) of solution candidates $X \in \mathbb{X}$ from the search space $\mathbb{X}$ iteratively [52, 5]. Their goal is to find the element(s) $X^\star \in \mathbb{X}$ for which the objective function $f : \mathbb{X} \mapsto \mathbb{R}$ takes on the optimal values. Evolutionary Algorithms which work on multiple such functions $F = \{f_1, f_2, .., f_n\}$ are called Multi-objective Evolutionary Algorithms (MOEAs) [16, 17].

All EAs proceed according to the schema depicted in Figure 3. First, an initial population of randomly configured individuals is created. Every iteration then starts with the evaluation of the objective functions on the individuals in the population. Based on their results, a relative fitness is assigned to each solution candidate in the population. These fitness values are the criteria on which selection algorithms operate to pick the most promising individuals for further investigation while discarding the less successful ones. The solution candidates which managed to enter the so called *mating pool* are then reproduced, i. e., combined via crossover or slightly changed by mutation operations. After this is done, the cycle starts again in the next generation.

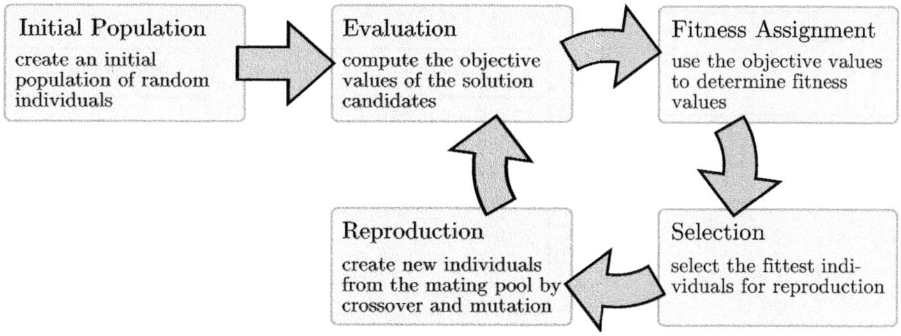

Fig. 3 The basic cycle of Evolutionary Algorithms

4.2 Search Space

When analyzing the problem structure outlined in Section 2.2, it becomes very obvious that standard encodings such as binary [27] or integer strings, matrixes, or real vectors cannot be used in the context of this very general logistics planning task. Although it might be possible to create a genotype-phenotype mapping capable of translating an integer string into a tuple x representing a valid tour, trying to encode a set X of a variable number of such tours in an integer string is not feasible. First, there are many substructures involved in a tour which have variable length such as the sets of orders $\underline{o}$ and swap bodies $\underline{b}$. Second, it would be practically impossible to ensure the required *physical soundness* of the tours given that the reproduction operations would randomly modify the integer strings.

In our work, we adhere to the premise that *all solution candidates must represent correct solutions* according to the specification given in Section 2.2 and *none of the search operations are allowed to violate this correctness.* A solution candidate $X \in \mathbb{X}$ does not necessarily contain a complete plan which manages to deliver all orders. Instead, partial solutions (again as demanded in Section 2.2) are admitted, too.

In order to achieve such a behavior, it is clear that all reproduction operations in the EA must have access to the complete set X of tuples x. Only then, they can check whether the modifications to be applied may impair the correctness of the plans. Therefore, the phenotypes are not encoded at all, but instead, they are the plan objects in their native representations as illustrated in Figure 4.

This figure holds the UML specification of the phenotypes in our planning system. The exactly same data structures are also used by the *in.west* middleware and graphical user interface. The location IDs (`startLocationID`, `endLocationID`) of the `Orders` and `Tours` are indices into a database. They are also used to obtain distances and times of travel between locations from a sparse distance matrix which can be updated asynchronously from different

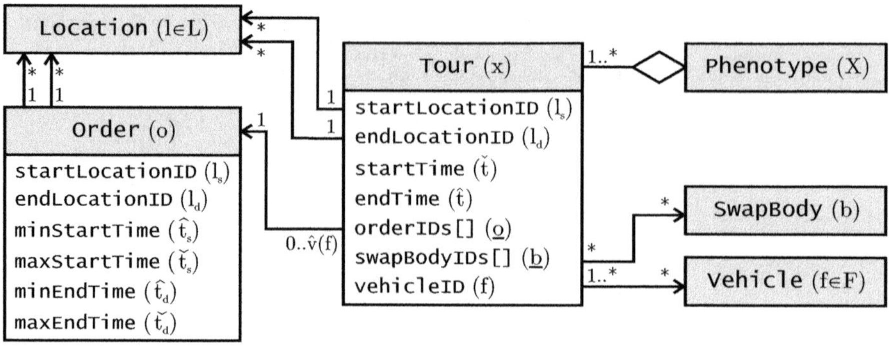

Fig. 4 The structure of the phenotypes X

information sources. The `orderIDs`, `swapBodyIDs`, and `vehicleIDs` are indices into a database as well.

4.3 Search Operations

By using this explicit representation, the search operations have full access to all the information in the freight plans. Standard crossover and mutation operators are, however, no longer applicable. Instead, intelligent operators have to be introduced which respect the correctness of the solution candidates.

For the *in.west* planning system, three crossover and sixteen mutation operations have been defined, each dealing with a specific constellation in the phenotypes and performing one distinct type of modification. During the evolution, individuals to be mutated are processed by a randomly picked operator. If the operator is not applicable because the individual does not belong to the corresponding constellation, another operator is tried. This is repeated, until either the individual is modified or all operators were tested. Two individuals to be combined with crossover are processed by a randomly selected operator as well.

Obviously, we cannot give detailed specifications on all twenty genetic operations [39] (including the initial individual creation) in this chapter. Instead, we will outline the mutation operators sketched in Figure 5 exemplarily.

The first operator Fig. 5(a) is applicable if there is at least one order which would not be delivered if the plan in the input phenotype X was carried out. This operator chooses randomly from all available means of transportation. *Available* in this context means "not involved in another tour for the time between the start and end times of the order". The freight transporters closer to the source of the order are picked with higher probability. Then, a swap body is allocated in the same manner. This process leads to between one and three new tours being added to the phenotype. If the transportation vehicle is a truck, a fourth tour is added which allows it to travel back to its starting

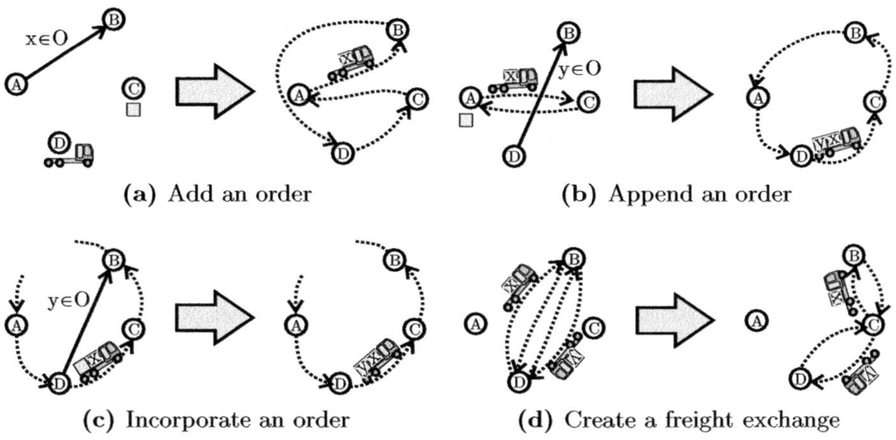

Fig. 5 Some mutation operators from the freight planning EA

point. This step is optional and is applied only if the system is configured to send all trucks back "home" after the end of their routes, as it is the case in *DHL*.

Fig. 5(b) illustrates one operator which tries to include an additional order o into an already existing set of related tours. If the truck driving these tours has space for another swap body, at least one free swap body b is available, and picking up o and b as well as delivering o is possible without violating the time constraints of the other transportation orders already involved in the set of tours, the order is included and the corresponding new tours are added to the plan.

The mutator sketched in Fig. 5(c) does the same if an additional order can be included in already existing tours because of available capacities in swap bodies. Such spare capacities occur from time to time since the containers are not left at the customers' locations after unloading the goods but transported back to the depots. For all operators which add new orders, swap bodies, or tours to the solution candidates, inverse operations which remove these elements are provided, too.

One exceptional operator is the "truck-meets-truck" mechanism. Often, two trucks are carrying out deliveries in opposite directions ($B \to D$ and $D \to B$ in Fig. 5(d)). The operator tries to find a location C which is close to both, B and D. If the time windows of the orders allow it, the two involved trucks can meet at this halting point C and exchange their freight. This way, the total distance that they have to drive can almost be halved from $4 * \overline{BD}$ to $2 * \overline{BC} + 2 * \overline{CD}$ where $\overline{BC} + \overline{CD} \approx \overline{BD}$.

The first recombination operator used in the *in.west* system copies all tours from the first parent and then adds all tours from the second parent in a way that does not lead to a violation of the solution's correctness. In this process, tours which belong together such as those created by the first mutator mentioned are kept together. A second crossover method tries to find sets of tours in the first parent which intersect with similar sets in the second parent and joins them into an offspring plan in the same way the truck-meets-truck mutator combines tours.

4.4 Objective Functions

The freight transportation planning process run by the EA is driven by a set F of three objective functions ($F = \{f_1, f_2, f_3\}$). These functions, all subject to minimization, are based on the requirements stated in Section 1 and are combined via Pareto comparisons [52, 16, 17] in the fitness assignment processes.

4.4.1 f_1: Order Delivery

One of the most important aspects of freight planning is to deliver as many orders as possible. Therefore, the first objective function $f_1(X)$ returns the number of orders which will not be delivered in a timely manner if the plan X was carried out. The optimum of f_1 is zero. Human operators need to hire external carriers for orders which cannot be delivered (due to insufficient resources, for instance).

4.4.2 f_2: Kilometers Driven

By using a sparse distance matrix stored in memory, the second objective function determines the total distance covered by all vehicles involved. Minimizing this distance will lead to less fuel consumption and thus, lower costs and lesser CO_2 production. The global optimum of this function is not known a priori and may not be discovered by the optimization process as well.

4.4.3 f_3: Full Utilization of the Capacities

The third objective function minimizes the spare capacities of the vehicles involved in tours. In other words, it considers the total volume left empty in the swap bodies on the road and the unused swap body slots of the trucks and trains. f_2 does not consider whether trucks are driving tours empty or loaded with empty containers. These aspects are handled by f_3 which again has the optimum zero.

5 Experiments

Because of the special requirements of the *in.west* project and the many constraints imposed on the corresponding optimization problem, the experimental results cannot be directly compared with other works. As we have shown in our discussion of related work in Section 3, none of the approaches in the vehicle routing literature are sufficiently similar to this scenario.

Hence, it was especially important to evaluate our freight planning system rigorously. We have therefore carried out a series of tests according to the full factorial design of experiments paradigm [8, 57]. These experiments (which we will discuss in Section 5.1) are based on a single, real-world set of orders. The results of additional experiments performed with different datasets are outlined in Section 5.2. All data used have been reconstructed from the actual order database of the project partner *DHL*, one of the largest logistics companies worldwide. This database is also the yardstick with which we have measured the utility of our system.

The experiments were conducted using a simplified distance matrix for both the EA and the original plans. Since the original plans did not involve trains, we deactivated the mutation operators which incorporate train tours into solution candidates too – otherwise the results would have been incomparable. Legal aspects like statutory idle periods of the truck drivers have not been considered in the reproduction operators either. However, only plans not violating these constraints were considered in the experimental evaluation.

5.1 *Full Factorial Tests*

EAs have a wide variety of parameters, ranging from the choice of subalgorithms (like those computing a fitness value from the vectors of objective values for each individual) to the mutation rate determining the fraction of the selected solution candidates which are to undergo mutation. The performance of an EA strongly depends on the configuration of these parameters. In different optimization problems, usually different configurations are beneficial and a setting finding optimal solutions in one application may lead to premature convergence to a local optimum in other scenarios. Because of the novelty of the presented approach for transportation planning, performing a large number of experiments with different settings of the EA was necessary in order to find the optimal configuration to be utilized in the *in.west* system in practice.

We, therefore, decided to conduct a full factorial experimental series, i. e., one where all possible combinations of settings of a set of configuration parameters are tested. As a basis for this series, we used a test case consisting of 183 orders reconstructed from one day in December 2007. The original freight plan X_o for these orders contained 159 tours which covered a total distance of $d = f_2(X_o) = 19\,109\,\text{km}$. The capacity of the vehicles involved was filled to 65.5%. The parameters examined in these experiments are listed in Table 1. These settings were varied in the experiments and each of the 192 possible configurations was tested ten times. All runs utilized a tournament selection scheme with five contestants and were granted 10 000 generations. The measurements collected are listed in Table 2.

Table 1 The configurations used in the full-factorial experiments

Param.	Setting and Meaning
ss	In every generation of the EA, new individuals are created by the reproduction operations. The parent individuals in the population are then either discarded (generational, $ss = 0$) or compete with their offspring (steady-state, $ss = 1$).
el	Elitist Evolutionary Algorithms keep an additional archive preserving the best solution candidates found ($el = 1$). Using elitism ensures that these solution candidates cannot be lost due to the randomness of the selection process. Turning off this feature ($el = 0$) may allow the EA to escape local optima easier.
ps	Allowing the EA to work with populations consisting of many individuals increases its chance of finding good solutions but also increases its runtime. Three different population sizes were tested: $ps \in \{200, 500, 1000\}$
fa	Either simple Pareto-Ranking [17] ($fa = 0$) or an extended assignment process ($fa = 1$, called *variety preserving* in [52]) with sharing was applied. Sharing [30, 22] decreases the fitness of individuals which are very similar to others in the population in order to force the EA to explore many different areas in the search space.
cp	The simple convergence prevention (SCP) method proposed in [52] was either used ($cp = 0.3$) or not ($cp = 0$). SCP is a clearing approach [38, 46] applied in the objective space which discards solution candidates with equal objective values with probability *cp*.
mr/*cr*	Different settings for the mutation rate $mr \in \{0.6, 0.8\}$ and the crossover rate $cr \in \{0.2, 0.4\}$ were tested. These rates do not necessarily sum up to 1, since individuals resulting from recombination may undergo mutation as well.

Table 2 The measurements taken during the experiments

Meas.	Meaning
ar	The number of runs which found plans that completely covered all orders.
at	The *median* number of generations needed by these runs until such plans were found.
gr	The number of runs which managed to find such plans which additionally were at least as good as the original freight plans.
gt	The *median* number of generations needed by these runs in order to find such plans.
et	The *median* number of generations after which f_2 did not improve by more than 1%, i. e., the point where the experiments could have been stopped without a significant loss in the quality of the results.
eτ	The *median* number of individual evaluations until this point.
d	The *median* value of f_2, i. e., the median distance covered.

Table 3 The best and the worst evaluation results in the full-factorial tests

#	mr	cr	cp	el	ps	ss	fa	ar	at	gr	gt	et	$e_{\mathcal{T}}$	d
1.	0.8	0.4	0.3	1	1000	1	1	10	341	10	609	3078	3 078 500	15 883 km
2.	0.6	0.2	0.3	0	1000	1	1	10	502	10	770	5746	5 746 500	15 908 km
3.	0.8	0.2	0.3	1	1000	1	1	10	360	10	626	4831	4 831 000	15 929 km
4.	0.6	0.4	0.3	0	1000	1	1	10	468	10	736	5934	5 934 000	15 970 km
5.	0.6	0.2	0.3	1	1000	1	1	10	429	10	713	6236	6 236 500	15 971 km
6.	0.8	0.2	0.3	0	1000	1	1	10	375	10	674	5466	5 466 000	16 003 km
7.	0.8	0.4	0.3	1	1000	1	0	10	370	10	610	5691	5 691 500	16 008 km
8.	0.8	0.2	0.3	0	1000	0	1	10	222	10	450	6186	6 186 500	16 018 km
9.	0.8	0.4	0	0	1000	0	1	10	220	10	463	4880	4 880 000	16 060 km
10.	0.8	0.2	0	1	1000	0	0	10	277	10	506	2862	2 862 500	16 071 km
11.	0.8	0.4	0.3	0	1000	1	0	10	412	10	734	5604	5 604 000	16 085 km
12.	0.8	0.2	0.3	1	1000	0	1	10	214	10	442	4770	4 770 500	16 093 km
13.	0.8	0.2	0.3	1	1000	1	0	10	468	10	673	4970	4 970 500	16 100 km
...	...	...	...	...	...	...	...	...	...	...	...	...	...	...
181.	0.8	0.2	0	0	200	1	0	10	1286	2	6756	6773	1 354 700	20 236 km
182.	0.6	0.2	0	0	500	1	0	10	1546	1	9279	9279	4 639 500	19 529 km
183.	0.8	0.4	0.3	0	200	0	0	10	993	0	∅	∅	∅	19 891 km
184.	0.8	0.4	0	0	200	0	0	10	721	0	∅	∅	∅	20 352 km
185.	0.6	0.2	0	0	200	1	0	10	6094	0	∅	∅	∅	23 709 km
186.	0.6	0.4	0	0	1000	1	0	0	∅	0	∅	∅	∅	∞
187.	0.8	0.4	0	0	1000	1	0	3	6191	0	∅	∅	∅	∞
188.	0.8	0.4	0	0	500	1	0	4	5598	0	∅	∅	∅	∞
189.	0.6	0.4	0	0	200	0	0	3	2847	0	∅	∅	∅	∞
190.	0.6	0.4	0	0	200	1	0	0	∅	0	∅	∅	∅	∞
191.	0.8	0.4	0	0	200	1	0	0	∅	0	∅	∅	∅	∞
192.	0.6	0.4	0	0	500	1	0	0	∅	0	∅	∅	∅	∞

Table 3 contains the thirteen best and the twelve worst configurations, sorted according to **gr**, **d**, and $\mathbf{e}_{\mathcal{T}}$. The best configuration managed to reduce the distance to be covered by over 3000 km (17%) consistently. Even the configuration ranked 170 (not in Table 3) saved almost 1100 km on an average. In total, 172 out of the 192 test series managed to surpass the original plans for the orders in the dataset in all ten runs and only ten configurations were unable to achieve this goal.

The experiments indicate that a *combination* of the highest tested population size ($ps = 1000$), steady-state and elitist population treatment, SCP with rejection probability $cp = 0.3$, a sharing-based fitness assignment process, a mutation rate of 80%, and a crossover rate of 40% is able to produce the best results. We additionally applied significance tests – the sign test [47, 52] and Wilcoxon's signed rank test [47, 56, 52] – in order to check whether there are settings of *single* parameters which generally have positive influence. On a significance level of $\alpha = 0.02$, we considered a tendency only if both (two-tailed) tests agreed. Applying the convergence prevention mechanism (SCP) [52], larger population sizes, variety preserving fitness assignment [52], elitism, and higher mutation and lower crossover rates have significantly positive influence in general.

Interestingly, the steady-state configurations lost in the significance tests against the generational ones, although the seven best-performing settings were steady-state. Here the utility of full factorial tests becomes obvious: steady-state population handling performed very well if (and only if) sharing and the SCP mechanism were applied, too. In the other cases, it led to premature convergence.

This behavior shows the following: transportation planning is a multimodal optimization problem with a probably rugged fitness landscape [55] or with local optima which are many search steps (applications of reproduction operators) apart. Hence, applying steady-state EAs for VRPs similar to the one described here can be beneficial, but only if diversity-preserving fitness assignment or selection algorithms are used in conjunction. Only then, the probability of premature convergence is kept low enough and different local optima and distant areas of the search space are explored sufficiently.

5.2 Tests with Multiple Datasets

We have run experiments with many other order datasets for which the actual freight plans used by the project partners were available. In all scenarios, our approach yielded an improvement which was never below 1%, usually above 5%, and for some days even exceeding 15%. Figure 6 illustrates the best f_2-values (the total kilometers) of the individuals with the most orders satisfied in the population for two typical example evolutions.

In both diagrams, the total distance first increases as the number of orders delivered by the solution candidates rises due to the pressure from f_1. At some point, plans which are able to deliver all orders evolved and f_1 is satisfied (minimized). Now, its corresponding dimension of the objective space begins to collapse, the influence of f_2 intensifies, and the total distances of the plans decrease. Soon afterwards, the efficiency of the original plans is surpassed. Finally, the populations of the EAs converge to a Pareto frontier and no further improvements occur. In Fig. 6(a), this limit was 54 993 km, an improvement of more than 8800 km or 13.8% compared to the original distance of 63 812 km.

Each point in the graph of f_2 in the diagrams represents one point in the Pareto frontier of the corresponding generation. Fig. 6(b) illustrates one additional graph for f_2: the best plans which can be created when at most 1% of the orders are outsourced. Compared to the transportation plan, including assignments for all orders which had a length of 79 464 km, these plans could reduce the distance to 74 436 km, i. e., another 7% of the overall distance could be saved. Thus, in this case, an overall reduction of around 7575 km is achieved in comparison to the original plan, which had a length of 82 013 km.

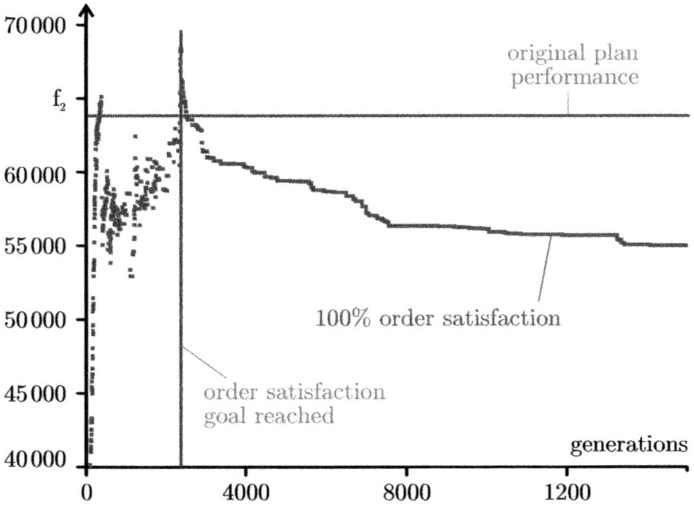

(a) For 642 orders (14% better)

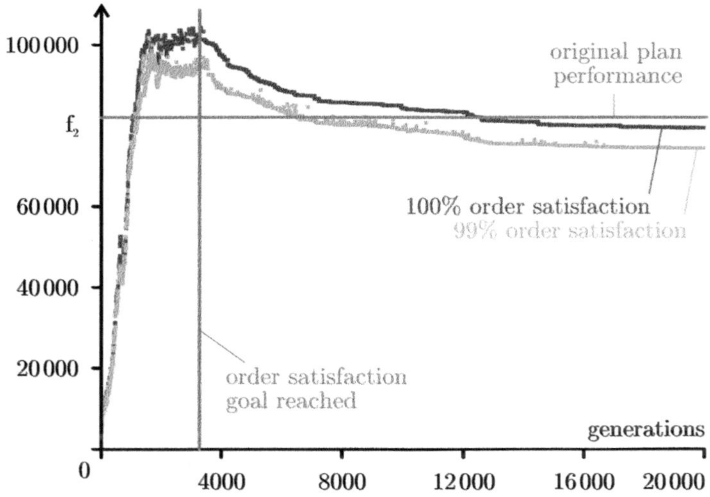

(b) For 1016 orders (3/10% better)

Fig. 6 Two examples for the freight plan evolution

5.3 Time Consumption

One run of the algorithm (prototypically implemented in Java) for the dataset used in the full factorial tests (Section 5.1) took around three hours. For sets with around 1000 orders it still easily fulfills the requirement of delivering the result within 24 hours. Runs with close to 2000 orders, however, take longer than one week (in our experiments, we used larger population sizes for them). Here, it should be pointed out that these measurements were taken on a single dual-core 2.6 GHz machine, which is only a fraction of the capacity available in the dedicated data centers of the project partners. It is well known that EAs can be efficiently parallelized and distributed over clusters [32, 53, 52]. The final implementation of the *in.west* system will incorporate distribution mechanisms and thus be able to deliver results in time for all situations.

6 Holistic Approach to Logistics

Logistics involve many aspects and the planning of efficient routes is only one of them. Customers and legislation [7], for instance, require traceability of the production data and therefore the goods on the road as well. Experts require more linkage between the traffic carriers and efficient distribution of traffic to cargo rail and water transportation.

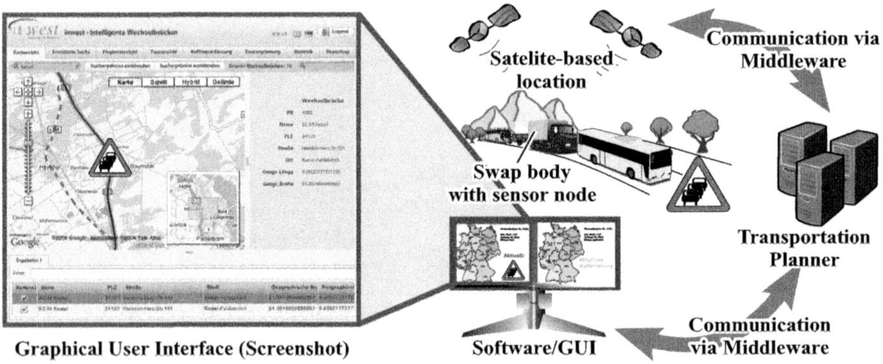

Fig. 7 An overview of the *in.west* system

Technologies like telematics are regarded as the enablers of smart logistics which optimize the physical value chain. Only a holistic approach combining intelligent planning and such technologies can solve the challenges [49, 13] faced by logistics companies. Therefore, the focus of the *in.west* project was to combine software, telematics, and business optimization approaches into one flexible and adaptive system sketched in Figure 7.

Data from telematics become input to the optimization system which suggests transportation plans to the human operator who selects or modifies these suggestions. The final plans then, in turn, are used to configure the telematic units. A combination of both allows the customers to track their deliveries and the operator to react to unforeseen situations. In such situations, traffic jams or accidents, for instance, the optimization component can again be used to make ad-hoc suggestions for resolutions.

6.1 The Project *in.west*

The prototype introduced in this chapter was provided in the context of the BMWi promoted project *in.west* and will be evaluated in field test in the third quarter of 2009. The analyses take place in the area of freight transportation in the business segment *courier-*, *express-* and *parcel services* of *DHL*, the market leader in this business. Today, the transport volume of this company constitutes a substantial portion of the traffic volume of the traffic carriers - road, ship, and rail. Hence, a significant decrease in the freight traffic caused by *DHL* might lead to a noticeable reduction in the freight traffic volume on German roads.

The goal of this project is to achieve this reduction by utilizing information and communication technologies on swap bodies, new approaches of planning, and novel control processes. The main objective is to design a decision support tool to assist the operator with suggestions for traffic reduction and a smart swap body telematic unit.

The requirements for the *in.west* software are various, since the needs of both the operators and the customers are to be satisfied. From their perspective, for example, a constant documentation of the transportation processes is necessary. The operators require that this documentation starts with the selection, movement, and employment of the swap bodies. From the view of the customer, only tracking and tracing of the containers on the road must be supported. With the web-based user interfaces we provide, a spontaneous check of the load condition, the status of the container, and information about the carrier is furthermore possible.

6.2 Smart Information and Communication Technology

All the information required by customers and operators on the status of the freight have to be obtained by the software system first. Therefore, *in.west* also features a hardware development project with the goal of designing a telematic unit. A device called *YellowBox* (illustrated in Figure 8) was developed which enables swap bodies to transmit real time geo positioning data of the containers to a software system. The basic functions of the device are location, communication, and identification. The data received from the *YellowBox* is processed by the software for planning and controlling the swap body in the logistic network.

The *YellowBox* consists of a main board, a location unit (GPS), a communication unit (GSM/GPRS) and a processor unit. Digital input and output interfaces ensure the scalability of the device, e.g., for load volume monitoring. Swap bodies do not offer reliable power sources for technical equipment like telematic systems. Thus, the *YellowBox* has been designed as a sensor node which uses battery current [40].

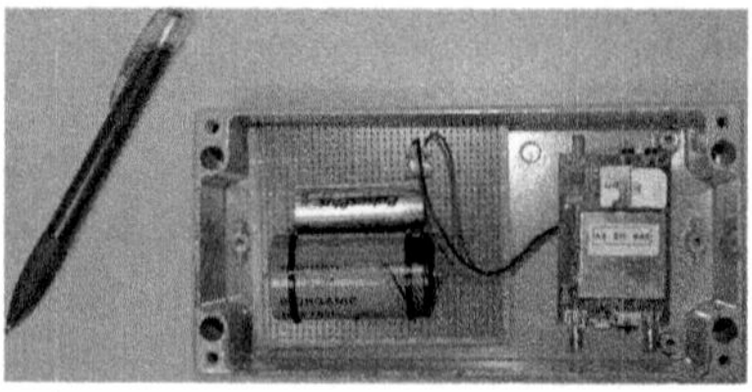

Fig. 8 The *YellowBox* – a mobile sensor node

One of the most crucial criteria for the application of these sensor nodes in practice was long battery duration and thus, low power consumption. The *YellowBox* is therefore turned on and off for certain time intervals. The software system automatically configures the device with the route along which it will be transported (and which has been evolved by the planner). In predefined time intervals, it can thus check whether or not it is "on the right track". Only if location deviations above a certain threshold are detected, it will notify the middleware. Also, if more than one swap body is to be transported by the same vehicle, only one of them needs to perform this notification. With these approaches, communication – one of the functions with the highest energy consumption – is effectively minimized.

7 Conclusions

In this chapter, we presented the *in.west* freight planning component which utilizes an Evolutionary Algorithm with intelligent reproduction operations for general transportation planning problems. The approach was tested rigorously on real-world data from the *in.west* partners and achieved excellent results. It has been integrated as a constituting part of the holistic *in.west* logistics software system.

We presented this work at the EvoTRANSLOG'09 workshop in Tübingen [54]. One point of the very fruitful discussion there was the question why we did not utilize heuristics to create some initial solutions for the EA. We intentionally left this for our future work for two reasons: First, we fear that creating such initial solutions may lead to a decrease of diversity in the population. In Section 5.1 we showed that diversity is a key to finding good solutions to this class of problem. Second, as can be seen in the diagrams provided in Figure 6, finding initial solutions where all orders are assigned to routes is not the time consuming part of the EA – optimizing them to plans with a low total distance is. Hence, incorporating measures for distribution and efficient parallelization may be a more promising addition to our approach. If a cluster of, for instance, twenty computers is available, we can assume that distribution according to client-server or island model schemes [34, 29, 53, 52] will allow us to decrease the runtime to at least one tenth of the current value. Nevertheless, testing the utility of heuristics for creating the initial population is on our agenda, too.

In the current phase, some of the functions of the component still work on a rather prototypical level. They will be updated in order to make the system ready for the field test in Fall 2009. We will therefore improve the support for parallelization and integrate components for distributing the computational load. As already pointed out, this is likely the best way to resolve the remaining timing issues for very large datasets. Additionally, features like online update of the distance matrix which is used both to compute f_2 and also to determe the time a truck needs to travel from one location to another, are planned.

The system will then be capable to *a)* perform planning for the whole orders of one day in advance, and *b)* update smaller portions of the plans online if traffic jams occur. It should be noted that even with such a system, the human operator cannot be replaced. There are always constraints and pieces of information which cannot be employed in even the most advanced automated optimization process. Hence, the solutions generated by our transportation planner are *suggestions* rather than doctrines. They are displayed to the operator and she may modify them according to her needs.

Looking forward to deploying this new system in the computer centers of the project partners, we are confident that *in.west* will fulfill their expectations.

Acknowledgements

The research work presented here has been sponsored by the German Federal Ministry of Economics and Technology. Also, we wish to thank the editors for their helpful comments and careful proofreading.

References

1. Alba, E., Dorronsoro, B.: Solving the vehicle routing problem by using cellular genetic algorithms. In: Gottlieb, J., Raidl, G.R. (eds.) EvoCOP 2004. LNCS, vol. 3004, pp. 11–20. Springer, Heidelberg (2004)
2. Alba, E., Dorronsoro, B.: Computing nine new best-so-far solutions for capacitated vrp with a cellular genetic algorithm. Information Processing Letters 98, 225–230 (2006)
3. Amberg, A., Domschke, W., Voß, S.: Multiple center capacitated arc routing problems: A tabu search algorithm using capacitated trees. European Journal of Operational Research (EJOR) 124(2), 360–376 (2000)
4. Augerat, P., Belenguer, J.M., Benavent, E., Corberán, A., Naddef, D., Rinaldi, G.: Computational results with a branch and cut code for the capacitated vehicle routing problem. Research Report 949-M, Universite Joseph Fourier, Grenoble, France (1995)
5. Bäck, T.: Evolutionary Algorithms in Theory and Practice: Evolution Strategies, Evolutionary Programming, Genetic Algorithms. Oxford University Press, Oxford (1996)

6. Badeau, P., Gendreau, M., Guertin, F., Potvin, J.-Y., Taillard, É.D.: A parallel tabu search heuristic for the vehicle routing problem with time windows. Transportation Research Part C: Emerging Technologies 5(2), 109–122 (1997)
7. van Betteray, K.: Gesetzliche und handelsspezifische anforderungen an die rückverfolgung. In: Vorträge des 7. VDEB-Infotags 2004, VDEB Verband IT-Mittelstand e.V, EU Verordnung 178/2002 (2004)
8. Box, G.E.P., Hunter, J.S., Hunter, W.G.: Statistics for Experimenters: Design, Innovation, and Discovery. John Wiley & Sons, Chichester (2005)
9. Bräysy, O.: Genetic algorithms for the vehicle routing problem with time windows. Arpakannus – Newsletter of the Finnish Artificial Intelligence Society (FAIS) 1, 33–38 (2001); Special issue on Bioinformatics and Genetic Algorithms
10. Bräysy, O., Gendreau, M.: Tabu search heuristics for the vehicle routing problem with time windows. TOP: An Official Journal of the Spanish Society of Statistics and Operations Research 10(2), 211–237 (2002)
11. Breedam, A.V.: An analysis of the behavior of heuristics for the vehicle routing problem for a selection of problems with vehicle-related, customer-related, and time-related constraints. Ph.D. thesis, University of Antwerp, RUCA, Belgium (1994)
12. Bullnheimer, B., Hartl, R.F., Strauss, C.: An improved ant system algorithm for the vehicle routing problem. Annals of Operations Research 89, 319–328 (1999)
13. Bundesministerium für Verkehr, Bau- und Stadtentwicklung: Verkehr in Zahlen 2006/2007. Deutscher Verkehrs-Verlag GmbH, Hamburg (2006)
14. Bundesministerium für Wirtschaft und Technologie: Mobilität und Verkehrstechnologien das 3. Verkehrsforschungsprogramm der Bundesregierung. BMWi, Öffentlichkeitsarbeit, Berlin, Germany (2008)
15. CEN/TC 119: Swap bodies – non-stackable swap bodies of class C – dimensions and general requirements. EN 284, CEN-CEN ELEC, Brussels, Belgium (2006)
16. Ceollo Coello, C.A.: A short tutorial on evolutionary multiobjective optimization. In: Zitzler, E., Deb, K., Thiele, L., Coello Coello, C.A., Corne, D.W. (eds.) EMO 2001. LNCS, vol. 1993, pp. 21–40. Springer, Heidelberg (2001)
17. Ceollo Coello, C.A., Lamont, G.B., van Veldhuizen, D.A.: Evolutionary Algorithms for Solving Multi-Objective Problems. In: Genetic and Evolutionary Computation, 2nd edn. (1st edn: 2002), vol. 5. Kluwer Academic Publishers, Springer (2007) doi:10.1007/978-0-387-36797-2
18. Christofides, N., Mingozzi, A., Toth, P.: The vehicle routing problem. In: Christofides, N., Mingozzi, A., Toth, P., Sandi, C. (eds.) Combinatorial Optimization, ch. 11, pp. 315–338. John Wiley & Sons, Chichester (1979)
19. Confessore, G., Galiano, G., Stecca, G.: An evolutionary algorithm for vehicle routing problem with real life constraints. In: Mitsuishi, M., Ueda, K., Kimura, F. (eds.) Manufacturing Systems and Technologies for the New Frontier – The 41st CIRP Conference on Manufacturing Systems, pp. 225–228. Springer, Heidelberg (2008)
20. Czech, Z.J., Czarnas, P.: Parallel simulated annealing for the vehicle routing problem with time windows. In: 10th Euromicro Workshop on Parallel, Distributed and Network-based Processing (PDP 2002), pp. 376–383. IEEE Computer Society, Los Alamitos (2002)
21. Dantzig, G.B., Ramser, J.H.: The truck dispatching problem. Management Science 6(1), 80–91 (1959)

22. Deb, K., Goldberg, D.E.: An investigation of niche and species formation in genetic function optimization. In: Schaffer, J.D. (ed.) Proceedings of the 3rd International Conference on Genetic Algorithms, pp. 42–50. Morgan Kaufmann Publishers Inc., San Francisco (1989)
23. Díaz, B.D.: Known best results (2007), `http://neo.lcc.uma.es/radi-aeb/WebVRP/results/BestResults.htm` (accessed 2007-12-28)
24. Doerner, K., Gronalt, M., Hartl, R.F., Reimann, M., Strauss, C., Stummer, M.: Savings ants for the vehicle routing problem. In: Cagnoni, S., Gottlieb, J., Hart, E., Middendorf, M., Raidl, G.R. (eds.) EvoIASP 2002, EvoWorkshops 2002, EvoSTIM 2002, EvoCOP 2002, and EvoPlan 2002. LNCS, vol. 2279, pp. 11–20. Springer, Heidelberg (2002)
25. Domschke, W.: Logistik, Rundreisen und Touren, fourth edn. Oldenbourgs Lehr- und Handbücher der Wirtschafts- u. Sozialwissenschaften. Oldenbourg Verlag (1997)
26. Glover, F.: Future paths for integer programming and links to artificial intelligence. Computers & Operations Research 13(5), 533–549 (1986)
27. Goldberg, D.E.: Genetic Algorithms in Search, Optimization, and Machine Learning. Addison-Wesley Longman Publishing Co., Inc., Boston (1989)
28. Golden, B., Wasil, E., Kelly, J., Chao, I.-M.: The impact of metaheuristics on solving the vehicle routing problem: Algorithms, problem sets, and computational results. In: Crainic, T.G., Laporte, G. (eds.) Teodor Gabriel Crainic and Gilbert Laporte, ch. 2. Center for Research on Transportation 25th Anniversary Series, 1971–1996, pp. 33–56. Kluwer/Springer, Boston/USA (1998)
29. Gorges-Schleuter, M.: Explicit parallelism of genetic algorithms through population structures. In: Schwefel, H.-P., Männer, R. (eds.) PPSN 1990. LNCS, vol. 496, pp. 150–159. Springer, Heidelberg (1991)
30. Holland, J.H.: Adaptation in Natural and Artificial Systems: An Introductory Analysis with Applications to Biology, Control, and Artificial Intelligence. The University of Michigan Press, Ann Arbor (1975); Reprinted by MIT Press, NetLibrary, Inc. (April 1992)
31. Jih, W., Hsu, J.Y.: Dynamic vehicle routing using hybrid genetic algorithms. In: IEEE International Conference on Robotics and Automation, pp. 453–458 (1999) doi: 10.1109/ROBOT.1999.770019
32. Luke, S., Panait, L., Balan, G., Paus, S., Skolicki, Z., Bassett, J., Hubley, R., Chircop, A.: Ecj: A java-based evolutionary computation research system (2006); Version 18, `http://cs.gmu.edu/~eclab/projects/ecj/` (accessed 2007-07-10)
33. Machado, P., Tavares, J., Pereira, F.B., Costa, E.J.F.: Vehicle routing problem: Doing it the evolutionary way. In: Langdon, W.B., Cantú-Paz, E., Mathias, K.E., Roy, R., Davis, D., Poli, R., Balakrishnan, K., Honavar, V., Rudolph, G., Wegener, J., Bull, L., Potter, M.A., Schultz, A.C., Miller, J.F., Burke, E.K., Jonoska, N. (eds.) GECCO 2002: Proceedings of the Genetic and Evolutionary Computation Conference, p. 690. Morgan Kaufmann Publishers Inc., San Francisco (2002)
34. Martin, W.N., Lienig, J., Cohoon, J.P.: Island (migration) models: Evolutionary algorithms based on punctuated equilibria. In: Bäck, T., Fogel, D.B., Michalewicz, Z. (eds.) Handbook of Evolutionary Computation, Computational Intelligence Library, ch. 6.3. Oxford University Press, Oxford (1997)

35. Ombuki-Berman, B.M., Hanshar, F.: Using genetic algorithms for multi-depot vehicle routing. In: Bio-inspired Algorithms for the Vehicle Routing Problem, pp. 77–99. Springer, Heidelberg (2009)
36. Pankratz, G., Krypczyk, V.: Benchmark data sets for dynamic vehicle routing problems (2007), `http://www.fernuni-hagen.de/WINF/inhfrm/benchmark_data.htm` (accessed 2008-10-27)
37. Pereira, F.B., Tavares, J. (eds.): Bio-inspired Algorithms for the Vehicle Routing Problem. SCI, vol. 161. Springer, Heidelberg (2009)
38. Pétrowski, A.: A clearing procedure as a niching method for genetic algorithms. In: Proceedings of IEEE International Conference on Evolutionary Computation, CEC 1996, pp. 798–803. IEEE Computer Society Press, Piscataway (1996)
39. Podlich, A.: Intelligente planung und optimierung des güterverkehrs auf straße und schiene mit evolutionären algorithmen. Master's thesis, University of Kassel, FB-16, Distributed Systems Group, Wilhelmshöher Allee 73, 34121 Kassel, Germany (2009)
40. Podlich, A., Weise, T., Menze, M., Gorldt, C.: Intelligente wechselbrückensteuerung für die logistik von morgen. In: Wagner, M., Hogrefe, D., Geihs, K., David, K. (eds.) First Workshop on Global Sensor Networks, GSN 2009 (2009); Electronic Communications of the EASST (ECASST), vol. 17, part Global Sensor Networks (GSN 2009), The European Association of Software Science and Technology (2009) ISSN 1863-2122
41. Potvin, J.-Y.: A review of bio-inspired algorithms for vehicle routing. In: Bio-inspired Algorithms for the Vehicle Routing Problem, pp. 1–34. Springer, Heidelberg (2009)
42. Radcliffe, N.J.: The algebra of genetic algorithms. Annals of Mathematics and Artificial Intelligence 10(4), 339–384 (1994)
43. Ralphs, T.: Vehicle routing data sets (2003), `http://www.coin-or.org/SYMPHONY/branchandcut/VRP/data/` (accessed 2009-04-08)
44. Ralphs, T.K., Kopman, L., Pulleyblank, W.R., Trotter, L.E.: On the capacitated vehicle routing problem. Mathematical Programming 94(2–3), 343–359 (2003)
45. von Randow, M.: Güterverkehr und logistik als tragende säule der wirtschaft zukunftssicher gestalten. In: Baumgarten, H. (ed.) Das Beste Der Logistik: Innovationen, Strategien, Umsetzungen. Bundesvereinigung Logistik (BVL), pp. 49–53. Springer, Heidelberg (2008)
46. Sareni, B., Krähenbühl, L.: Fitness sharing and niching methods revisited. IEEE Transactions on Evolutionary Computation 2(3), 97–106 (1998)
47. Siegel, S., Castellan Jr., N.J.: Nonparametric Statistics for The Behavioral Sciences. Humanities/Social Sciences/Languages. McGraw-Hill, New York (1988)
48. Sigurjónsson, K.: Taboo search based metaheuristic for solving multiple depot vrppd with intermediary depots. Master's thesis, Informatics and Mathematical Modelling, IMM, Technical University of Denmark, DTU (2008), `http://orbit.dtu.dk/getResource?recordId=224453&objectId=1&versionId=1` (accessed 2009-04-09)
49. Steierwald, G., Künne, H.D., Vogt, W.: Stadtverkehrsplanung: Grundlagen, Methoden, Ziele, 2., neu bearbeitete und erweiterte auflage edn. Springer, Berlin (2005)
50. Taillard, É.D.: Parallel iterative search methods for vehicle routing problems. Networks 23(8), 661–673 (1993)

51. Thangiah, S.R.: Vehicle routing with time windows using genetic algorithms. In: Practical Handbook of Genetic Algorithms: New Frontiers, pp. 253–277. CRC, Boca Raton (1995)
52. Weise, T.: Global Optimization Algorithms – Theory and Application, 2nd edn (2009), `http://www.it-weise.de/` (accessed 2009-07-14)
53. Weise, T., Geihs, K.: DGPF – An Adaptable Framework for Distributed Multi-Objective Search Algorithms Applied to the Genetic Programming of Sensor Networks. In: Filipič, B., Šilc, J. (eds.) Proceedings of the Second International Conference on Bioinspired Optimization Methods and their Applications (BIOMA 2006), pp. 157–166. Jožef Stefan Institute (2006)
54. Weise, T., Podlich, A., Reinhard, K., Gorldt, C., Geihs, K.: Evolutionary freight transportation planning. In: Giacobini, M., et al. (eds.) EvoWorkshops 2009. LNCS, vol. 5484, pp. 768–777. Springer, Heidelberg (2009)
55. Weise, T., Zapf, M., Chiong, R., Nebro, A.J.: Why is optimization difficult? In: Chiong, R. (ed.) Nature-Inspired Algorithms for Optimisation, ch. 1. SCI, vol. 193, pp. 1–50. Springer, Heidelberg (2009)
56. Wilcoxon, F.: Individual comparisons by ranking methods. Biometrics Bulletin 1(6), 80–83 (1945)
57. Yates, F.: The Design and Analysis of Factorial Experiments. Imperial Bureau of Soil Science, Commonwealth Agricultural Bureaux (1937); Tech. Comm. No. 35
58. Zhu, K.Q.: A diversity-controlling adaptive genetic algorithm for the vehicle routing problem with time windows. In: 15th IEEE International Conference on Tools with Artificial Intelligence, pp. 176–183. IEEE Computer Society Press, Los Alamitos (2003)

A Genetic Algorithm with Priority Rules for Solving Job-Shop Scheduling Problems

S.M. Kamrul Hasan, Ruhul Sarker, Daryl Essam, and David Cornforth

Abstract. The Job-Shop Scheduling Problem (JSSP) is one of the most difficult NP-hard combinatorial optimization problems. In this chapter, we consider JSSPs with an objective of minimizing makespan while satisfying a number of hard constraints. First, we develop a genetic algorithm (GA) based approach for solving JSSPs. We then introduce a number of priority rules to improve the performance of GA, such as partial re-ordering, gap reduction, and restricted swapping. The addition of these rules results in a new hybrid GA algorithm that is clearly superior to other well-known algorithms appearing in the literature. Results show that this new algorithm obtained optimal solutions for 27 out of 40 benchmark problems. It thus makes a significantly new contribution to the research into solving JSSPs.

1 Introduction

In this chapter, we present a new hybrid algorithm for solving the Job-Shop Scheduling Problems (JSSPs) that is demonstrably superior to other well-known algorithms. The JSSP is a common problem in the manufacturing industry. A classical JSSP involves a combination of N jobs and M machines. Each job consists of a set of operations that has to be processed on a set of known machines, where each operation has a known processing time. A schedule is a complete set of operations required by a job, to be performed on different machines in a given order. In addition, the process may need to satisfy other constraints. The total time between the start of the first

S.M. Kamrul Hasan and Ruhul Sarker and Daryl Essam
School of IT&EE, University of New South Wales at the Australian Defence Force Academy, Northcott Drive, Canberra, ACT 2600, Australia
e-mail: {kamrul, r.sarker, d.essam}@adfa.edu.au

David Cornforth
Division of Energy Technology, Commonwealth Scientific and Industrial Research Organization, Murray Dwyer Circuit, Mayfield West, NSW 2304, Australia
e-mail: david.cornforth@csiro.au

R. Chiong, S. Dhakal (Eds.): Nat. Intel. for Sched., Plan. and Pack. Prob., SCI 250, pp. 55–88.
springerlink.com

operation and the end of the last operation, is termed as the *makespan*. Makespan minimisation is widely used as an objective in solving JSSPs [2, 8, 16, 37, 47, 50, 51]. A feasible schedule contains no conflicts such as (i) no more than one operation of any job can be executed simultaneously and (ii) no machine can process more than one operation at the same time. The schedules are generated on the basis of a predefined sequence of machines and the given order of job operations.

The JSSPs are widely acknowledged as one of the most difficult NP-complete problems [22, 23, 35]. They are also well-known for their practical applications in many manufacturing industries. Over the last few decades, a good number of algorithms have been developed to solve JSSPs. However, no single algorithm can solve all kinds of JSSPs optimally (or near optimally) within a reasonable time limit. Thus, there is scope to analyze the difficulties of JSSPs as well as to design improved algorithms that may be able to solve them.

In this work, we start by examining the performance of the traditional GA (TGA) for solving JSSPs. Each individual represents a particular schedule and the individuals are represented by binary chromosomes. After reproduction, any infeasible individual is repaired to make it feasible. The phenotype representation of the problem is a matrix of $N \times M$ numbers where each row represents the sequence of jobs for a given machine. We apply both genotype and phenotype representations to analyze the schedules. The binary genotype is effective for the simple crossover and mutation techniques. After analyzing the traditional GA solutions, we realize that the solutions could be improved further by applying simple rules or local searches. So, we introduce three new priority rules to improve the performance of traditional GA, namely: partial reordering (PR), gap reduction (GR), and restricted swapping (RS). These priority rules are integrated as a component of TGA. The actions of these rules will be accepted if and only if they improve the solution. The details of these priority rules are discussed in a section later in this chapter. We also implement our GA incorporating different combinations of these priority rules. For ease of explanation, in this chapter, we designate these as PR with GA, GR with GA and GR plus RS with GA, as PR-GA, GR-GA, and GR-RS-GA respectively. To test the performance of our proposed algorithms, we solve 40 of the benchmark problems originally presented in Lawrence [33]. The proposed priority rules improved the performance of traditional GAs for solving JSSPs. Among the GAs with priority rules, GR-RS-GA was the best performing algorithm. It obtained optimal solutions for 27 out of 40 test problems. The overall performance of GR-RS-GA is better than many key JSSP algorithms that appear in the literature. The current version of our algorithm is much refined than our earlier version. The initial and intermediate developments of the algorithm, with limited experimentation, can be found in Hasan *et al.* [28, 29, 30].

The rest of this chapter is organized as follows. After the introduction, a brief review of approaches for solving JSSPs is provided. Section 3 defines the JSSP considered in this research. Section 4 discusses traditional GA approaches for solving JSSPs, including chromosome representations used for JSSPs, and ways of handling infeasibility in JSSPs. Section 5 introduces new priority rules for improving the performance of traditional GA. Section 6 presents the proposed algorithms and implementation aspects. Section 7 shows the experimental results and the necessary statistical analysis used to measure the performance of the algorithms. Finally, the conclusions and comments on future research are presented.

2 Solving JSSP: A Brief Review

Scheduling is a very old and widely accepted combinatorial optimization problem. Conway [14] in late 60s, Baker [5] in mid 70s, and French [21] in early 80s. showed many different ways of solving various scheduling problems, which were frequently used in later periods. Job-shop scheduling is one of those challenging optimization problems. A JSSP consists of N jobs, where each job is composed of a finite number of operations. There are M machines, where each machine is capable of executing a set of operations O_k, where k is the machine index. The size of the solution space for such a problem is $(n_1)!(n_2)!(n_3)...(n_k)...(n_{M-1})!(n_M)$, where n_k is the number of operations executable by machine k. For equal number of operations in each machine, this is equal to $(N!)^M$. Of course, many solutions are infeasible, and more than one optimal solutions may exist. As the number of alternative solutions grows at a much faster rate than the number of jobs and the number of machines, it is infeasible to evaluate all solutions (i.e., complete enumeration) even for a reasonably sized practical JSSP. The feasible solutions can be classified as semi-active, active, and non-delay schedules [44]. The set of non-delay schedules is a complete subset of the set of active schedules, where the active set itself is a complete subset of the set of semi-active schedules [27]. In the semi-active schedules, no operation can be locally shifted to the left, where in the active schedules, no left shift is possible either locally or globally. These two kinds of schedules may contain machine delay. Solutions having zero machine delay time are termed as the non-delay schedules. In our algorithms, we force the solutions to be in the non-delay scheduling region.

In the early stages, Akers and Friedman [3], and Giffler and Thompson [24] explored only a subset of the alternative solutions in order to suggest acceptable schedules. Although such an approach was computationally expensive, it could solve the problems much quicker than a human. Later, the branch-and-bound (B&B) algorithm became very popular for solving JSSPs. It uses the concept of omitting a subset of solutions comprising those, that were out of bounds [4, 10, 12]. Among them, Carlier and Pinson [12] solved a 10×10 JSSP optimally for the first time, a problem that was proposed in 1963 by Muth and Thompson [36]. They considered the $N\times M$ JSSP as M one-machine problems and evaluated the best preemptive solution for each machine. Their algorithm relaxed the constraints on all other machines except the one under consideration. The concept of converting a M machines problem to a one-machine problem was also found in Emmons [19] and Carlier [11]. As the complexity of this algorithm is directly dependent on the number of machines, it is not computationally efficient for large scale problems.

Although the above algorithms can achieve optimum or near optimum makespan, they are computationally expensive, infeasible for large problems, even with current computational power. For this reason, numerous heuristic and meta-heuristic approaches have been proposed in the last few decades. These approaches do not guarantee optimality, but provide a good quality solution within a reasonable period of time. Examples of such approaches applied to JSSPs are genetic algorithm (GA) [7, 16, 18, 38, 51], tabu search (TS) [6, 17], shifting bottleneck (SB) [2, 15], greedy randomized adaptive search procedure (GRASP) [8], and simulated annealing (SA) [32]. The vast majority of this work focuses on the static JSSP, where there are a fixed number of machines and jobs. However, many practical problems can be flexible in terms of the flexibility of constraints, availability of jobs, etc., and this can make solutions even more difficult to obtain. For example, Kacem *et al.* [31] demonstrated a localization method and an evolutionary approach, while Zribi *et al.* [53] proposed a hierarchical method for solving flexible JSSPs (FJSSPs).

Over the last few decades, a substantial amount of work aimed at solving JSSPs by using genetic algorithms and hybrid genetic algorithms has been reported. Lawrence [34] explained how a genetic algorithm can be applied for solving JSSPs. It is very common to improve the performance of the GA by incorporating different search and heuristic techniques, and this approach is readily applied to solving the JSSP using a GA. For example, after applying the crossover operator, Shigenobu *et al.* [43] used the G&T method to build active schedules (i.e., schedules having no unnecessary machine idle time). The G&T method, which is named after Giffler and Thompson [24], is an enumeration technique that explores only a subset of the feasible schedules. These feasible schedules are recognized as active schedules. The active schedules in turn convert to a subset of optimal schedules. However, the G&T method ensures only the activeness, and not the optimality, as finding the optimal solution for larger problems is expensive. Park *et al.* [41] used the G&T method to generate a set of good and diverse initial solutions. Croce *et al.* [16] focused on the activeness of the schedules and proposed a genetic algorithm based approach for solving JSSPs.

Aarts *et al.* [1] provided a comparative analysis of different methods, such as multi-stage iterative improvement (MSII), threshold acceptance (TA), simulated annealing (SA) and GA with neighborhood search, that can be useful for solving JSSPs. Ombuki and Ventresca [38] reported two different approaches for solving JSSPs. The first approach is based on a simple GA with a task assignment scheme (TAS), where a task is assigned to the earliest free machine, taking preemption, concurrency and machine idle time into account. TAS works well for those solutions where only one operation waits for each machine at any instance. Otherwise some priority-rules may be used to improve the performance. In the second approach, the authors incorporated a local search (LS) mutator to improve the quality of solutions obtained by GA with TAS, though they reported that it does not guarantee any improvement. The local search looks for the best makespan by swapping each consecutive pair of a solution. It is computationally very expensive and does not work when more than one swapping is necessary. For this reason, they hybridized the genetic approach by replacing its local search by a simple tabu search proposed by Glover [25]. It is still expensive in terms of computation and memory, and is not good enough to identify required multiple swapping. Gonalves *et al.* [27] also applied two-exchange local search over the genetic algorithms. The search allows up to two swaps at once to improve the fitness. The authors considered the set of active and non-delay schedules in two of their approaches. They also proposed parameterized active schedules which may have a machine delay up to a certain threshold. Tsai and Lin [46] applied a single-swap local search which tries to swap each and every consecutive pair of a selected solution and accepts the new solution if it improves the fitness. But in major cases, more than one swap is necessary to improve the fitness.

Xing *et al.* [49] implemented an adaptive GA where the reproduction parameters are adapted by the algorithm itself. They proposed to use variable crossover and mutation probability which are calculated by an exponential function. The function is a factor of the best and average fitness of the current generation. Yang *et al.* [52] proposed a similar approach for solving JSSPs. They considered the current number of generation as another factor to calculate the reproduction parameters. As the generation progresses, the rate of mutation is increased. Genetic programming (GP) was integrated with GA for solving JSSPs by Werner *et al.* [48]. This technique has higher time complexity as for a single generation of GP, GA runs for a hundreds of generations.

The Shifting Bottleneck (SB) approach, which is derived from the concept of one-machine scheduling, has been using by some researchers for solving JSSP [11, 12, 19]. It starts by arranging the machines according to a specific order, then identifies the first bottleneck machine and schedules it optimally. Then, it selects the next machine in the order and updates the starting time of the jobs that have already been scheduled. The main purpose of this technique is to identify the best order of the machines. The most frequently used strategy is to rearrange the machines according to the criticalness of the machines as identified by the longest processing time. Carlier and Pinson [12] proposed to apply B&B for one machine schedules, which is effective only for independent jobs. However there might be a path that exists between two operations of a job that creates a dependency. In that case, Dauzere-Peres and Lasserre [15] proposed to increase the release date of some unselected jobs to reduce the waiting time between the dependent jobs. Adams *et al.* [2] also focused on the importance of the appropriate ordering of machines in the SB heuristic for JSSP. The main demerit of this heuristic is that it considers only the local information, i.e., only the status of the current and previously considered machines, which may not be effective for all cases.

Feo *et al.* [20] proposed a metaheuristic method, known as greedy randomized adaptive search procedure (GRASP) which was used later by Binato *et al.* [8], for solving JSSPs, this consists of two phases: construction where feasible solutions are built, and local search where the neighborhood solutions are explored. In the construction phase, the authors proposed to maintain a restricted candidature list (RCL) consisting of all possible schedulable operations, to select an appropriate operation to schedule. Different probability distributions are found for this selection in Bresina [9]. The authors proposed to select the operation which gives the minimum increment of schedule time from that instance. This technique may not work in all cases, as it reduces the schedule time for one machine and may delay some operations in other machines. They proposed a local search that identifies the longest path in the disjunctive graph and swaps the critical paths to improve the makespan. GRASP has a problem that it does not take into account any information from the previous iterations. To address this, the authors proposed an intensification technique which keeps track of the elite solutions (e.g., having better fitness value) and includes new solutions in the record if they are better than the worst from the elite list. They also applied the proximate optimality principle (POP) to avoid the error in scheduling early in the construction process in such a way that may lead to errors in the following operations. The new approach proposed in this chapter overcomes these shortcomings by utilising heuristics that effectively comprise a local search technique, while maintaining an elite group.

3 Problem Definition

The standard job-shop scheduling problem makes the following assumptions: Each job consists of a finite number of operations.

- The processing time for each operation using a particular machine is defined.
- There is a pre-defined sequence of operations that has to be maintained to complete each job.
- Delivery times of the products are undefined.
- There is no setup or tardiness cost.
- A machine can process only one job at a time.
- Each job is performed on each machine only once.
- No machine can deal with more than one type of task.
- The system cannot be interrupted until each operation of each job is finished.
- No machine can halt a job and start another job before finishing the previous one.
- Each and every machine has full efficiency.

The objective of the problem is the minimization of the maximum time taken to complete each and every operation, while satisfying the machining constraints and the required operational sequence of each job. In this research, we develop three different algorithms for solving JSSPs. These algorithms are briefly discussed in the next three subsections of the next section.

4 Job-Shop Scheduling with Genetic Algorithm

In this chapter, we consider the minimization of makespan as the objective of JSSPs. According to the problem definition, the sequence of machine use (this is also the sequence of operations) by each job is given. In this case, if we know either the start or end time of each operation, we can calculate the makespan for each job and hence generate the whole schedule. In JSSPs, the main aim is to find the sequence of jobs to be operated on each machine that minimizes the overall makespan. The chromosome representation is an important issue in solving JSSPs using GAs. We discuss the representation aspect in the next section.

4.1 Chromosome Representation

In solving JSSPs using GAs, the chromosome of each individual usually comprises the schedule. Chromosomes can be represented by binary, integer or real numbers. Some popular representations for solving JSSPs are: operation based, job based, preference-list based, priority-rule based, and job pair-relationship based representations [42]. A list of representations commonly used for JSSPs can be found in the survey of Cheng *et al.* [13]. In the operation-based representation, each gene stands for a sequence of operations where every integer numbers in a gene represents a job ID. The first occurrence of a job ID in a chromosome stands for the first operation of that job in a defined machine. Suppose for a 3×3 JSSP; a chromosome may be represented as ||211323321|| where the first number '2' means the first operation of job $j_2(O_{21})$, similarly the second number '1' indicates the first operation of job $j_1(O_{11})$, the third number '1' indicates the second operation (as second appearance of the number '1') of job $j_1(O_{12})$ and so on. Then the operations can be read from the chromosome as $O_{21}O_{11}O_{12}$... and so on, where O_{ji} represents the ith operation of the job j [42]. As the job processing time is known, it is possible to construct the complete schedule from the chromosome information.

In the job-based representation, both the machine sequences and job sequences are necessary to represent a solution. Here, the first job is scheduled first. Moreover, the sequence of operations of each job is determined from the machine sequence. In this representation, there should be two strings of same length; one is the chromosome and another is the machine sequences. Each number in the chromosome represents a machine ID and the numbers in machine sequence represent the job ID. For example, if the chromosome is ||322112313|| and the sequence of machines is ||213122133||, the number 2 (which is job 2) occurs three times in the chromosome. Here, the corresponding values in the machine field are 1, 3 and 2. That means that the sequence of the job j_2 is $m_1m_3m_2$.

In the preference list-based representation, the chromosome represents the preferences of each job. There are M sub-chromosomes in each chromosome where each sub-chromosome represents the preference of the jobs on that machine. Suppose, if the chromosome looks like ||312123321||, it means that the first preferential jobs on machines m_1, m_2 and m_3 are jobs j_3, j_1 and j_3 respectively.

We select the job pair-relationship based representation for the genotype, as in [37, 39, 50, 51], due to the flexibility of applying genetic operators to it. In this representation, a chromosome is symbolized by a binary string, where each bit stands for the order of a job pair (u, v) for a particular machine m.

For a chromosome C_i

$$C_i(m,u,v) = \begin{cases} 1 & \text{if the job } j_u \text{ preceeds the job } j_v \text{ in machine } m \\ 0 & \text{otherwise} \end{cases} \tag{1}$$

This means that in the chromosome representing the individual i, if the relevant bit for job j_u and job j_v for machine m is 1, j_u must be processed before j_v in machine m. The job having the maximum number of 1s is the highest priority job for that machine. The length of each chromosome is

$$l = M \times (N-1) \times N/2 \tag{2}$$

where N stands for the number of jobs, and M for the number of machines. Thus, l is the number of pairs formed by a job with any other job. This binary string acts as the genotype of individuals. It is possible to construct a phenotype which is the job sequence for each machine. This construction is described in Table 1. This representation is helpful if the conventional crossover and mutation techniques are used. We use this representation for the flexibility of applying simple reproduction operators. More representations can be found in the survey [13]. As this chromosome does not contribute to the evaluation process, it does not affect the speed of evaluation. Like some other crossovers, such as partially matched crossover (PMX) and order crossover (OX) that operate only on the phenotypes, the operation-based or job-based representation can be used instead of the binary job-pair relation based representation. We also use the constructed phenotype as the chromosome on which to apply some other heuristic operators that are discussed later.

In this algorithm, we map the phenotype directly from the binary string, i.e., the chromosome, and perform the simple two-point crossover and mutation on it. For this, the crossover points are selected randomly. After applying the operators, as these reproduction operators may produce an infeasible solution [37, 50, 51], we perform the following repairing techniques: local and global harmonization, in order to make this solution feasible. The solutions that remain feasible or unaffected by the reproduction operations need not to be involved in this repairing. We also apply heuristic operators on the constructed phenotype, and these are discussed later.

4.2 Local Harmonization

This is the technique of constructing (which can be recognized as decoding) the phenotype (i.e., the sequence of operations for each machine) from the binary genotype. M tables are formed from a chromosome of length l as described in Eq. (2). Each of the tables is of $N \times N$ size which reflects the relationship between the corresponding jobs of every job pair which contains only binary values. The job having the maximum number of 1s' represents the most preferred job having the highest priority score. These jobs are then rearranged according to their own priorities. Table 1 shows the way to construct the phenotype from the genotype by applying local harmonization.

Table 1.a represents the binary chromosome (for a 3 jobs and 3 machines problem) where each bit represents the preference of one job with respect to another job in the corresponding machine. The third row shows the machine pairs in a given order. The second row indicates the order of the machines for the first job of the pair shown in the third row. The first bit is 1, which means that job j_1 will appear before job j_2 in machine m_1. Table 1.b.1, Table 1.b.2 and Table 1.b.3 represent the

Table 1 Construction of the phenotype from the Binary genotype and Pre-defined Sequences

1	0	1	0	0	1	1	0	0
m_1	m_3	m_2	m_1	m_3	m_2	m_2	m_1	m_3
$j_1 - j_2$			$j_1 - j_3$			$j_2 - j_3$		

1.a

	j_1	j_2	j_3	S
j_1	*	1	0	1
j_2	0	*	0	0
j_3	1	1	*	2

1.b.1

	j_1	j_2	j_3	S
j_1	*	1	1	2
j_2	0	*	1	1
j_3	0	0	*	0

1.b.2

	j_1	j_2	j_3	S
j_1	*	0	0	0
j_2	1	*	0	1
j_3	1	1	*	2

1.b.3

	j_{o1}	j_{o2}	j_{o3}
j_1	m_1	m_3	m_2
j_2	m_2	m_1	m_3
j_3	m_1	m_3	m_2

1.c

	m_{t1}	m_{t2}	m_{t3}
m_1	j_3	j_1	j_2
m_2	j_1	j_2	j_3
m_3	j_3	j_2	j_1

1.d

job pair based relationship in machines m_1, m_2 and m_3 respectively, as mentioned in Eq. (1). In Table 1.b.1, the '1' in cell $j_1 - j_2$ indicates that job j_1 will appear before job j_2 in machine m_1. Similarly, the '0' in cell $j_1 - j_3$ indicates that job j_1 will not appear before job j_3 in machine m_1. In the same Table 1.b, column S represents the priority of each job, which is the row sum of all the 1s for the job presented in each row. A higher number represents a higher priority because it precedes all other jobs. So for machine m_1, job j_3 has the highest priority. If more than one job has equal priority in a given machine, a repairing technique modifies the order of these jobs to introduce different priorities. Consider a situation where the order of jobs for a given machine is $j_1 - j_2$, $j_2 - j_3$, and $j_3 - j_1$. This will provide $S = 1$ for all jobs in that machine. By swapping the content of cells $j_1 - j_3$, and $j_3 - j_1$, it would provide $S = 2$, 1 and 0 for jobs j_1, j_2, and j_3 respectively.

Table 1.c shows the pre-defined operational sequence of each job. In this table, j_{o1}, j_{o2}, and j_{o3} represent the first, second and third operation for a given job. According to the priorities found from Table 1.b, Table 1.d is generated, which is the phenotype or schedule. For example, the sequence of m_1 is $j_3 j_1 j_2$, because in Table 1.b.1, j_3 has the highest priority and j_2 is the lowest priority job.

In Table 1.d, the top row (m_{t1}, m_{t2}, and m_{t3}) represents the first, second and third task on a given machine. For example, considering Table 1.d, the first task of machine m_1 is to process the first task of job j_3.

4.3 Global Harmonization

In a solution space of $(N!)^M$ size, only a small percentage may contain the feasible solutions. The solutions mapped from the chromosome do not guarantee feasibility.

Global harmonization is a repairing technique for changing infeasible solutions into feasible solutions. Suppose that the job j_3 must process its first, second and third operations on machines m_3, m_2, and m_1 respectively; and the job j_1 must process its first, second and third operations on machines m_1, m_3, and m_2 respectively. Further assume that an individual solution (or chromosome) indicates that j_3 is scheduled first on machine m_1 to process its first operation and job j_1 thereafter. Such a schedule is infeasible as it violates the defined sequence of operations for job j_3. In this case, swapping the places between job j_1 with job j_3 on machine m_1 would allow job j_1 to have its first operation on m_1 as required and it may provide an opportunity to job j_3 to visit m_3, and m_2 before visiting m_1 as per its order. Usually, the process identifies the violations sequentially and performs the swap one by one until the entire schedule is feasible. In this case, there is a possibility that some jobs swapped earlier in the process are required to be swapped back to their original position to make the entire schedule feasible. This technique is useful not only for the binary representations, but also for the job-based or operation based representation. Further details on the use of global harmonization with GAs for solving JSSPs can be found in [37, 50, 51].

In our proposed algorithm, we consider multiple repairs to narrow down the deadlock frequency. As soon as a deadlock occurs, the algorithm identifies at most one operation from each job that can be scheduled immediately. Starting from the first operation, the algorithm identifies the corresponding machine of the operation and swaps the tasks in that machine so that at least the selected task prevents deadlock the next time. For N jobs, the risk of getting into deadlock will be removed for at least N operations.

After performing global harmonization, we obtain a population of feasible solutions. We then calculate the makespan of all the feasible individuals and rank them based on their fitness values. We then apply genetic operators to generate the next population. We continue this process until the stopping criteria are satisfied.

5 Priority Rules and JSSPs

As reported in the literature, different priority rules are imposed in conjunction with GAs to improve the JSSP solution. Dorndorf and Pesch [18] proposed twelve different priority rules for achieving better solutions for JSSPs. However they suggested choosing only one of these rules while evaluating the chromosome. They also applied the popular shifting bottleneck heuristic proposed by Adams *et al.* [2] for solving JSSP. This heuristic ingeniously divides the scheduling problem into a set of single machine optimization and re-optimization problems. It selects machines identified as bottlenecks one by one. After the addition of a new machine, all previously established sequences are re-optimized. However these algorithms were implemented while evaluating the individuals in GA and generating the complete schedule. In this section, we introduce a number of new priority rules. We propose to use these rules after the fitness evaluation as the process requires analyzing the individual solutions from the preceding generation. The rules are briefly discussed below.

5.1 Paritial Reordering (PR)

In the first rule, we identify the machine (m_k) which is the deciding factor for the makespan in phenotype p (i.e., the last machine on which a job is executed) and the last job (j_k) that is to be processed by the machine m_k. The machine m_k can be termed as the bottleneck machine in the chromosome under consideration. Then we find the machine (say m) required by the first operation of job j_k. The re-ordering rule then suggests that the first operation of job j_k must be the first task on machine m if it is not currently scheduled. If we move the job j_k from its current l-th position to the first position, we may need to push some other jobs currently scheduled on machine m to the right. In addition, it may provide an opportunity to shift some jobs to the left on other machines. The overall process helps to reduce the makespan for some chromosomes. Algorithm 1 and Algorithm 2 in the Appendix describe this re-ordering process.

The following explains the re-ordering process with a simple example. In Fig. 1(a), the makespan is the completion time of job j_3 on machine m_1. That means machine

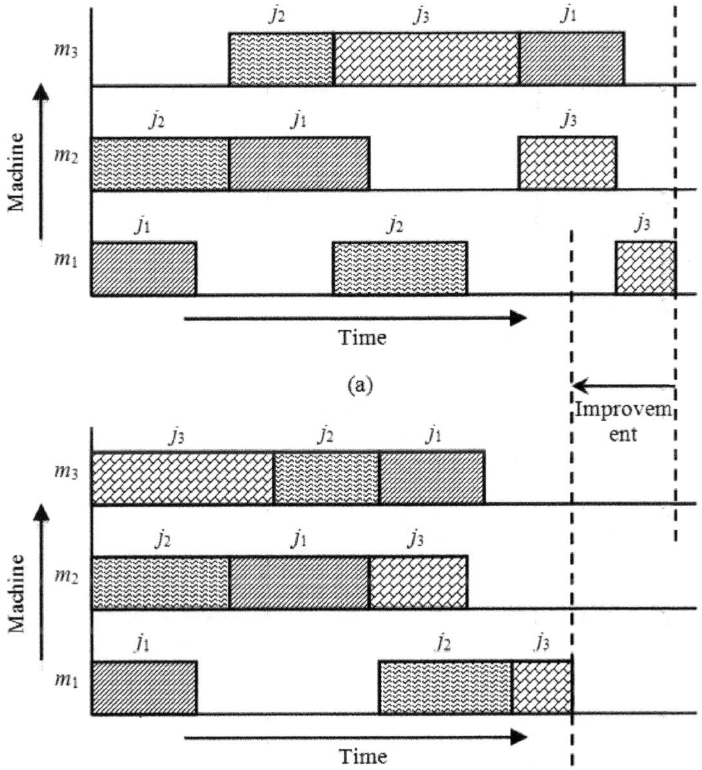

Fig. 1 Gantt chart of the solution: (a) before applying the partial reordering, (b) after applying partial reordering and reevaluation

m_1 is the bottleneck machine. Here, job j_3 requires machine m_3 for its first operation. If we move j_3 from its current position to the first operation of machine m_3, it is necessary to shift job j_2 to the right for a feasible schedule on machine m_3. These changes create an opportunity for the jobs j_1 on m_3, j_3 on m_2 and j_3 on m_1 to be shifted towards the left without violating the operational sequences. As can be seen in Fig. 1(b), the resulting chromosome is able to improve its makespan. The change of makespan is indicated by the dotted lines. Algorithm 2 in the Appendix also shows how the partial reordering can be done.

5.2 Gap Reduction (GR)

After each generation, the generated phenotype usually leaves some gaps between the jobs. Sometimes, these gaps are necessary to satisfy the precedence constraints.

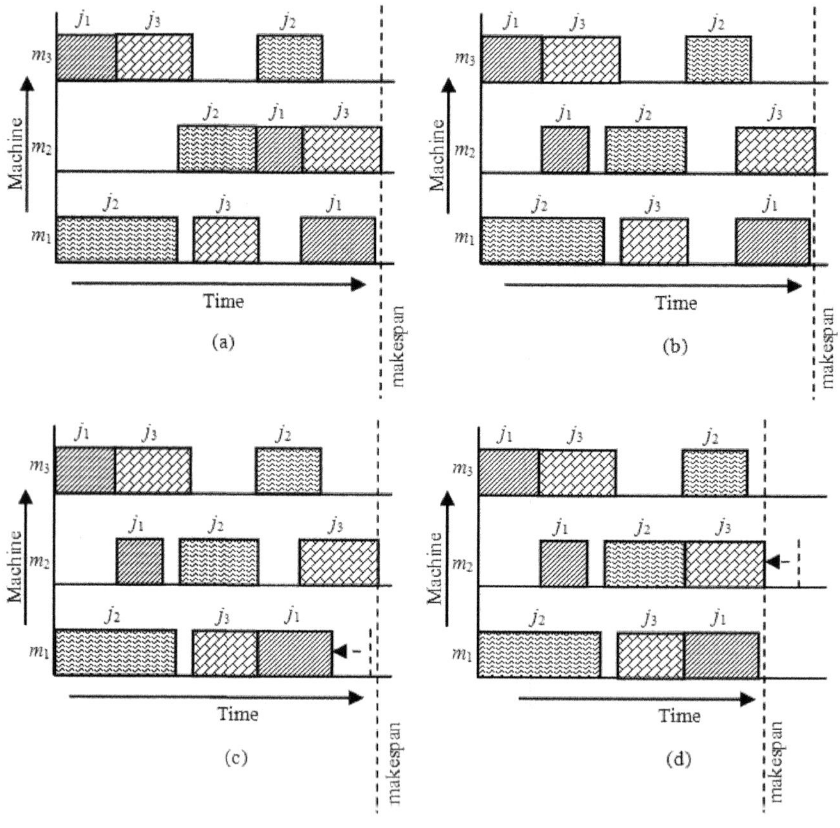

Fig. 2 Two steps of a partial Gantt chart while building the schedule from the phenotype for a 3×3 job-shop scheduling problem. The X axis represents the execution time and the Y axis represents the machines.

However, in some cases, a gap could be removed or reduced by replacing the gap with a job on the right side of the gap. For a given machine, this is like swapping between a gap from the left and a job from the right of a schedule. In addition, a gap may be removed or reduced by simply moving a job on the right-hand side of the gap leftwards. This process would help developing a compact schedule from the left and continuing up to the last job for each machine. Of course, it must ensure no conflict or no infeasibility before accepting the move.

Thus, the rule must identify the gaps in each machine, and the candidate jobs which can be placed in those gaps, without violating the constraints and without increasing the makespan. The same process is carried out for any possible leftwards shift of jobs of the schedule.

The gap reduction rule, with swapping between gaps and jobs, is explained using a simple example. A simple instance of a schedule is shown in Fig. 2(a). In the phenotype p, j_1 follows j_2 in machine m_2, however, job j_1 can be placed before j_2, as shown in Fig. 2(b), due to the presence of an unused gap before j_2. A swap between this gap and job j_1 would allow the processing of j_1 on m_2 earlier than the time shown in Fig. 2(a). This swapping of j_1 on m_2 creates an opportunity to move this job to the left on machine m_3 (see Fig. 2(c)). Finally, j_3 on m_2 can also be moved to the left, which ultimately reduces the makespan as shown in Fig. 2(d). Algorithm 3 in the Appendix gives the step by step instructions of the GR algorithm.

5.3 *Restricted Swapping (RS)*

For a given machine, the restricted swapping rule allows swapping between the adjacent jobs if and only if the resulting schedule is feasible. This process is carried out only for the job which takes the longest time to complete.

Suppose that job j takes the longest time for completion for the phenotype p. This algorithm starts from the last operation of j in p and checks with the immediate predecessor operation whether these two are swappable or not. The necessary conditions for swapping are: none of the operations can start before the finishing time of the immediate predecessor operation of that corresponding job, and both operations have to finish before the start of the immediate successive operations of the corresponding jobs. Interestingly, the algorithm does not violate the feasibility of the solution. It may change the makespan if any of the operations are the last operation of the corresponding machine, but it will also give an alternative solution which may improve the fitness of the solution in successive generations, when the phenotype is rescheduled. The details of this algorithm are described in Algorithm 4 in the Appendix.

This swapping can be seen from Fig. 3(a) and (b). The makespan in Fig. 3(b) is improved due to the swapping between jobs j_3, and j_2 in machine m_3. This swapping is not permissible by GR, because there is not enough gap for j_3 in front of j_2. However, one such swapping may create the scope for another such swapping in the next bottleneck machine. This process also allows swapping between two randomly selected individuals. This is done for a few individuals only. As the complexity of

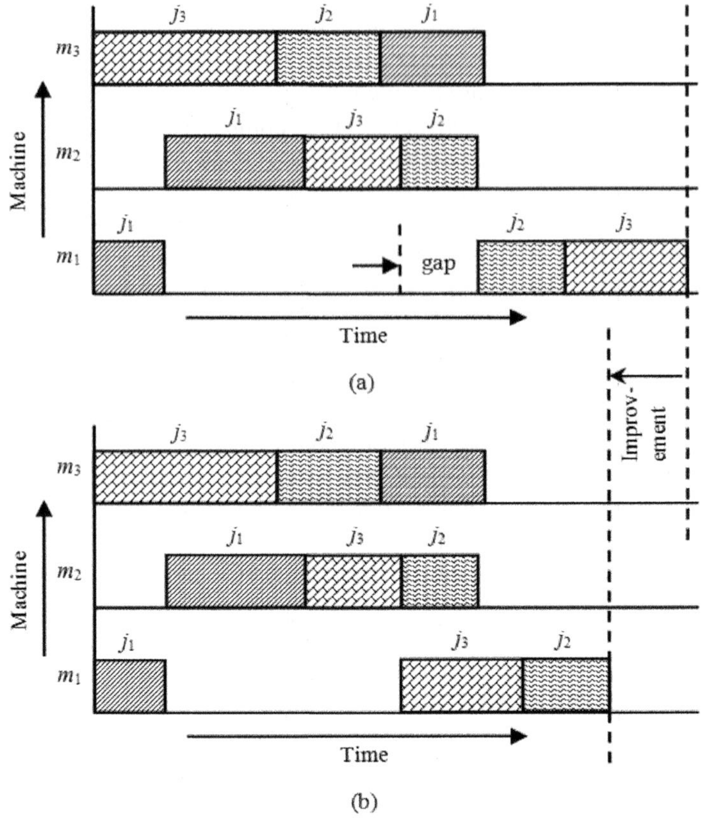

Fig. 3 Gantt chart of the solution: (a) before applying the restricted swapping, (b) after applying restricted swapping and reevaluation.

this algorithm is simply of order N, it does not affect the overall computational complexity much.

6 Implementation

For the initial implementation of TGA, we generate a set of random individuals. Each individual is represented by a binary chromosome. We use the job-pair relationship based representation as Nakano and Yamada [37] and Paredis *et al.* [40] successfully used this representation to solve job-shop scheduling problems and reported the effectiveness of the representation. We use the simple two-point crossover and bit flip mutation as reproduction operators. We have carried out a set of experiments with different crossover and mutation rates to analyze the robustness of the algorithm. After the successful implementation of the TGA, we introduce the priority rules, as discussed in the last section, to TGA as follows:

- Partial re-ordering rule with TGA (PR-GA)
- Gap reduction rule with TGA (GR-GA) and
- Gap reduction and restricted swapping rule with TGA (GR-RS-GA)

For ease of explanation, we describe the steps of GR-RS-GA below:

Let R_c and R_m be the selection probabilities for two-point crossover and bit-flip mutation respectively. $P'(t)$ is the set of current individuals at time t and $P(t)$ is the evaluated set of individuals which is the set of individuals repaired using local and global harmonization at time t. K is the total number of individuals in each generation. s is an index which indicates a particular individual in the current population.

1. Initialize $P'(t)$ as a random population $P'(t=0)$ of size K, where each random individual is a bit string of length l.
2. Repeat until some stopping criteria are met
   a. Set $t := t+1$ and s :=NULL
   b. Evaluate $P(t)$ from $P'(t-1)$ by the following steps;
      i. Decode each individual p by using the job-based decoding with the local harmonization and global harmonization methods to repair illegal bit strings.
      ii. Generate the complete schedule with the starting and ending time of each operation by applying the gap reduction rule (GR) and calculate the objective function f of p.
      iii. Go to `step 2(b)i` until every individual is evaluated.
      iv. Rank the individuals according to the fitness values from higher to lower fitness value.
      v. Apply elitism; i.e., preserve the solution having the best fitness value in the current generation so that it can survive at least until to the next generation.
   c. Apply the restricted swapping rule (RS) on some of the individuals selected in a random manner.
   d. Go to `step 3` if the stopping criteria are met.
   e. Modify $P(t)$ using the following steps;
      i. Select the current individual p from $P(t)$ and select a random number R between 0 and 1.
      ii. If $R \leqslant R_c$ then
         A. If s =NULL
            - Save the location of p into s.
            - Go to `step 2(e)i`.

            [End of `step 2(e)iiA` If]

B. Select randomly one individual p_1 from the top 15% of the population and two individuals from the rest. Play a tournament between the last two and choose the winning individual w. Apply two-point crossover between p_1 and w; generate p_1 and w. Replace p with p_1 and content of s with w. Set s with NULL.

C. Else if $R > R_c$ and $R \leqslant (R_c + R_m)$ then randomly select one individual p_1 from $P(t)$ and apply bit-flip mutation. Replace p with p_1.

D. Else continue.

[End of `step 2(e)ii` If]

iii. Reassign the $P'(t)$ by $P(t)$ to initialize the new generation preserving the best solution as elite.

[End of `step 2a` Loop]

3. Save the best solution among all the feasible solutions.

[End of Algorithm]

Sometimes the actions of the genetic operators may direct the good individuals to less attractive regions in the search space. In this case, the elitism would ensure the survival of the best individuals [26, 27]. We apply elitism in each generation to preserve the best solution found so far and also to inherit the elite individuals more often than the rest.

During the crossover operation, we use the tournament selection that chooses one individual from the elite class of individuals (i.e., the top 15%) and two individuals from the rest. Increasing that rate reduces the quality of solutions. On the other hand, reducing the rate initiates a quicker but premature convergence. This selection then plays a tournament between the last two and performs crossover between the winner and the elite individual. As we apply a single selection process for both the reproduction processes, the probability of selecting an individual multiple times is low, but not zero. We rank the individuals on the basis of their fitness values, and a high selection pressure on the better individuals may contribute to premature convergence. Consequently, we consider the situation where 50% or more of the elite class are the same solution. In this case, their offspring will be quite similar after some generations. To counter this, when this occurs, a higher mutation rate will be used to help diversify the population. We set the population size to 2500 and the number of generations to 1000. In our approach, the GR rule is used as a part of evaluation. That means GR is applied to every individual. On the other hand, we apply PR and RS to only 5% of randomly selected individuals in every generation. Because of the role of GR in the evaluation process, it is not possible to apply it as an additional component like PR or RS. Moreover, PR and RS are effective on feasible individuals which prohibit using these rules before evaluation. To test the performance of our proposed algorithms, we have solved the 40 benchmark problems designed by Lawrence [34] and have compared our results with several existing algorithms. The problems range from 10×5 to 30×10 and 15×15 where $N \times M$ represents N jobs and M machines.

7 Result and Analysis

The results for the benchmark problems were obtained by executing the algorithms on a personal computer. All results are based on 30 independent runs with different random seeds. To select the appropriate set of parameters, we have performed several experiments varying the reproduction parameters (crossover and mutation). The results presented below are based on the best parameter set. The details of parametric analysis are discussed later in the chapter. These results and parameters are tabulated in Table 2-Table 7.

Table 2 Comparing our Four Algorithms

No. of Problems	Algorithm	Optimal Found	ARD (%)	SDRD (%)	Fitness Eval. (10^3)
40 (la01-la40)	TGA	15	3.591	4.165	664.90
	PR-GA	16	3.503	4.192	660.86
	GR-GA	23	1.360	2.250	356.41
	GR-RS-GA	27	0.968	1.656	388.58

Table 2 compares the performance of our four algorithms (TGA, PR-GA, GR-GA, and GR-RS-GA) in terms of the percentage average relative deviation (ARD) from the best result published in literature, the standard deviation of the percentage relative deviation (SDRD), and the total number of evaluations required. We do not consider the CPU time as a unit of measurement due to the fact that we have conducted our experiments using different platforms and the unavailability of such information in the literature. From Table 2, it is clear that the performance of GR-GA is better than both PR-GA and TGA. The addition of RS to GR-GA, which is known as GR-RS-GA, has clearly enhanced the performance of the algorithm. Out of the 40 test problems, both GR-GA and GR-RS-GA obtained exact optimal solutions for 23 problems. In addition, GR-RS-GA obtained optimal solutions for 4 more problems and substantially improved solutions for 10 problems. In general, these two algorithms converge quickly as can be seen from the average number of generations.

To analyze the individual contribution of the PR, RS and GR, we have experimented on a sample of five problems (la21-la25) with the same set of parameters in the same computing environment. For these problems, the individual percentage improvements of PR, GR and RS over TGA after 100, 250 and 1000 generations

Table 3 Individual Contribution of the Priority Rules after 100, 250 and 1000 Generations

No. of Problems	Algorithm	% improvement from TGA		
		100	250	1000
5 (la21-la25)	PR-GA	1.33	1.12	0.38
	RS-GA	0.89	1.05	0.18
	GR-GA	8.92	7.73	6.31

are reported in Table 3. To measure this, we calculate the improvement of the g-th generation from the $(g-1)$-th generation up to the G-th generation where G is 100, 250 and 1000 respectively.

The result in Table 3 is the average of the improvements in percentage scale. Although all three priority rules have a positive effect, GR's contribution is significantly higher than the other two rules and is consistent over many generations.

Interestingly, the improvement rapidly decreases in the case of GR compared to PR and RS. The reason for this is that, GR-GA starts with a set of good initial solutions, for example, an 18.17% improvement compared to TGA for the problems la21-la25. This is why, the effects in each generation decrease simultaneously.

To observe the contribution more closely, we measure the improvement due to the individual rule in every generation in the first 100 generations. A sample comparison of the fitness values for our three algorithms in the first 100 generations is shown in Fig. 4. It is clear from the figures that the improvement rates of TGA, PR-GA and RS-GA are higher than GR-GA, but GR-GA provides better fitness in all the tested problems. As JSSP is a minimization problem, GR-GA outperforms the others every case.

PR considers only the bottleneck job, whereas GR is applied to all individuals. The process of GR eventually makes most of the changes performed by PR over some (or many) generations. We identify a number of individuals where PR could make a positive contribution. We apply GR and PR on those individuals, to compare their relative contribution. For the five problems we consider over 1000 generations, we observe that GR made a 9.13% more improvement than PR. It must be noted here that GR is able to make all the changes which PR does. That means PR cannot make an extra contribution over GR. As a result, the inclusion of PR with GR does not help to improve the performance of the algorithm. That is why we do not present other possible variants, such as PR-RS-GA and GR-RS-PR-GA.

Both PR and RS were applied only to 5% of the individuals. The role of RS is mainly to increase the diversity. A higher rate of PR and RS does not provide significant benefit either in terms of quality of solution or computational time. We have experimented with varying the rate of PR and RS individually from 5% to 25% and tabulated the percentage relative improvement from TGA in Table 4. From Table 4, it is clear that the increase of the rate of applying PR and RS does not improve the quality of the solutions. Moreover, it was found from experiments that it takes extra time to converge.

Table 5 presents the percentage of relative deviation from the best known solution, for 40 test problems, for our four algorithms. Table 6 shows the same results for a number of well-known algorithms appearing in the literature. The first two columns of Table 5 represent the problem instances and the size of the problems. These columns are followed by the average relative deviation (ARD) of the best fitness found from four of our algorithms compared to the best fitness found from the literature in percentage scale.

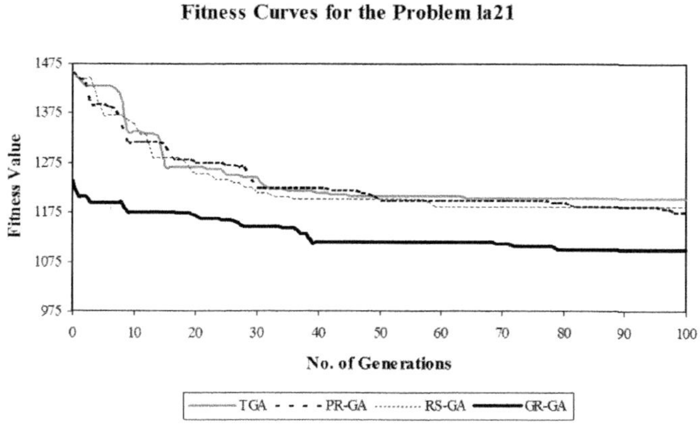

(a)

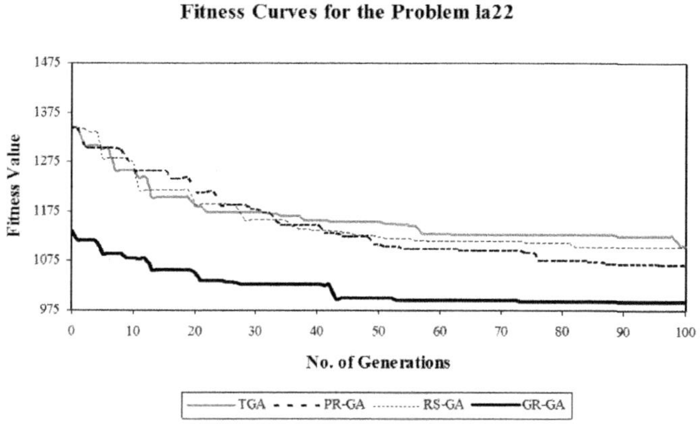

(b)

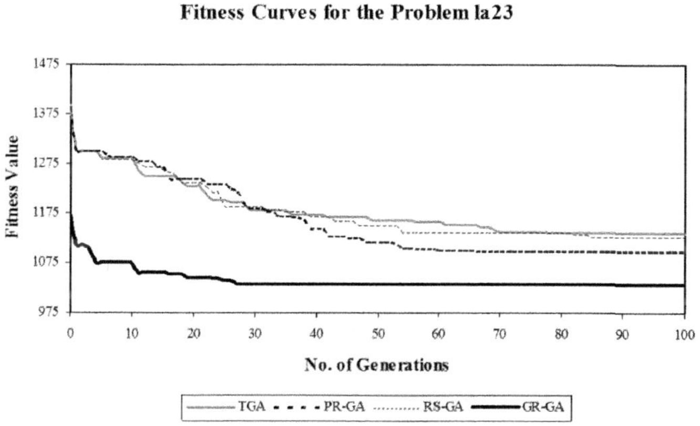

(c)

Fig. 4 Fitness curve for the problem la21–la25 up to the first 100 generations

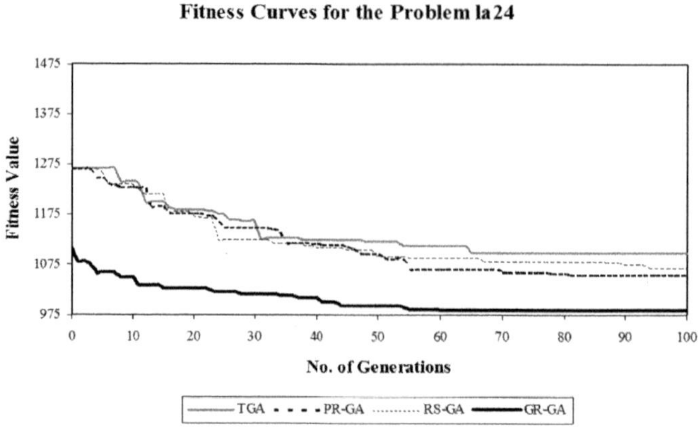

(d)

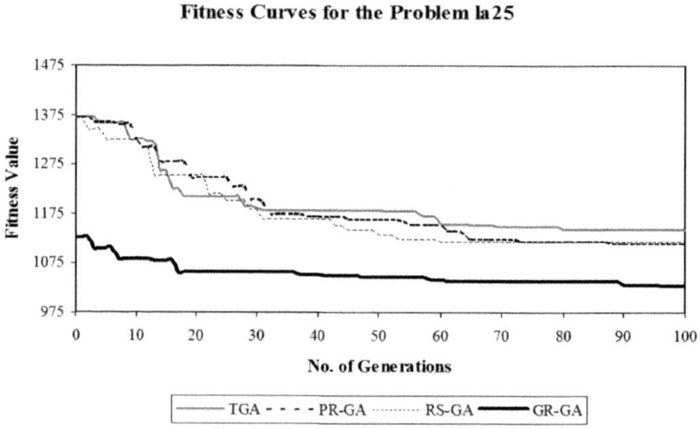

(e)

Fig. 4 (*continued*)

Table 6 starts with the column of problem instance and each column next to that represents the ARD in percentage of some other algorithms found in the literature. Here, we consider our four algorithms (TGA, PR-GA, GR-GA, and GR-RS-GA), local search GA [27, 38], GA with genetic programming [48], GRASP [8], normal GA and shifting-bottleneck GA [18], local search GA [1], GA [16] and shifting bottleneck heuristic [2]. The details of these algorithms were discussed in earlier sections of this chapter.

As shown in Table 5 and Table 6, for most of the test problems, our proposed GR-RS-GA performs better than other algorithms in terms of the quality of solutions. To compare the overall performance of these algorithms, we calculate the average of relative deviation (ARD) for the test problems and the standard deviation of the relative deviations (SDRD), and present them in Table 7. In Table 7, we compare the overall performance with only our GR-RS-GA. As different authors used different

Table 4 Percentage Relative Improvement of the Five Problems (la21-la25)

Changing	5%	10%	15%	20%	25%
PR	5.26	4.94	0.17	2.80	1.37
RS	4.20	4.95	2.22	3.02	3.03

Table 5 Comparison of the Average Percentage Relative Deviations from the Best Result Found in Literature

Problem	Size	GR-RS-GA	GR-GA	PR-GA	TGA	Problem	Size	GR-RS-GA	GR-GA	PR-GA	TGA
la01	10×5	0.00	0.00	0.15	0.15	la21	15×10	3.15	4.11	4.21	4.97
la02	10×5	0.00	0.00	0.00	0.00	la22	15×10	3.56	3.88	6.26	6.36
la03	10×5	0.00	0.00	3.35	3.35	la23	15×10	0.00	0.00	1.07	1.07
la04	10×5	0.00	0.00	2.71	2.71	la24	15×10	2.57	4.71	5.45	5.24
la05	10×5	0.00	0.00	0.00	0.00	la25	15×10	1.43	1.43	10.24	10.24
la06	15×5	0.00	0.00	0.00	0.00	la26	20×10	0.00	0.16	6.98	6.32
la07	15×5	0.00	0.00	0.00	0.00	la27	20×10	4.13	5.10	7.53	7.53
la08	15×5	0.00	0.00	0.00	0.00	la28	20×10	1.64	2.22	6.25	6.66
la09	15×5	0.00	0.00	0.00	0.00	la29	20×10	5.53	6.91	9.51	9.33
la10	15×5	0.00	0.00	0.00	0.00	la30	20×10	0.00	0.00	0.59	1.62
la11	20×5	0.00	0.00	0.00	0.00	la31	30×10	0.00	0.00	0.00	0.00
la12	20×5	0.00	0.00	0.00	0.00	la32	30×10	0.00	0.00	0.00	0.00
la13	20×5	0.00	0.00	0.00	0.00	la33	30×10	0.00	0.00	0.00	0.00
la14	20×5	0.00	0.00	0.00	0.00	la34	30×10	0.00	0.00	1.10	1.57
la15	20×5	0.00	0.00	0.00	0.00	la35	30×10	0.00	0.00	0.00	0.53
la16	10×10	0.00	0.11	5.19	5.19	la36	15×15	3.08	3.86	9.54	9.46
la17	10×10	0.00	0.00	0.13	1.28	la37	15×15	3.22	4.94	10.52	11.74
la18	10×10	0.00	1.53	1.53	1.53	la38	15×15	5.85	9.03	14.30	14.30
la19	10×10	0.00	0.95	6.41	5.70	la39	15×15	1.54	2.43	8.84	8.52
la20	10×10	0.55	0.55	7.21	7.21	la40	15×15	2.45	2.45	11.05	11.05

numbers of problems, we compare the results based on the number of test problems solved by others in Table 5. For example, as Ombuki and Ventresca [38] solved 25 (problems la26-la40) out of the 40 test problems considered in this research, we calculate ARD and SDRD for these 25 problems to make a fairer comparison.

As in Table 7, for 10 selected problems, in terms of the quality of the solutions, GR-RS-GA outperforms Croce *et al.* [16]. For 24 selected problems, GR-RS-GA also outperforms Adams *et al.* (SB II) algorithm [2]. For 25 test problems (la26-la40), our proposed GR-RS-GA is very competitive with SBGA (60) but much better than Ombuki and Ventresca [47]. When considering all 40 test problems, our GR-RS-GA clearly outperforms all the algorithms compared in Table 5.

Table 6 Comparison of the Percentage Relative Deviations from the Best Results with that of Other Authors

Problem	Goncalves *et al.*	Werner *et al.*	Aarts *et al.*		Ombuki & Ventresca	Dorndorf & Pesch			Croce *et al.*	Binato *et al.*	Adams *et al.*	
			GLS1	GLS2		PGA	SBGA-1	SBGA-2			SB-I	SB-II
la01	0.00		0.00	0.00		0.00	0.00		0.00	0.00	0.00	
la02	1.50		1.98	0.61		3.97	1.68		1.68	0.00	9.92	2.14
la03	1.16		2.68	2.01		3.85	1.17		11.56	1.17	4.36	1.34
la04	0.00		1.53	0.68		5.08	0.00			0.00	1.19	0.51
la05	0.00		0.00	0.00		0.00	0.00			0.00	0.00	
la06	0.00		0.00	0.00		0.00	0.00		0.00	0.00	0.00	
la07	0.00		0.00	0.00		0.00	0.00			0.00	0.00	
la08	0.00		0.00	0.00		0.00	0.00			0.00	0.58	0.00
la09	0.00		0.00	0.00		0.00	0.00			0.00	0.00	
la10	0.00		0.00	0.00		0.00	0.00			0.00	0.10	
la11	0.00		0.00	0.00		0.00	0.00		0.00	0.00	0.00	
la12	0.00		0.00	0.00		0.00	0.00			0.00	0.00	
la13	0.00		0.00	0.00		0.00	0.00			0.00	0.00	
la14	0.00		0.00	0.00		0.00	0.00			0.00	0.00	
la15	0.00		0.00	0.00		2.49	0.00			0.00	0.00	
la16	2.88	3.60	3.39	3.39	1.48	6.67	1.69	1.69	3.60	0.11	8.04	3.49
la17	1.01	2.42	0.89	0.89	1.02	3.19	0.38	0.00		0.00	1.53	0.38
la18	0.82	3.42	0.94	1.18	1.06	8.02	0.00	0.00		0.00	5.07	1.30
la19	1.06	4.87	2.49	2.02	2.14	4.51	2.49	0.71		0.00	3.92	2.14
la20	2.59	4.10	1.22	1.55	0.55	2.88	1.00	0.89		0.55	2.44	1.33
la21	3.06	14.53	3.63	3.73	6.50	8.89	2.68	2.68	4.88	4.30	12.05	3.63
la22	2.42		2.91	1.83	6.69	7.66	0.86	0.97		3.56	12.19	1.83
la23	0.00		0.00	0.00	0.29	3.88	0.00	0.00		0.00	2.81	0.00
la24	3.61	11.23	3.74	4.92	10.37	8.45	2.67	2.35		4.60	6.95	4.39
la25	3.55	12.59	3.99	3.38	7.16	3.79	3.17	3.07		5.22	7.27	4.09
la26	0.00		1.81	1.48	7.31	4.93	0.08	0.00	1.07	4.35	7.06	0.49
la27	3.67	17.00	5.91	5.26	9.31	11.58	3.00	2.75		6.88	7.29	4.53
la28	2.72		5.35	4.03	7.89	9.13	1.97	2.06		6.33	3.29	2.80
la29	4.06	20.22	11.50	8.90	13.31	15.47	4.06	4.58		11.75	11.84	7.09
la30	0.00		3.47	2.29	7.08	4.13	0.00	0.00		0.96	3.54	0.00
la31	0.00		0.00	0.00	0.00	0.00	0.00	0.00	0.00	0.00	0.00	
la32	0.00		0.00	0.00	0.00	0.00	0.00	0.00		0.00	0.00	
la33	0.00		0.00	0.00	1.51	0.00	0.00	0.00		0.00	0.00	
la34	0.00		0.93	0.52	3.66	0.00	0.00	0.00		1.86	0.00	
la35	0.00		0.32	0.11	3.71	0.00	0.00	0.00		0.00	0.00	
la36	2.69		4.42	3.39	7.10	8.28	3.86	3.86	2.92	5.21	6.55	2.92
la37	2.78		3.72	3.79	8.59	7.23	6.23	3.51		4.29	6.30	1.86
la38	4.47	17.64	7.44	7.27	13.88	8.36	4.60	3.76		5.94	7.02	4.93
la39	1.36		3.73	3.73	12.81	9.57	3.97	3.57		4.62	7.14	3.24
la40	2.40	20.54	4.17	3.11	8.27	8.10	4.26	2.45		3.03	8.51	3.85

Table 7 Comparing the Algorithms Based on Average Relative Deviations and Standard Deviation of Average Relative Deviations

No. of Problems	Test Problems	Author	Algorithm	ARD(%)	SDRD(%)
40	la01 la40	Our Proposed	GR-RS-GA	0.97	1.66
		Gonalves *et al.*	Non-delay	1.20	1.48
		Aarts *et al.*	GLS1	4.00	4.09
		Aarts *et al.*	GLS2	2.05	2.53
		Dorndorf & Pesche	PGA	1.75	2.20
		Dorndorf & Pesche	SBGA (40)	1.25	1.72
		Binato *et al.*	-	1.87	2.78
		Adams *et al.*	SB I	3.67	3.98
25	la16 la40	Our Proposed	GR-RS-GA	1.55	1.88
		Ombuki and Ventresca	-	5.67	4.38
		Dorndorf & Pesch	SBGA (60)	1.56	1.58
24	Selected (see Table 6)	Our Proposed	GR-RS-GA	1.68	1.86
		Adams *et al.*	SB II	2.43	1.85
12	Selected (see Table 6)	Our Proposed	GR-RS-GA	2.14	2.17
		Werner *et al.*	GC	11.01	7.02
10	Selected (see Table 6)	Our Proposed	GR-RS-GA	0.62	1.31
		Croce *et al.*	-	2.57	3.60

To get a clear view of the performance of our three algorithms over the traditional GA, we perform a statistical significance test for each of these algorithms against the traditional GA. We use the student's t-test [45] where the t-values are calculated from the average and standard deviation of 30 independent runs for each problem. The values are normally distributed. The results of the test are tabulated in Table 8.

We derive nine levels of significance, to judge the performance of PR-GA, GR-GA, and GR-RS-GA over the TGA, using the critical t-values 1.311 (which is for 80% confidence level), 2.045 (for 95% confidence level), and 2.756 (for 99% confidence level). We define the significance level S as follows.

$$S = \begin{cases} ++++ & t_i \geq 2.756 \\ +++ & 2.045 \leq t_i < 2.756 \\ ++ & 1.311 \leq t_i < 2.045 \\ + & 0 < t_i < 1.311 \\ = & t_i = 0 \\ - & -1.311 < t_i < 0 \\ -- & -2.045 < t_i \leq -1.311 \\ --- & -2.756 < t_i \leq -2.045 \\ ---- & t_i \leq -2.756 \end{cases} \tag{3}$$

Table 8 Statistical Significance Test (Student's t-Test) Result of GR-RS-GA Compared to the TGA, PR-GA, and GR-GA

Problem	*t*-value			Significance			Problem	*t*-value			Significance		
	PR-GA	GR-GA	GR-RS-GA	PR-GA	GR-GA	GR-RS-GA		PR-GA	GR-GA	GR-RS-GA	PR-GA	GR-GA	GR-RS-GA
la01	-0.04	6.72	6.68	-	++++	++++	la21	1.02	9.15	11.00	+	++++	++++
la02	-1.53	3.59	1.95	- -	++++	++	la22	1.00	13.06	11.46	+	++++	++++
la03	-0.43	4.85	4.87	-	++++	++++	la23	1.09	14.04	14.05	+	++++	++++
la04	-0.12	25.31	22.74	-	++++	++++	la24	0.11	4.91	5.03	+	++++	++++
la05	0.00	0.00	0.00	=	=	=	la25	0.53	25.58	20.07	+	++++	++++
la06	0.00	0.00	0.00	=	=	=	la26	-1.16	20.60	20.05	-	++++	++++
la07	1.19	8.33	8.33	+	++++	++++	la27	0.11	12.16	12.54	+	++++	++++
la08	0.00	0.00	0.00	=	=	=	la28	-0.48	22.74	21.76	-	++++	++++
la09	0.00	0.00	0.00	=	=	=	la29	-0.62	12.76	13.86	-	++++	++++
la10	0.00	0.00	0.00	=	=	=	la30	0.54	16.02	15.36	+	++++	++++
la11	0.00	0.00	0.00	=	=	=	la31	-1.00	0.00	0.00	-	=	=
la12	0.00	0.00	0.00	=	=	=	la32	0.61	5.42	5.42	+	++++	++++
la13	0.00	0.00	0.00	=	=	=	la33	0.91	4.88	4.88	+	++++	++++
la14	0.00	0.00	0.00	=	=	=	la34	-2.56	20.53	20.53	- - -	++++	++++
la15	-1.85	10.66	10.63	- -	++++	++++	la35	1.35	10.40	10.48	++	++++	++++
la16	-0.10	11.23	10.27	-	++++	++++	la36	0.09	18.83	18.15	+	++++	++++
la17	0.97	17.39	16.70	+	++++	++++	la37	-0.05	27.24	26.16	-	++++	++++
la18	-0.27	12.91	12.94	-	++++	++++	la38	0.08	29.09	25.28	+	++++	++++
la19	-0.29	31.49	27.70	-	++++	++++	la39	-0.51	19.01	19.03	-	++++	++++
la20	-0.30	29.66	29.64	-	++++	++++	la40	-1.26	26.15	21.19	-	++++	++++

Better
++++ Extremely Significant
+++ High Significant
++ Significant
+ Slightly Significant
= Equal

Worse
- - - - Extremely Significant
- - - High Significant
- - Significant
- Slightly Significant

It is clear from Table 8 that GR-GA and GR-RS-GA are extremely better than traditional GA as these two algorithms made extremely significant improvement over TGA in 30 and 29 problems respectively. Also both the algorithms are either better or equally good for rest of the problems. Although the algorithm PR-GA is not extremely better than TGA for any problem, it is either slightly, or significantly better, or equal to TGA for most of the test problems. The above analysis supports the fact that the priority rules improve the performance of traditional GA significantly.

7.1 Parameter Analysis

In GA, different reproduction parameters are used. We have performed experiments with different combinations of parameters to identify the appropriate set of parameters and their effect on solutions. A higher selection pressure on better individuals, with a higher rate of crossover, contributes towards diversity reduction and hence the solutions converge prematurely. In JSSPs, when a large portion of the population converges to particular solutions, the probability of solution improvement reduces because the rate of selecting the same solution increases. Therefore, it is important to find an appropriate rate for crossover and mutation. Three sets of experiments were carried out as follows:

- **Experiment 1:** Crossover varied from 0.60 to 0.95 with an increment of 0.05, and mutation from 0.35 to 0.00 with a reduction of 0.05. However, the crossover rate plus the mutation rate must be equal to 0.95. The detailed combinations are shown as 8 sets in Table 9.
- **Experiment 2:** Experiments were conducted by varying crossover while keeping mutation fixed. The value of mutation was taken from the best set of experiment 1.
- **Experiment 3:** Experiments were conducted by varying mutation while keeping crossover fixed. The value of crossover was taken from the best value of experiment 2.

Table 9 Combination of Different Reproduction Parameters

	Set 1	Set 2	Set 3	Set 4	Set 5	Set 6	Set 7	Set 8
crossover	0.60	0.65	0.70	0.75	0.80	0.85	0.90	0.95
mutation	0.35	0.30	0.25	0.20	0.15	0.10	0.05	0.00

The detailed results for different combinations are tabulated and graphically shown in this section. Initially, we have fixed the sum of crossover rate and mutation rate approximately equal to 1.0. The combinations are tabulated in Table 9. Performing further experiments with more parameters would not provide any benefit, but would rather consume a significant amount of computational time.

The two curves in Fig. 5 represent how the quality of solutions varies with the changing crossover and mutation rates. For the parameter set 2, our algorithm provides the best solution. We show a cumulative measurement of the average and standard deviation of the relative deviations in Fig. 5 (bottom). Here, the parameter set 2 still performs well. We multiply ARD with STDev as both of them are better when minimized. Figures (bottom) in the next three figures represent the combined impact of both the ARD and STDev.

In Fig. 5, the algorithm performs better for the parameter set 2 where the crossover rate is 0.65 and the mutation rate is 0.30. In the second set of experiments, we vary the crossover rate from 0.70 to 0.35 with a step size of 0.05 while fixing the mutation at 0.30.

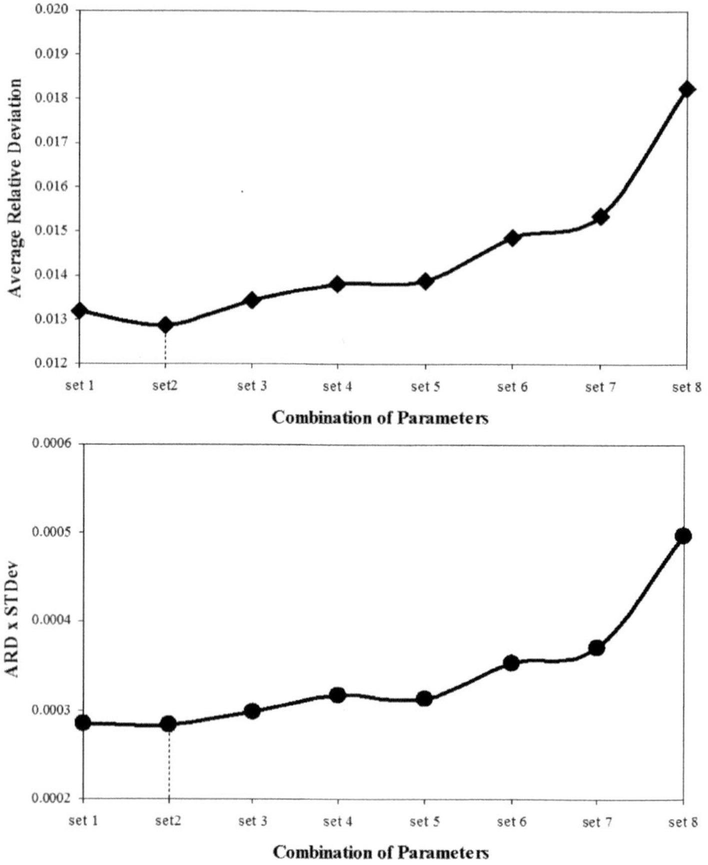

Fig. 5 Product of average relative deviation (ARD) and standard deviation with respect to different parameter combinations tabulated in Table 9.

Fig. 6 presents the outcome of the second set of experiments which shows that the crossover rate of 0.45 are the best with a mutation rate of 0.30. The products of ARD and standard deviation of ARD for a crossover rate of 0.45 are slightly better.

In the third set of experiments, we fix the crossover rate at 0.45 and varied the mutation rate from 0.5-0.0. The third set of experiments showed the effect of mutation when the crossover rate was fixed as the best crossover rate found from the first set of experiments. Fig. 7 shows that the algorithm performed better as soon as the mutation rate was increased, and the algorithm performs better even for the mutation rate of 0.35, which is almost the best mutation rate found in the first set of experiments.

It can be concluded from the above experiments that higher mutation rates and lower crossover rates perform better for this combination of problems and the algorithm. It can be noted that the average relative deviations in Fig. 5 to Fig. 7 are around 1.30%, whereas it is 0.97% in Table 7. This is due to the fact that the results presented in Table 5 and Table 7 are based on the individual parameter tuning, and the results provided in Fig. 5 to Fig. 7 are based on the same parameters for all problems.

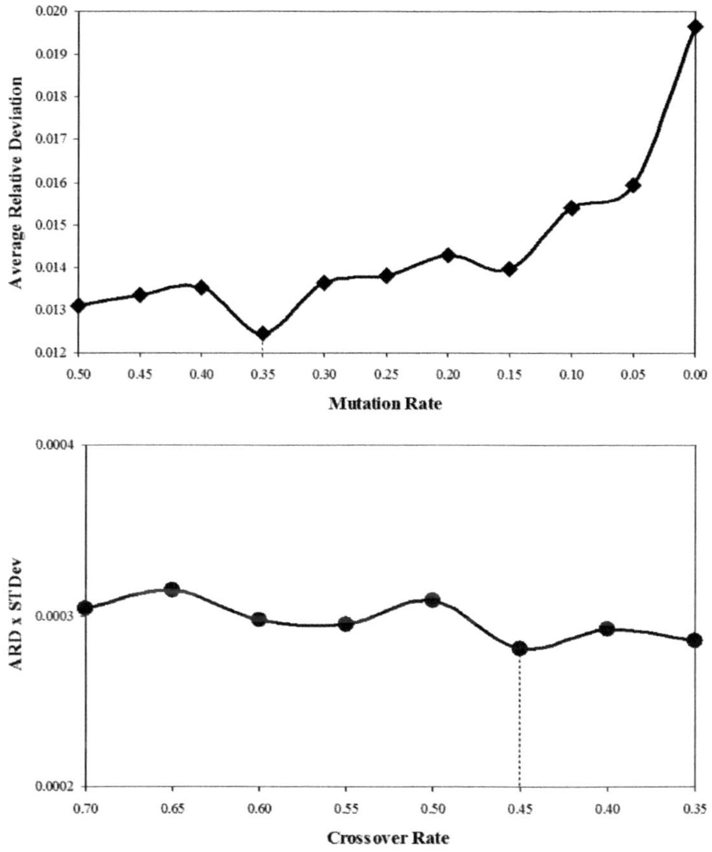

Fig. 6 Average relative deviation (ARD) and the product of ARD and standard deviation based on fixed mutation and variable crossover rate.

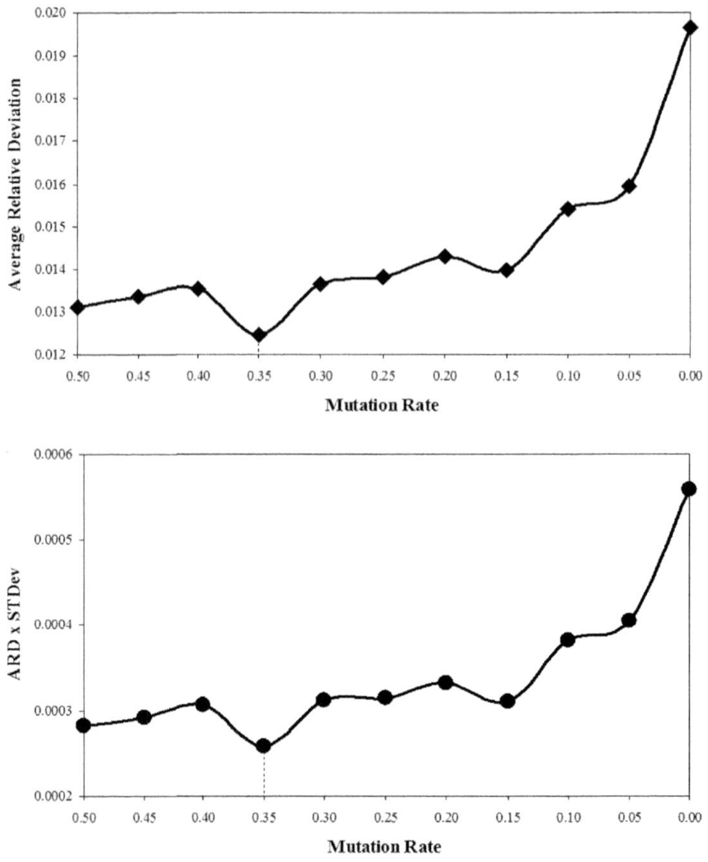

Fig. 7 Average relative deviation (ARD) and the product of ARD and standard deviation based on fixed crossover and variable mutation rate.

8 Conclusion

Although JSSP is a very old and popular problem, no algorithm can yet assure the optimal solution for all test problems, specifically for the larger problems appearing in the literature. GAs are gaining popularity day by day due to their effectiveness in solving optimization problems within a reasonable time period. In this chapter, we presented genetic algorithm based approaches to solve job-shop scheduling problems. After developing a traditional GA with different kind of operations, we designed and applied three priority rules. Combinations of these rules help to improve the performance of the GA. We solved 40 benchmark problems and compared our results with well-known algorithms appearing in the literature. Our algorithm GR-RS-GA clearly outperforms all other algorithms considered in this chapter. We also provided sensitivity analysis of parameters and experimented with different parameters and algorithms for analyzing their contributions. Although our algorithm performed well, we feel that the algorithm requires further work to ensure consistent performance for a wide range of practical JSSPs. We have experimented with different sizes of the test problems varying from 10×5 to 30×10. To justify the robustness of the algorithms, we would like to conduct experiments on large-scale problems with higher complexities. The real life job-shop scheduling problems may not be as straight forward as we considered here. The problems may involve different kinds of uncertainties and constraints. Therefore, we would also like to extend our research by introducing situations like machine breakdown, dynamic job arrival, machine addition and removal, due date restrictions and others. For the case of machine breakdown, we would like to consider two different scenarios: (i) the breakdown information is known in advance, and (ii) the breakdown happened while the schedule is on process. The later case is more practical in terms of the real-life problems. Regarding the dynamic job arrival, our objective is to re-optimize the remaining operations along with the newly arrived job as a separate sub-problem. Machine addition and removal require reorganizing the operations related to the new/affected machine(s). Changing the due date might be similar to changing the priority of an existing job. Setting the preferred due dates may relax or tighten the remaining operations after re-optimization. Finally, the proposed algorithms are significant contributions to the research into solving JSSPs.

References

1. Aarts, E.H.L., Van Laarhoven, P.J.M., Lenstra, J.K., Ulder, N.L.J.: A computational study of local search algorithms for job shop scheduling. ORSA Journal on Computing 6(2), 118–125 (1994)
2. Adams, J., Balas, E., Zawack, D.: The shifting bottleneck procedure for job shop scheduling. Management Science 34(3), 391–401 (1988)
3. Akers Jr., S.B., Friedman, J.: A non-numerical approach to production scheduling problems. Journal of the Operations Research Society of America 3(4), 429–442 (1955)
4. Ashour, S., Hiremath, S.R.: A branch-and-bound approach to the job-shop scheduling problem. International Journal of Production Research 11(1), 47–58 (1973)
5. Baker, K.R.: Introduction to sequencing and scheduling. Wiley, New York (1974)
6. Barnes, J.W., Chambers, J.B.: Solving the job shop scheduling problem with tabu search. IIE Transactions 27(2), 257–263 (1995)
7. Biegel, J.E., Davern, J.J.: Genetic algorithms and job shop scheduling. Computers & Industrial Engineering 19(1-4), 81–91 (1990)
8. Binato, S., Hery, W.J., Loewenstern, D.M., Resende, M.G.C.: A grasp for job shop scheduling. In: Ribeiro, C.C., Hansen, P. (eds.) Essays and surveys on metaheuristics, pp. 58–79. Kluwer Academic Publishers, Boston (2001)
9. Bresina, J.L.: Heuristic-biased stochastic sampling. In: 13th National Conference on Artificial Intelligence, vol. 1, pp. 271–278. CSA Illumina, Portland (1996)
10. Brucker, P., Jurisch, B., Sievers, B.: A branch and bound algorithm for the job-shop scheduling problem. Discrete Applied Mathematics 49(1-3), 107–127 (1994)
11. Carlier, J.: The one-machine sequencing problem. European Journal of Operational Research 11(1), 42–47 (1982)

12. Carlier, J., Pinson, E.: An algorithm for solving the job-shop problem. Management Science 35(2), 164–176 (1989)
13. Cheng, R., Gen, M., Tsujimura, Y.: A tutorial survey of job-shop scheduling problems using genetic algorithms–i. representation. Computers & Industrial Engineering 30(4), 983–997 (1996)
14. Conway, R.W., Maxwell, W.L., Miller, L.W.: Theory of scheduling. Addison-Wesley Pub. Co., Reading (1967)
15. Dauzere-Peres, S., Lasserre, J.B.: A modified shifting bottleneck procedure for job-shop scheduling. International Journal of Production Research 31(4), 923–932 (1993)
16. Della-Croce, F., Tadei, R., Volta, G.: A genetic algorithm for the job shop problem. Computers & Operations Research 22(1), 15–24 (1995)
17. Dell'Amico, M., Trubian, M.: Applying tabu search to the job-shop scheduling problem. Annals of Operations Research 41(3), 231–252 (1993)
18. Dorndorf, U., Pesch, E.: Evolution based learning in a job shop scheduling environment. Computers & Operations Research 22(1), 25–40 (1995)
19. Emmons, H.: One-machine sequencing to minimize certain functions of job tardiness. Operations Research 17(4), 701–715 (1969)
20. Feo, T.A., Resende, M.G.C.: A probabilistic heuristic for a computationally difficult set covering problem. Operations Research Letters 8(2), 67–71 (1989)
21. French, S.: Sequencing and scheduling: an introduction to the mathematics of the job-shop. Ellis Horwood series in mathematics and its applications. E. Horwood; Wiley, Chichester, White Sussex (1982)
22. Garey, M.R., Johnson, D.S.: Computers and intractability: a guide to the theory of NP-completeness. Freeman, W. H., San Francisco (1979)
23. Garey, M.R., Johnson, D.S., Sethi, R.: The complexity of flowshop and jobshop scheduling. Mathematics of Operations Research 1(2), 117–129 (1976)
24. Giffler, B., Thompson, G.L.: Algorithms for solving production-scheduling problems. Operations Research 8(4), 487–503 (1960)
25. Glover, F.: Tabu search – part i. ORSA Journal on Computing 1(3), 190 (1989)
26. Goldberg, D.E.: Genetic algorithms in search, optimization, and machine learning. Addison-Wesley Pub. Co., Reading (1989)
27. Goncalves, J.F., de Magalhaes, M., Jorge, J., Resende, M.G.C.: A hybrid genetic algorithm for the job shop scheduling problem. European Journal of Operational Research 167(1), 77–95 (2005)
28. Hasan, S.M.K., Sarker, R., Cornforth, D.: Hybrid genetic algorithm for solving job-shop scheduling problem. In: 6th IEEE/ACIS International Conference on Computer and Information Science, pp. 519–524. IEEE Computer Society Press, Melbourne (2007)
29. Hasan, S.M.K., Sarker, R., Cornforth, D.: Modified genetic algorithm for job-shop scheduling: A gap-utilization technique. In: IEEE Congress on Evolutionary Computation, pp. 3804–3811. IEEE Computer Society Press, Singapore (2007)
30. Hasan, S.M.K., Sarker, R., Cornforth, D.: Ga with priority rules for solving job-shop scheduling problems. In: IEEE World Congress on Computational Intelligence, pp. 1913–1920. IEEE Computer Society Press, Hong Kong (2008)
31. Kacem, I., Hammadi, S., Borne, P.: Approach by localization and multiobjective evolutionary optimization for flexible job-shop scheduling problems. IEEE Transactions on Systems, Man, and Cybernetics, Part C: Applications and Reviews 32(1), 1–13 (2002)
32. van Laarhoven, P.J.M., Aarts, E.H.L., Lenstra, J.K.: Job shop scheduling by simulated annealing. Operations Research 40(1), 113–125 (1992)

33. Lawrence, D.: Resource constrained project scheduling: An experimental investigation of heuristic scheduling techniques. Tech. rep., Graduate School of Industrial Administration, Carnegie-Mellon University (1984)
34. Lawrence, D.: Job shop scheduling with genetic algorithms. In: First International Conference on Genetic Algorithms, pp. 136–140. Lawrence Erlbaum Associates, Inc., Mahwah (1985)
35. Lenstra, J.K., Rinnooy Kan, A.H.G.: Computational complexity of discrete optimization problems. Annals of Discrete Mathematics 4, 121–140 (1979)
36. Muth, J.F., Thompson, G.L.: Industrial scheduling. Prentice-Hall international series in management. Prentice-Hall, Englewood Cliffs (1963)
37. Nakano, R., Yamada, T.: Conventional genetic algorithm for job shop problems. In: Belew, Booker (eds.) Fourth International Conference on Genetic Algorithms, pp. 474–479. Morgan Kaufmann, San Francisco (1991)
38. Ombuki, B.M., Ventresca, M.: Local search genetic algorithms for the job shop scheduling problem. Applied Intelligence 21(1), 99–109 (2004)
39. Paredis, J.: Handbook of evolutionary computation. In: Parallel Problem Solving from Nature 2. Institute of Physics Publishing and Oxford University Press, Brussels (1992)
40. Paredis, J.: Exploiting constraints as background knowledge for evolutionary algorithms. In: Back, T., Fogel, D., Michalewicz, Z. (eds.) Handbook of Evolutionary Computation, pp. G1.2:1–6. Institute of Physics Publishing and Oxford University Press, Bristol, New York (1997)
41. Park, B.J., Choi, H.R., Kim, H.S.: A hybrid genetic algorithm for the job shop scheduling problems. Computers & Industrial Engineering 45(4), 597–613 (2003)
42. Ponnambalam, S.G., Aravindan, P., Rao, P.S.: Comparative evaluation of genetic algorithms for job-shop scheduling. Production Planning & Control 12(6), 560–674 (2001)
43. Shigenobu, K., Isao, O., Masayuki, Y.: An efficient genetic algorithm for job shop scheduling problems. In: Eshelman, L.J. (ed.) 6th International Conference on Genetic Algorithms, pp. 506–511. Morgan Kaufmann Publishers Inc., Pittsburgh (1995)
44. Sprecher, A., Kolisch, R., Drexl, A.: Semi-active, active, and non-delay schedules for the resource-constrained project scheduling problem. European Journal of Operational Research 80(1), 94–102 (1995)
45. Student: The probable error of a mean. Biometrika 6(1), 1–25 (1908)
46. Tsai, C.F., Lin, F.C.: A new hybrid heuristic technique for solving job-shop scheduling problem. In: Proceedings of the Second IEEE International Workshop on Intelligent Data Acquisition and Advanced Computing Systems: Technology and Applications, 2003., pp. 53–58 (2003)
47. Wang, W., Brunn, P.: An effective genetic algorithm for job shop scheduling. Proceedings of the Institution of Mechanical Engineers - Part B: Journal of Engineering Manufacture 214(4), 293–300 (2000)
48. Werner, J.C., Aydin, M.E., Fogarty, T.C.: Evolving genetic algorithm for job shop scheduling problems. In: Adaptive Computing in Design and Manufacture, Plymouth, UK (2000)
49. Xing, Y., Chen, Z., Sun, J., Hu, L.: An improved adaptive genetic algorithm for job-shop scheduling problem. In: Third International Conference on Natural Computation, Haikou, China, vol. 4, pp. 287–291 (2007)
50. Yamada, T.: Studies on metaheuristics for jobshop and flowshop scheduling problems. Ph.D. thesis, Kyoto University (2003)
51. Yamada, T., Nakano, R.: Genetic algorithms for job-shop scheduling problems. In: Modern Heuristic for Decision Support, UNICOM seminar, London, pp. 67–81 (1997)

52. Yang, G., Lu, Y., Li, R.W., Han, J.: Adaptive genetic algorithms for the job-shop scheduling problems. In: 7th World Congress on Intelligent Control and Automation, pp. 4501–4505. IEEE Computer Society Press, Dalian (2008)
53. Zribi, N., Kacem, I., Kamel, A.E., Borne, P.A.B.P.: Assignment and scheduling in flexible job-shops by hierarchical optimization. IEEE Transactions on Systems, Man, and Cybernetics, Part C: Applications and Reviews 37(4), 652–661 (2007)

Appendix

Algorithm 1. Algorithm to find out the Bottleneck Job

Let $Q_p(m,k)$ be the kth job in machine m for the phenotype p and $C(m,n)$ is the finishing time of nth operation of machine m; where m varies from 1 to M and n varies from 1 to N. *getBottleneckJob* is a function that returns the job which is takes the maximum time in the schedule.
getBottleneckJob (void)

1. Set $m := 1$ and $max := -1$
2. Repeat while $m \leq M$
   a. If $max < C(m,N)$ then
      i. Set $max := C(m,N)$
      ii. Set $j := Q_p(m,N)$
      [End of step 2a If]
   b. Set $m := m+1$

   [End of step 2 Loop]
3. Return j

[End of Algorithm]

Algorithm 2. Algorithm for the Partial Reordering Technique (PR)

Let $D(j,m)$ be the mth machine in job j in the predefined machine sequence D. $O_p(j,m)$ is the order of job j in machine m for the phenotype p.

1. Set $j :=$ *getBottleneckJob* and $m := D(j,1)$
2. Set $k := O_p(j,m)$
3. Repeat until $k > 1$
   a. Swap between $C_p(m,k)$ and $C_p(m,k-1)$
   b. Set $k := k-1$

   [End of step 3 Loop]

[End of Algorithm]

Algorithm 3. Algorithm for the Gap-Reduction Technique (GR)

Let p be the phenotype of an individual i, M and N are the total number of machines and jobs respectively. S and C is the set of starting and finishing times of all the operations respectively of those that have already been scheduled. $T(j,m)$ is the execution time of the current operation of job j in machine m. $Q_p(m,k)$ is the kth operation of machine m for phenotype p. $mFront(m)$ represents the front operation of machine m, $jFront(j)$ is the machine where schedulable operation of job j will be processed. $jBusy(j)$ and $mBusy(m)$ are the busy time for job j and machine m respectively. $max(m,n)$ returns the m or n which is the maximum.

1. Set $m := 1$ and $mFront(1:M) := 0$
2. Repeat until all operations are scheduled
   a. Set $Loc := mFront(m)$ and $jID := Q_p(m,Loc)$
   b. If $jFront(jID) = m$ then
      i. Set $flag := 1$ and $k := 1$
      ii. Repeat until $k \leq Loc$
         A. Set $X := max(C(m,k-1), jBusy(jID))$
         B. Set $G := S(m,k) - X$
         C. If $G \geq T(jID,m)$
            - Set $Loc := k$
            - Go to Step F

            [End of Step b If]
         D. Set $k := k+1$

         [End of Step ii Loop]

      Else Set flag:=flag+1

      [End of Step B If]
   c. Set $j_1 := 1$
   d. Repeat while $j_1 \leq J$
      i. Set $mID := jFront(j_1)$
      ii. Find the location h of j_1 in machine mID
      iii. Put j_1 in the front position and do 1-bit right shift from location $mFront(mID)$ to h.
      iv. Set $j_1 := j_1 + 1$

      [End of Step D Loop]
   e. Go to Step A
   f. Place jID at the position Loc
   g. Set $S(m,Loc) := X$
   h. Set C (m,Loc):= S (m,Loc)+T(jID,m)
   i. Set m:=(m+1) mod M

   [End of Step 2 Loop]

[End of Algorithm]

Algorithm 4. Algorithm for the Restricted Swapping Technique (RS)

Let $Q_p(m,k)$ be the kth job in machine m and $O_p(j,m)$ is the order of job j in machine m particularly for the phenotype p. $nonConflict(m,i,j)$ is a function that returns true if the ending time of immediate predecessor operation of j does not overlap with the modified starting time of the same job in machine m and the starting time of the immediate following operation of job j does not conflict with the ending time of the same job in machine m.

1. Set $j' := getBottleneckJob$ and $k := N-1$
2. Repeat while $k \geq 1$
   a. Set $m := S(j',k)$
   b. If $O_p(j',m) \neq 1$ then
      i. Set $j" := Q_p(m,(O_p(j',m)-1))$
      ii. If $nonConflict(m,j',j")$=true
         - Swap j' with $j"$ in phenotype p
         - Go to `step 2c`

         [End of `step 2(b)ii` If]

      [End of `step 2b` If]
   c. Set $k := k-1$

   [End of `step 2` loop]

[End of Algorithm]

An Estimation of Distribution Algorithm for Flowshop Scheduling with Limited Buffers

Mansour Eddaly, Bassem Jarboui, Patrick Siarry, and Abdelwaheb Rebaï

Abstract. Most of the works that address the flowshop scheduling problem presume unlimited buffers between successive machines. However, with the advent of new technologies in the manufacturing systems, limited capacity storage between machines has become profitable. Aimed at makespan minimization, the flowshop scheduling problem with buffer constraints is NP-hard in the strong sense. Therefore, several approximate algorithms have been proposed in the literature. In this chapter, we propose an Estimation of Distribution Algorithm for solving a flowshop scheduling problem with buffer constraints. The main characteristics of the problem, such as the order of jobs and similar blocks of jobs in the sequence, are taken into account while building the probabilistic model. In order to enrich the search procedure of the algorithm, a skewed variable neighbourhood search algorithm is embedded into it, restricted by a calculated probability which depends on the quality of the created offspring. The computational results show that our algorithm outperforms genetic algorithm and particle swarm algorithm, and can obtain several optimal solutions in a short time.

1 Introduction

In a classical flowshop scheduling problem (FSP), each job j ($j = 1,2,...,n$) must be processed on every machine i ($i = 1,2,...,m$) and all the jobs have to pass through all the machines following the same route. In such problems, an infinite buffer size between every two successive machines is assumed. Due to the modernization of

Mansour Eddaly, Bassem Jarboui, and Abdelwaheb Rebaï
FSEGS, route de l'aéroport km 4.5, B.P. No. 1088, Sfax 3018, Tunisie
e-mail: `eddaly.mansour@gmail.com`, `bassem_jarboui@yahoo.fr`, `abdelwaheb.rebai@fsegs.rnu.tn`

Patrick Siarry
LiSSi, Université de Paris 12, 61 avenue du Général de Gaulle, 94010 Créteil, France
e-mail: `siarry@univ-paris12.fr`

R. Chiong, S. Dhakal (Eds.): Nat. Intel. for Sched., Plan. and Pack. Prob., SCI 250, pp. 89–110.
springerlink.com

manufacturing technologies and the emergence of just-in-time manufacturing and Kanban control systems, which maintain a limited in-process inventory, the constraint of intermediate buffers with a limited capacity between every two successive machines is introduced. Therefore, the study of flowshop scheduling with limited buffer storage has become attractive for many researchers [28]. In real world applications, the FSP with limited buffers may occur in industrial production, e.g. petrochemical processing industries and cell manufacturing [31], as well as in computer architecture [13]. Papadimitriou and Kanellakis [21] have proved that this problem is strongly NP-hard with respect to the makespan criterion. Therefore, many approximate algorithms have been developed in the literature for solving the FSP with limited buffers.

With the makespan criterion, Dutta and Cunninghan [5] and Reddi [26] have proposed a dynamic programming technique aimed at finding an optimal solution for the problem with two or more machines successively. Leisten [13] has analysed the importance of the limited storage restrictions on the performance of the production systems. Moreover, he has presented several constructive heuristics to solve this variant of FSP. Nowicki [20] has proposed a tabu search approach (TS) by using a non-trivial generalization of the block elimination properties known for the classical FSP. Qian et al. [25] have proposed a differential evolution algorithm (DE) where they developed a rule to convert the continuous version of DE individuals to job permutations. Also, they applied a local search procedure based on the insertion of a part of individuals at each generation. In [31] a genetic algorithm (GA) is adapted to solve the problem where multiple crossover and mutation operators are used simultaneously. Moreover, a decision probability is used to control the utilization of genetic mutation and local search. Liu et al. [14] have presented a particle swarm optimization (PSO) algorithm. They have proposed a new encoding scheme allowing the transition from the continuous values of the PSO to job permutations. A local search procedure and a simulated annealing algorithm (SA) are introduced into the algorithm in order to enrich the search procedure and obtain solutions with good diversity.

In 1996, a new class of evolutionary algorithms was introduced by Mühlenbein and Paaß [19], called Estimation of Distribution Algorithm (EDA). In order to generate a new individual, EDA uses a probabilistic model learned from a population of individuals. Therefore, a distribution of probability is estimated from the selected candidates of the initial population and then new offspring are generated according to this distribution. Finally, according to the quality of the new individual, it is decided whether it will survive in the next generation. Concerning its applications, EDA was devoted to solving several combinatorial optimization problems. A detailed survey of these applications can be found in Larrañaga and Lozano [12], Pelikan [22], and Lozano et al. [15].

In this chapter, we propose an EDA for solving the FSP with limited buffers while minimizing the makespan criterion. Aimed at enhancing the quality of the solution of EDA, it is recommended to use a local search algorithm [15]. The Skewed Variable Neighborhood Search (SVNS) [7] is embedded to the EDA for improving its performance.

2 Problem Definition

In a FSP with limited buffers, there is a set of n jobs to be processed through a set of m machines with the same sequence on all machines and without interruption in the processing (no pre-emption). Let $p_{i[j]}$ be the processing time for the job at position j in the sequence s on the machine i, where $j = 1,2,...,n$, $i = 1,2,...,m$, $s = \{s_1, s_2, ..., s_n\}$ and p_{ij} are fixed, known in advance and have non-negative values. The intermediate storage constraint implies that between each successive pair of machines $(i-1,i)$ a buffer with limited size B_i of unfinished jobs can temporarily stay. Let $D_{i[j]}$ denotes the departure time (starting time) of the job at the jth position in the sequence on the machine i.

The completion time of the last executed job, also known as the makespan ($C_{\max}$) can be found through the recursive expression of the departure time as follows:

$$D_{1[1]} = 0$$
$$D_{1[j]} = D_{1[j-1]} + p_{1[j-1]},\ j = 2,3,...,n$$
$$D_{i[j]} = \max\left(D_{i-1[j]} + p_{i-1[j]}, D_{i[j-1]} + p_{i[j-1]}, D_{i+1[j-B_{i+1}-1]}\right)\ i = 2,...,m, j = 2,...,n$$

Thus,

$$C_{\max} = D_{n[m]} + p_{n[m]}$$

Following this definition, two cases can be discussed. If $B_i = 0$, the limited buffer constraint disappears and the problem will be transformed into the blocking flowshop scheduling problem [28]. Otherwise, if $B_i \geq n-1$, the problem under consideration becomes the classical FSP. Nowicki [20] has presented a detailed analysis of the limited buffer FSP based on a graph representation model.

3 An Overview of Estimation of Distribution Algorithms (EDAs)

The framework of the basic EDA [19] can be presented as follows: starting with a randomly generated initial population, one selects a subpopulation of M parent individuals through a selection method based on the fitness function. Next, one estimates the probability of distribution of the selected parents with a probabilistic model. Then, one generates new offspring, according to the estimated probability distribution. Finally, some individuals in the current population are replaced with newly generated offspring. These steps are repeated until a stopping criterion is met.

Three classes of EDA were developed, according to the chosen probabilistic model. The first class consists of models which do not take into account the dependencies between variables of candidate solutions, i.e. all variables are independent. The second class assumes at most two-order dependencies between these variables and the last class assumes multiple dependencies between the variables.

3.1 EDAs with No Dependencies

3.1.1 Population-Based Incremental Learning Algorithm (PBIL)

The PBIL algorithm was introduced by Baluja [1]. It involves transforming the population of candidate solutions into a probability vector $p(x) = \{p(x_1), p(x_2), ..., p(x_n)\}$ where $p(x_i)$ denotes the probability of a "1" in the ith position of solution bits. In other words, this algorithm checks the proportion of ones in each individual chromosome. All positions are initially equally likely, that is all the probability values are set to 0.5. In this way, the highest level of diversity will be found [1]. In each iteration, after transforming the population into a probability vector, the PBIL algorithm generates P_1 solutions according to the current probability. Then, each component of probability vector, $p(x_i)$, is updated according to the following:

$$p(x_i) = p(x_i) + \alpha(x_i^{best} - p(x_i))$$

where $\alpha \in [0,1]$ is the so-called learning rate and x_i^{best} is the ith bit in the best solution found so far. Using this update rule, the value of $p(x_i)$ increases when x_i^{best} is equal to "1" and decreases otherwise. Furthermore, the values in the probability vector move towards the fittest of the generated solutions. The aim of PBIL is to actively create a probability vector which, with high probability, represents a population of highly evaluated solution vectors. It is worthy to note that each bit is examined independently and thus no interaction is considered.

3.1.2 Compact Genetic Algorithm (CGA)

CGA was proposed by Harik et al. [8]. Similar to PBIL algorithm, CGA creates a probability vector from the population of solutions and all vector values are initialized to 0.5. The update rule used in this algorithm is equivalent to a steady-state tournament selection. So, the procedure is as follows: at each iteration, two individuals are selected at random from the population. These individuals are compared with respect to the fitness values. Then, the probability vector is updated, following this formula:

$$p(x_i) = \begin{cases} p(x_i) + \frac{1}{P} & \text{if } x_i^{best} = 1 \text{ and } x_i^{worst} = 0; \\ p(x_i) - \frac{1}{P} & \text{if } x_i^{best} = 0 \text{ and } x_i^{worst} = 1; \\ p(x_i) & \text{otherwise.} \end{cases}$$

where x_i^{best} and x_i^{worst} denote the ith bit in the best and the worst solutions respectively, among the two selected solutions and P is the population size. According to this method of updating the probability, the CGA replaces one candidate solution by another one using a population of size P.

Two points of difference between PBIL and CGA can be found. Firstly, the update rule used by CGA depends only on the size of the population. Secondly, the CGA is more profitable in terms of memory requirements. The update phase in CGA requires a constant size of $\frac{1}{P}$ whereas the update rule values in PBIL can take an arbitrary value in the range [0,1] and the number of values that can be stored in an element of the vector is infinite.

3.1.3 Univariate Marginal Distribution Algorithm (UMDA)

The UMDA was proposed by Mühlenbein et al. [19]. This algorithm works differently compared to PBIL and CGA. The UMDA does not transform the population into a population vector. After generating a population of individuals, UMDA selects some individuals according to their fitness values. Then, it constructs a probability vector through the selected solution. Formally, the probabilistic model used in UMDA can be written as follows:

$$p(x) = \prod_{i=1}^{n} p(x_i).$$

Finally, it generates new solutions based on the built probabilistic model. The process is repeated until a termination criterion is met. Although UMDA has different framework than CGA and PBIL, the performances of these algorithms are similar [22].

3.2 *EDAs with Two-Order Dependencies*

3.2.1 Mutual Information Maximization for Input Clustering (MIMIC)

MIMIC was proposed by De Bonet et al. [4]. It considers pairwise interactions between the variables in its distribution of probability.

Let $p(x) = p(x_1/x_2,...,x_n)p(x_2/x_3,...,x_n)...p(x_{n-1}/x_n)p(x_n)$ denote the true joint probability, and $\hat{p}_\pi(x) = p(x_{i1}/x_{i2})p(x_{i2}/x_{i3})...p(x_{in-1}/x_{in})p(x_{in})$ where $\pi\{i_1,i_2,...,n\}$ is a permutation of the numbers between 1 and n and $\hat{p}_\pi(x)$ is estimated by the marginal and conditional relative frequencies of the corresponding variables within the subset of selected individuals. The aim of MIMIC algorithm is to generate a permutation π^* that minimizes the divergence between $\hat{p}_\pi(x)$ and $p(x)$. This probabilistic model has the framework of a chain. In a sequential way, all positions, except for the first position of the chain, are conditionally dependent on the previous position in the chain. Starting with an initial random population, the MIMIC algorithm selects a set of promising solutions. After that, it computes the marginal and conditional probabilities. Next, MIMIC uses a greedy algorithm, based on Kullback-Leibler information divergence, to maximize mutual information between the adjacent positions in the chain.

3.2.2 Combining Optimizers with Mutual Information Trees (COMIT)

COMIT algorithm, proposed by Baluja and Davies [2], also considers pairwise interactions between variables of the problem. Instead of using the shape of a chain in the probabilistic model, as in MIMIC, this algorithm attempts to present the probability vector through a tree distribution. The probabilistic model, constructed here, can be written as follows:

$$p(x) = \prod_{i=1}^{n} p(x_i/x_j).$$

where i is a descendant node from j. If i is a root node then $p(x_i/x_j) = p(x_i)$. The objective is to find a tree that maximizes mutual information between parent nodes and their descendants. In [2], it is shown that COMIT outperforms MIMIC algorithm.

3.2.3 Bivariate Marginal Distribution Algorithm (BMDA)

BMDA was introduced by Pelikan and Mühlenbein [24]. It constitutes a kind of a generalization of both the previous algorithms and UMDA. For constructing the probability distribution, this algorithm uses a set of disjoint tree distributions. The criterion used for measuring the dependencies between variables is Pearson's chi-square statistics using the following:

$$\chi_{ij}^2 = M \times \sum_{x_i, x_j} \frac{(p(x_i, x_j) - p(x_i)p(x_j))^2}{p(x_i)p(x_j)}$$

3.3 EDAs with Multiple Dependencies

3.3.1 Factorized Distribution Algorithm (FDA)

FDA was proposed by Mühlenbein and Mahnig [17]. The method involves using an additive decomposition of function (ADF), to decompose the fitness function of the problem and a factorization method, based on Boltzmann distribution which is used to factor the distribution into consistent, conditional and marginal distributions. In such a way, N individuals are generated according to uniform distribution as in UMDA. Then, the Boltzmann selection is used to select M individuals from those generated previously. In addition, the conditional probabilities are estimated from the selected individuals. Finally, M new individuals are generated according to these probabilities. The process is repeated until a termination criterion is met.

3.3.2 The Extended Compact Genetic Algorithm (ECGA)

The ECGA [8] is an extension inspired from CGA. It works by partitioning the variables into different partitions by using the so called Marginal Product Model (MPM). This model does not estimate the conditional probabilities as in FDA. However, it includes both univariate marginal distribution as well as multivariate marginal distribution. The form of probability distribution used in MPM can be presented as:

$$p(x) = \sum_{i \in \Omega} p(x_i)$$

where Ω is the set of alleles included in the same partitions. Based on MPM, ECGA uses a greedy procedure in order to obtain the best partitions. At the beginning, MPM assumes that all variables are independent, i.e. all partitions are composed of one element. Then, at each iteration, it merges two partitions together with respect to the Bayesian Information criterion. Finally, when no further possible improvement is obtained, the model used is retained.

3.3.3 Bayesian Optimization Algorithm (BOA)

In [23], the BOA was developed. This algorithm considers multiple interactions among the variables. At first, an initial random population is generated. Secondly, a selected method is chosen to select the promising solutions. Thirdly, a Bayesian Network is constructed from a probabilistic model based on the selected individuals. Formally, an acyclic Bayesian network with directed edges is used in this algorithm. The estimated joint probability can be written as follows:

$$p(x) = \prod_{i=0}^{n-1} p(x_i / \Phi_{x_i})$$

where Φ_{x_i} is the set of nodes from which there exists an edge to x_i. In such way, a variable is sampled after all its parents have already been sampled. In order to evaluate the quality of the constructed network, a metric named Bayesian Dirichlet metric is used by Pelikan et al. [23]. Finally, new offspring are generated according to this network.

3.3.4 Learning Factorised Distribution Algorithm (LFDA)

LFDA [18] extends the FDA by using the Bayesian network. Instead of using the ADF structure, as in FDA, LFDA uses the Bayesian network framework to build the probabilistic model. In fact, this algorithm does not need a priori information about the structure of the problem. Therefore, LFDA performs similarly to BOA with the exception of the metric used . In this context, LFDA uses a Bayesian Information Criterion (BIC) in order to assess the network's quality.

3.3.5 Estimation of Bayesian Network Algorithm (EBNA)

EBNA was proposed by Etxeberria and Larrañaga [6]. Also, these authors exploit the Bayesian network structure in estimating the joint probability. This algorithm follows the same process as BOA and LFDA. In [11] three different variants of EBNA have been proposed where they differ in the metrics used in the evaluation step. The first one, called EBNA$_P C$, uses Chi square tests to detect the conditional independencies. The BIC criterion is devoted to the second variant EBNA$_{BIC}$. The last one is called EBNA$_{k_2+pen}$ which uses the marginal likelihood in order to construct the Bayesian network and a penalty is added to the metric for reducing the complexity of the algorithm.

4 Estimation of Distribution Algorithm (EDA) for the Flowshop Scheduling Problem with Limited Buffer

EDA is a new evolutionary algorithm proposed by Mühlenbein and Paaß in 1996 [19]. Instead of recombination and mutation, EDA generates new individuals with respect to a probabilistic model, learned from the population of parents. To the best of our knowledge, in the literature, there is no application of this algorithm in the scheduling context, with the exception of the work of Jarboui et al. [10]. In this section, we discuss, in detail, our proposed EDA for solving the FSP with respect to makespan criterion.

4.1 Encoding Solutions and Initial Population

For encoding the solution, we use the well-known representation scheme for the permutation FSP, which is the permutation of n jobs, where the jth number in the permutation denotes the job located at position j. For the purpose of diversity, the initial population of P individuals is generated randomly.

4.2 Selection

The selection procedure adopted in our algorithm consists of two phases. First, the individuals of the initial population are sorted according to their objective function value. Second, M individuals are selected from the subset of 20% of the best individuals from the sorted list.

4.3 Probabilistic Model: Construction of New Solution

The choice of the probabilistic model is closely related to the performance of the EDA [15]. So, it is recommended to carefully construct this model. In this chapter, the probabilistic model is built while taking into account the specific characteristics of the problem under consideration, as in [10]. Therefore, both the order of jobs in the sequence and the similar blocks of jobs presented in the selected parents are considered. In order to construct a new sequence, Jarboui et al. [10] have proposed a probability π_{jk} for selecting a job j at position k:

$$\pi_{jk} = \frac{\eta_{jk} \times \mu_{j[k-1]}}{\sum_{l \in \Omega_k} \eta_{lk} \times \mu_{l[k-1]}}$$

with η_{jk} denotes the number of times job j that appears before or at the position k in the subset of the selected sequences plus a given constant δ_1. This parameter indicates the importance of the order of jobs in the sequence. $\mu_{j[k-1]}$ denotes the

```
EDA Algorithm
Initialize the population P using random permutation;
loop
    select M individuals from 20% of the best individuals in the population;
    select s0 at random from the M selected individuals;
    for k = 1,2,...,n
        update Ω_k using the q first job not already scheduled from s0;
        compute the probability π_jk for each job j ∈ Ω_k;
        s_current,k receives the job selected from Ω_k following the computed probability;
    end for
    if (the fitness of newly created induvidual s_current is better than the worst individual in the
        population) then
        replace the latter by s_current;
    end if
until (stopping criterion is reached)
```

Fig. 1 EDA procedure

number of times job j comes immediately after the job at the position $k-1$ in the new individual from the sequences of the selected parents. This parameter is used to highlight the importance of similar blocks in the sequence. Ω_k represents the set of jobs not already scheduled until position k. According to this model, the task of constructing new offspring has a complexity order of $O(n^2)$. Although this model provides a good solution quality, it consumes high CPU times, especially when the size of the problem increases. In this work, we propose an improvement to the model of Jarboui et al. [10] by reducing its complexity. First, a sequence of jobs s_0 is selected, at random, among the M selected parents. Then, the probabilistic model is built like the one described above while modifying the set Ω_k into Ω_{kq}. The latter set is composed of q jobs not already scheduled, following their order in s_0, until position k. So, while comparing the model proposed here and the one proposed by Jarboui al. [10], we find that the phase of construction of new individual requires a linear time $O(n)$ in this proposed model. Moreover, this formulation grants more intensification to the algorithm. The pseudo-code of the EDA is given in Fig. 1.

4.4 *Skewed Variable Neighborhood Search (SVNS) Algorithm*

SVNS algorithm is an extension inspired by the original variable neighborhood search algorithm proposed by Hansen and Mladenovič [7]. In order to escape from the local optima, this algorithm is added to our EDA after creating a new solution. Firstly, to increase the performance of the SVNS, we restrict the application of the algorithm to only a part of the individuals [29]. To do this, we define a calculated probability p^c, as in [10], which decides if the new individual will be subject to an application of the SVNS algorithm. This probability depends on the quality of the created individual, as follows:

$$p^c = \max\left\{\exp\left(\frac{RD}{\alpha}\right), \varepsilon\right\},$$

where

$$RD = \left(\frac{f(s_{current}) - f(s_{best})}{f(s_{best})}\right)$$

and $s_{current}$ and s_{best} denote the sequences of the new offspring and the best individual found so far, respectively. This probability depends on the distance separating the current solution and the best solution. Therefore, the best obtained solutions have a high probability of being subject to an application of SVNS algorithm.

A random number in the range $[0,1]$ is drawn for each individual. If this number is less than or equal to p^c, then the considered individual will be eligible to be subjected to the SVNS algorithm. This algorithm consists of two structures of neighbourhood ($k = 1,2$). The first structure leads to the selection of two distinct positions at random from $s_{current}$ and the permutation of the jobs located on those positions. The second one consists of selecting, randomly, a job j from $s_{current}$ and inserting it at a randomly selected position. After setting k to 1, the shaking phase generates a sequence s_1 according to the kth neighborhood. This phase is introduced into the SVNS algorithm in order to avoid cycling which might occur if any deterministic rule was used.

Then, two local search methods are employed. The first procedure performs all possible swap moves in the sequence s_1. The obtained local optimum is denoted by s_2 (see Fig. 2). After that, the second local search procedure starts performing all insertion moves to s_2. Let s_3 be the resulting local optimum (see Fig. 3). If $f(s_3) < f(s_2)$, we replace s_1 by s_3, and we return to the first local search procedure. The process is repeated until no improvement is possible. Then we compare the objective

Swap Local Search

```
swap_local_search()
Let s0 be the input solution
Let sbest be the output solution
sbest = s0;
set i = 1;
loop
    for j = i+1, i+2, ..., n
        create a new solution by exchanging the jobs at position i and j in the best sequence;
        if (the fitness of the new solution is better than the sbest) then
            update sbest;
            break;
        end if
    endfor
    i = i+1;
    if (i > n-1) set i = 1;
until (no possible improvement)
```

Fig. 2 Swap local search procedure

function values of s_3 (the local optimum) and $s_{current}$ (the incumbent solution). If s_3 is better than $s_{current}$, then the latter is updated and we go to the shaking phase with $k = 1$. Else, we define a distance function $\rho(s_3, s_{current})$ as a discrete function, as follows:

$$\rho(s_3, s_{current}) = \frac{\sum_{j=1}^{n} \left| pos_3^j - pos_{current}^j \right|}{n}$$

where pos_3^j denotes the position of the job j in s_3 and $pos_{current}^j$ denotes the position of the job j in the incumbent sequence $s_{current}$. So, $\rho(s_3, s^{current})$ represents the position lags between the obtained local optimum and the incumbent solution. The calculation of this distance function is illustrated by the example below. Let $s_3 = \{6,4,3,1,2,5\}$ and $s_{current} = \{3,1,6,4,2,5\}$, we define by pos_3 and $pos_{current}$ two n-dimensional vectors, where pos_3^j and $pos_{current}^j$ indicate the positions of the job j in s_3 and $s_{current}$, respectively. So, $pos_3 = \{4,5,3,2,6,1\}$ and $pos_{current} = \{2,5,1,4,6,3\}$ and consequently:

$$\rho(s_3, s^{current}) = \frac{|4-2| + |5-5| + |3-1| + |2-4| + |6-6| + |1-3|}{6} = 1.33$$

Also, we define a parameter β: when $f(s_3) - f(s_{current}) < \beta \times \rho(s_3, s_{current})$, then we accept the deterioration of the current solution and we replace $s_{current}$ by s_3 and thereafter k is reinitialized to 1. Otherwise, k is incremented to 2 and we return to the shaking process. Finally, the whole procedure is repeated until a maximum number of iterations ($iter_{\max}$) is reached. Fig. 4 describes the pseudo-code of the SVNS algorithm.

Insert Local Search

Let s_0 be the input solution
Let s_{best} be the output solution
$s_{best} = s_0$;
set $i = 1$;
loop
  for $j = 1, 2, ..., n$ and $j \neq i$
    create a new solution by inserting the job at position i into position j in the best sequence;
    if (the fitness of the new solution is better than the s_{best}) **then**
      update s_{best};
      break;
    end if
  endfor
  $i = i + 1$;
  if $(i > n)$ set $i = 1$;
until (no possible improvement)

Fig. 3 Insert local search procedure

```
SVNS Algorithm
Let s0 be the input solution
Let s_best be the output solution
s_best = s0;
s_current = s0;
loop
    set k = 1;
    do
        s1 = s_current;
        if (k = 1) then
            select i and j at random and insert the job at position i into position j in s1;
        else
            select i and j at random and exchange the jobs at position i and j in s1;
        endif
        do
            find the optimum local s2 by applying swap local search to s1;
            find the optimum local s3 by applying insert local search to s2;
            if (s3 is better than s2) then
                s1 = s3;
            endif
        while (s3 is better than s2)
        if (s3 is better than s_best) then
            s_best = s3;
            s_current = s3;
            k = 1;
        else if (f(s3) - f(s_current) < β × ρ(s3, s_current)) then
            s_current = s3;
            k = 1;
        else
            k = k + 1;
        endif
    while(k ≤ 2)
until(stopping criterion is reached)
```

Fig. 4 SVNS procedure

4.5 *Replacement and Stopping Criterion*

In our algorithm, we compare the new individual with the worst individual in the current population. If the first one is better than the latter and the sequence of the offspring is unique, then the worst individual is removed from the population and is replaced by the new individual. In the literature, various stopping criteria have been proposed, such as the maximum number of generations, bound of time, the maximum number of iterations without improvement, etc. In our work, we set a maximal computational time. The pseudo-code of the whole hybrid algorithm is provided in Fig. 5.

Hybrid EDA

Initialize the population P using random permutation;
loop
  select M individuals from 20% of best individuals in the population;
  select s_0 at random from the M selected individuals;
  for $k = 1,...,n$
    update Ω_k using the q first job not already scheduled from s_0;
    compute the probability π_{jk} for each job $j \in \Omega_k$;
    $s_{current,k}$ receives the job selected from Ω_k following the computed probability;
  end for
  compute p^c;
  if $(rand \leq p^c)$ **then** // *rand* denotes a random value uniformly distributed in the range $[0,1]$
    Apply the SVNS procedure to improve $s_{current}$;
  end if
  if (the fitness of new created induvidual $s_{current}$ is better than the worst individual in the population) **then**
    replace the latter by $s_{current}$;
  end if
until (stopping criterion is reached)

Fig. 5 EDA with SVNS procedure

5 Experimental Results and Comparative Study

The algorithm was coded in C++ programming language. All experiments with EDA for FSP with limited buffers, with respect to the makespan criterion, were run in Windows XP on a desktop PC with Intel Pentium IV, 3.2 GHz processor and 512 MB memory. EDA was tested for 29 instances distributed as follows: 8 instances of Carlier [3] and 21 instances of Reeves [27]. All results were obtained after 20 replications for each instance.

The corresponding parameters of EDA were set, experimentally, to the following values: $P = 20$, $\delta_1 = \delta_2 = 4/n$, the number of the selected parents $M = 3$, the number of generated offspring $O = 3$. Numerically, $p^c = 0.5$ leads to the acceptance of a sequence with a makespan relatively superior by 1% to the best value of the makespan found so far. So, thereafter we determined p^c according to this formula: $p^c = \max\left\{\exp\left(\frac{RD}{\alpha}\right), \varepsilon\right\}$, with $\varepsilon = 0.01$. The maximum number of iterations of VNS algorithm $(iter_{\max})$ was set at 3 and the coefficient β was fixed to 2. The performance measure employed in our numerical study was the average relative percentage deviation in the makespan:

$$\Delta_{avg} = \frac{\sum_{i=1}^{R}\left(\frac{Heu_i - LB}{LB}\right)}{R} \times 100,$$

Heu_i is the objective value of the solution found by our algorithm at the ith replication and LB (Lower Bound) is the best known value of makespan when the size of the buffer is infinite.

The comparative approaches used in this study consist of the representative methods employed for solving this problem in the literature, including Hybrid Genetic Algorithm (HGA) of Wang et al. [31], Hybrid Particle Swarm Optimization (HPSO) of Liu et al. [14] and (PSOvns) of Tasgetiren al. [30] .

5.1 *Results of the Proposed EDA under Zero and Infinite Buffer Sizes*

Table 1 and Table 2 summarize the results obtained by our EDA against HGA and HPSO for the extreme cases of the problem. In Table 1, the problem with no buffer,

Table 1 Computational results under $B_i = 0$

		HGA		HPSO		EDA	
instances	*LB*	$\Delta_{\min}$	Δ_{avg}	$\Delta_{\min}$	Δ_{avg}	$\Delta_{\min}$	Δ_{avg}
Car1	7038	5.27	6.56	5.27	6.46	5.27	5.27
Car2	7166	8.39	10.2	5.14	7.75	4.70	4.70
Car3	7312	10.2	11.06	9.6	10.02	9.60	9.60
Car4	8003	10.41	12.44	9.18	11.18	9.18	9.18
Car5	7720	6.45	7.19	6.45	7.51	6.45	6.45
Car6	8505	4.33	4.71	4.33	4.64	4.33	4.33
Car7	6590	3.00	3.00	3.01	3.01	3.00	3.00
Car8	8366	2.62	3.41	2.62	3.18	2.62	2.62
Rec1	1247	15.8	17.8	13.87	15.77	13.71	13.76
Rec3	1109	15.78	17.77	14.43	15.95	12.98	13.66
Rec5	1242	16.59	18.39	14.9	16.52	14.09	14.13
Rec7	1566	13.22	14.26	12.13	13.28	10.34	10.51
Rec9	1537	13.27	14.53	12.17	13.59	11.19	11.23
Rec11	1431	12.23	13.73	11.39	12.49	10.97	11.02
Rec13	1930	9.64	10.24	9.48	10.32	9.02	9.20
Rec15	1950	7.69	9.62	7.03	8.85	6.41	6.52
Rec17	1902	10.04	11.52	9.46	10.88	9.46	9.46
Rec19	2093	16.96	18.11	14.96	16.74	14.48	15.12
Rec21	2017	19.53	20.67	17.8	18.94	16.51	16.99
Rec23	2011	16.61	18.34	15.12	17.12	13.77	14.37
Rec25	2513	15.68	17.6	13.33	14.77	11.98	12.46
Rec27	2373	14.83	15.75	14.08	15.18	12.85	13.78
Rec29	2287	13.29	14.21	13.21	13.9	12.46	13.15
Rec31	3045	24.6	26.49	20.99	22.83	20.36	21.18
Rec33	3114	25.56	26.4	22.19	23.34	20.84	21.66
Rec35	3277	18.71	19.66	15.17	16.39	13.70	14.74
Rec37	4951	25.77	26.87	23.77	24.64	23.13	24.55
Rec39	5087	25.56	26.5	22.33	23.35	21.37	22.39
Rec41	4960	29.01	30.02	26.27	27.42	25.95	26.38
average		14.17	15.42	12.75	14.00	12.09	12.46

Table 2 Computational results under B_i = infinite

instances	LB	HGA $\Delta_{\min}$	HGA Δ_{avg}	HPSO $\Delta_{\min}$	HPSO Δ_{avg}	EDA $\Delta_{\min}$	EDA Δ_{avg}
Car1	7038	0.00	0.00	0.00	0.00	0.00	0.00
Car2	7166	0.00	0.59	0.00	0.00	0.00	0.00
Car3	7312	0.00	0.86	0.00	0.79	0.00	0.00
Car4	8003	0.00	0.00	0.00	0.00	0.00	0.00
Car5	7720	0.00	0.00	0.00	0.59	0.00	0.00
Car6	8505	0.00	0.57	0.00	0.61	0.00	0.00
Car7	6590	0.00	0.00	0.00	0.00	0.00	0.00
Car8	8366	0.00	0.26	0.00	0.03	0.00	0.00
Rec1	1247	0.00	1.36	0.00	0.41	0.00	0.04
Rec3	1109	0.00	1.35	0.18	0.30	0.00	0.00
Rec5	1242	0.00	0.40	0.24	0.29	0.24	0.24
Rec7	1566	0.00	1.91	0.70	1.66	0.00	0.00
Rec9	1537	0.78	2.26	0.00	1.54	0.00	0.00
Rec11	1431	0.00	3.09	0.00	1.20	0.00	0.00
Rec13	1930	0.73	2.08	0.21	1.25	0.00	0.14
Rec15	1950	0.82	1.66	0.67	1.36	0.00	0.24
Rec17	1902	1.37	3.36	0.00	2.33	0.00	0.06
Rec19	2093	1.29	2.85	0.67	1.35	0.29	0.51
Rec21	2017	1.44	2.50	1.44	1.61	1.44	1.44
Rec23	2011	1.39	3.47	0.90	1.84	0.45	0.52
Rec25	2513	2.51	3.69	1.11	2.42	0.52	0.94
Rec27	2373	1.39	2.80	0.55	1.83	0.42	0.99
Rec29	2287	3.10	3.92	1.01	3.05	0.26	0.89
Rec31	3045	3.19	3.88	1.38	2.34	0.49	1.67
Rec33	3114	0.83	2.08	0.00	0.78	0.13	0.64
Rec35	3277	0.00	0.21	0.00	0.01	0.00	0.00
Rec37	4951	3.80	4.77	2.26	3.03	1.66	2.99
Rec39	5087	2.95	3.62	1.47	2.11	1.22	1.88
Rec41	4960	4.92	5.53	2.74	3.48	2.42	3.24
average		1.05	2.04	0.54	1.25	0.33	0.57

i.e. the blocking flowshop scheduling problem, is addressed. It is clear that EDA is better than HGA and HPSO with respect to both δ_{avg} and the minimum relative percentage deviations of best found solutions, $\delta_{\min}$. Therefore, we observe that, in average for all instances, δ_{avg} provided by our algorithm (12,46 %) is less than $\delta_{\min}$ provided by the compared algorithms (14,17% for HGA and 12,75% for HPSO). When the size of the buffer is infinite, i.e. for the classical FSP, the EDA again outperforms the other approaches in terms of δ_{avg} and $\delta_{\min}$ (Table 2). Moreover, for Carlier instances [3], our EDA seems to be able to reach all LB values for all replications in short CPU times (Table 6). Globally, the EDA reaches 17 LB values out of 29 and 14 LB values are obtained by HGA and HPSO.

Table 3 Computational results under $B_i = 1$

		HGA		PSOvns		HPSO		EDA	
instances	*LB*	$\Delta_{\min}$	Δ_{avg}	$\Delta_{\min}$	Δ_{avg}	$\Delta_{\min}$	Δ_{avg}	$\Delta_{\min}$	Δ_{avg}
Car1	7038	0.00	0.00	0.00	0.00	0.00	0.00	0.00	0.00
Car2	7166	0.00	0.88	0.00	0.15	0.00	0.00	0.00	0.00
Car3	7312	0.00	1.05	0.00	0.84	0.00	0.91	0.00	0.00
Car4	8003	0.00	0.00	0.00	0.04	0.00	0.00	0.00	0.00
Car5	7720	0.00	0.17	0.00	0.23	0.00	0.22	0.00	0.00
Car6	8505	0.00	0.57	0.00	0.27	0.00	0.50	0.00	0.00
Car7	6590	0.00	0.00	0.00	0.04	0.00	0.00	0.00	0.00
Car8	8366	0.00	0.78	0.00	0.03	0.00	0.00	0.00	0.00
Rec1	1247	1.84	3.34	0.96	2.13	0.32	1.04	0.16	0.18
Rec3	1109	1.98	2.88	1.44	2.99	0.54	1.84	0.18	0.25
Rec5	1242	0.56	1.38	1.45	2.37	0.40	1.39	0.24	0.30
Rec7	1566	1.79	3.51	1.15	1.76	0.13	2.26	0.00	0.11
Rec9	1537	1.76	2.96	1.63	2.62	0.98	1.86	0.33	0.46
Rec11	1431	2.24	4.97	1.33	2.53	0.14	1.65	0.14	0.19
Rec13	1930	1.97	2.50	1.40	2.41	0.78	1.67	0.36	0.58
Rec15	1950	1.18	1.64	1.13	2.29	1.08	1.45	0.00	0.34
Rec17	1902	2.52	4.28	2.79	3.70	1.05	2.38	0.89	0.89
Rec19	2093	3.11	4.27	3.20	3.96	2.10	2.88	1.15	1.55
Rec21	2017	2.88	4.59	2.83	4.34	1.69	3.16	1.64	1.85
Rec23	2011	4.92	6.63	3.33	4.40	2.54	3.88	1.49	2.04
Rec25	2513	3.90	5.48	3.42	4.88	2.71	3.89	1.63	2.27
Rec27	2373	3.50	4.66	3.88	4.54	2.40	3.39	1.69	2.01
Rec29	2287	3.76	3.99	3.41	5.45	2.93	3.69	1.75	2.28
Rec31	3045	6.37	7.28	4.99	5.90	3.55	4.75	2.50	3.38
Rec33	3114	4.34	5.42	3.82	4.98	2.99	3.80	1.86	2.68
Rec35	3277	1.77	2.82	0.76	1.42	0.00	0.45	0.00	0.09
Rec37	4951	8.14	8.73	8.14	8.97	6.14	7.18	5.82	6.70
Rec39	5087	6.17	6.84	5.98	6.80	4.40	5.42	3.87	4.57
Rec41	4960	8.06	8.74	7.92	9.00	6.25	7.03	6.03	6.60
average		2.51	3.46	2.24	3.07	1.49	2.30	1.09	1.36

5.2 *Results of the Proposed EDA under Different Buffer Sizes*

Even for the buffer sizes of 1, 2 and 4 where the results are depicted in Tables 3, 4 and 5 respectively, our proposed algorithm remains the best one in terms of the quality of the solutions. Moreover, it is shown that while the buffer sizes increase the deviations become close to zero. Therefore, the performance of the EDA increases when the problem presents fewer constraints and approaches the classical FSP. Also, it can be seen that the differences between $\delta_{\min}$ and δ_{avg} are very small, which implies the high robustness of our algorithm in a consistent manner (Table 6).

Table 4 Computational results under $B_i = 2$

		HGA		PSOvns		HPSO		EDA	
instances	*LB*	$\Delta_{\min}$	Δ_{avg}	$\Delta_{\min}$	Δ_{avg}	$\Delta_{\min}$	Δ_{avg}	$\Delta_{\min}$	Δ_{avg}
Car1	7038	0.00	0.00	0.00	0.00	0.00	0.00	0.00	0.00
Car2	7166	0.00	0.59	0.00	0.15	0.00	0.00	0.00	0.00
Car3	7312	0.00	0.93	0.00	0.89	0.00	0.68	0.00	0.00
Car4	8003	0.00	0.00	0.00	0.01	0.00	0.00	0.00	0.00
Car5	7720	0.00	0.04	0.00	0.35	0.00	0.66	0.00	0.00
Car6	8505	0.00	0.53	0.00	0.27	0.00	0.69	0.00	0.00
Car7	6590	0.00	0.00	0.00	0.35	0.00	0.00	0.00	0.00
Car8	8366	0.00	0.52	0.00	0.03	0.00	0.03	0.00	0.00
Rec1	1247	0.00	1.32	0.16	0.81	0.16	0.46	0.00	0.05
Rec3	1109	0.72	1.74	0.18	0.58	0.00	0.32	0.00	0.00
Rec5	1242	0.24	0.51	0.24	0.87	0.24	0.35	0.24	0.24
Rec7	1566	1.15	2.69	1.15	1.45	0.00	1.90	0.00	0.00
Rec9	1537	1.43	2.37	0.00	1.59	0.00	1.43	0.00	0.00
Rec11	1431	0.28	3.13	0.14	2.28	0.00	1.45	0.00	0.00
Rec13	1930	1.35	2.22	1.24	2.44	0.10	1.24	0.00	0.21
Rec15	1950	1.18	1.72	0.97	1.97	0.56	1.48	0.05	0.31
Rec17	1902	2.37	4.21	1.05	3.12	0.42	3.18	0.00	0.02
Rec19	2093	1.58	3.07	1.48	2.71	0.62	1.43	0.29	0.69
Rec21	2017	2.53	3.03	1.64	2.22	1.64	1.70	0.59	1.39
Rec23	2011	2.88	4.06	1.39	2.80	0.85	1.90	0.40	0.52
Rec25	2513	2.83	3.73	2.71	3.61	0.80	2.38	0.76	1.21
Rec27	2373	1.81	2.97	2.15	3.39	1.56	2.35	0.55	1.07
Rec29	2287	3.85	3.97	3.24	4.37	1.27	2.58	0.79	1.07
Rec31	3045	2.66	3.44	3.02	3.96	1.61	2.43	0.53	1.69
Rec33	3114	1.73	2.63	1.32	1.91	0.64	1.17	0.22	0.71
Rec35	3277	0.00	0.76	0.00	0.55	0.00	0.05	0.00	0.00
Rec37	4951	4.87	5.88	4.93	6.01	2.97	3.79	2.54	3.52
Rec39	5087	4.42	5.40	3.54	4.22	1.71	2.69	1.30	2.13
Rec41	4960	5.34	5.90	5.18	5.74	3.19	3.76	2.72	3.73
average		1.49	2.32	1.23	2.02	0.63	1.38	0.38	0.64

5.3 Comparative Results Based on Unilateral Paired t-Test

In order to compare the performance of our algorithms, we have used the unilateral paired t-test procedure [16] at the 99% significance level. This procedure consists of testing the differences in two samples of observations within pairs. Let μ_A and μ_B denote the true average relative percent deviations of the algorithms A and B respectively. The tested hypotheses are:

$$\begin{cases} H_0 : \mu_A - \mu_B = 0 \\ H_1 : \mu_A - \mu_B < 0 \end{cases}$$

Table 5 Computational results under $B_i = 4$

		HGA		PSOvns		HPSO		EDA	
instances	*LB*	$\Delta_{\min}$	Δ_{avg}	$\Delta_{\min}$	Δ_{avg}	$\Delta_{\min}$	Δ_{avg}	$\Delta_{\min}$	Δ_{avg}
Car1	7038	0.00	0.00	0.00	0.00	0.00	0.00	0.00	0.00
Car2	7166	0.00	0.89	0.00	0.15	0.00	0.00	0.00	0.00
Car3	7312	0.00	1.11	0.00	0.55	0.00	1.00	0.00	0.00
Car4	8003	0.00	0.00	0.00	0.00	0.00	0.00	0.00	0.00
Car5	7720	0.00	0.19	0.00	0.47	0.00	0.85	0.00	0.00
Car6	8505	0.00	0.50	0.00	0.27	0.00	0.42	0.00	0.00
Car7	6590	0.00	0.00	0.00	0.04	0.00	0.00	0.00	0.00
Car8	8366	0.00	0.48	0.00	0.07	0.00	0.09	0.00	0.00
Rec1	1247	0.00	1.53	0.16	0.54	0.16	0.42	0.00	0.05
Rec3	1109	0.18	1.36	0.09	0.50	0.00	0.36	0.00	0.00
Rec5	1242	0.24	0.60	0.24	0.80	0.24	0.38	0.24	0.24
Rec7	1566	0.00	2.17	0.89	1.60	0.38	1.87	0.00	0.00
Rec9	1537	0.98	2.26	0.39	1.55	0.00	1.40	0.00	0.00
Rec11	1431	0.00	3.08	0.49	2.25	0.00	0.82	0.00	0.00
Rec13	1930	1.14	2.15	0.99	2.43	0.26	1.45	0.00	0.11
Rec15	1950	1.18	1.65	0.92	2.39	0.72	1.67	0.00	0.23
Rec17	1902	2.58	4.26	1.84	3.00	0.00	2.43	0.00	0.06
Rec19	2093	1.29	2.96	1.96	2.73	0.67	1.58	0.29	0.56
Rec21	2017	1.64	2.42	1.49	2.18	1.44	1.62	0.55	1.38
Rec23	2011	1.39	3.69	0.95	2.42	0.60	2.14	0.45	0.51
Rec25	2513	2.43	3.46	2.47	3.51	0.72	2.25	0.48	0.97
Rec27	2373	1.98	3.02	1.43	2.89	0.97	1.84	0.72	0.96
Rec29	2287	2.62	3.90	2.36	4.40	1.05	3.08	0.48	0.86
Rec31	3045	3.02	3.71	2.96	3.79	1.18	2.21	0.99	1.74
Rec33	3114	0.77	1.94	0.93	1.63	0.35	0.83	0.00	0.64
Rec35	3277	0.00	0.29	0.00	0.24	0.00	0.01	0.00	0.00
Rec37	4951	4.20	5.01	4.83	5.44	2.65	3.22	2.18	3.16
Rec39	5087	3.09	3.65	2.99	4.01	1.34	2.24	1.14	1.85
Rec41	4960	4.82	5.53	4.84	5.31	2.70	3.46	2.20	3.36
average		1.16	2.13	1.15	1.90	0.53	1.30	0.33	0.58

H_0 implies that the average relative percent deviations of the two algorithms are similar, while H_1 implies that the average relative percent deviations of the algorithm A are lower than the ones for the algorithm B. The obtained results are given in Table 7.

For different buffer sizes, the statistical tests prove that the negative difference between EDA and all other algorithms is meaningful at the significance level of 0.99. Therefore, the effectiveness of our algorithm against the representative approaches from the literature for this problem is proved.

Table 6 Average CPU times of EDA over 20 replications (in seconds)

instances	buffer 0	buffer 1	buffer 2	buffer 4	infinite
car1	0.03	0.00	0.00	0.00	0.00
car2	0.01	0.00	0.00	0.00	0.00
car3	0.00	0.01	0.01	0.02	0.01
car4	0.05	0.00	0.00	0.00	0.00
car5	0.00	0.00	0.02	0.02	0.02
car6	0.00	0.00	0.00	0.00	0.00
car7	0.00	0.00	0.00	0.00	0.00
car8	0.00	0.00	0.00	0.00	0.00
Rec1	0.61	0.55	0.38	0.41	0.29
Rec3	0.46	0.60	0.25	0.26	0.24
Rec5	0.55	0.51	0.01	0.01	0.01
Rec7	0.97	1.43	0.63	0.75	0.67
Rec9	0.98	1.33	0.27	0.17	0.17
Rec11	0.94	0.91	0.15	0.14	0.16
Rec13	1.83	2.01	1.85	1.66	1.51
Rec15	1.90	1.93	2.17	1.95	2.39
Rec17	1.53	1.04	1.34	1.77	1.29
Rec19	2.03	2.04	2.10	2.07	2.10
Rec21	2.09	2.27	1.50	0.72	0.59
Rec23	2.77	2.52	1.60	2.23	2.48
Rec25	3.29	2.66	3.36	2.59	3.41
Rec27	2.88	3.15	3.39	2.01	2.98
Rec29	3.61	3.64	3.03	2.95	3.14
Rec31	3.90	4.01	3.85	3.71	4.01
Rec33	3.72	4.06	3.54	2.21	2.06
Rec35	4.38	3.37	1.20	0.78	0.79
Rec37	16.45	15.00	15.96	14.49	14.96
Rec39	15.92	15.27	16.19	14.03	15.63
Rec41	14.10	15.46	13.87	14.47	15.92
average	2.93	2.89	2.64	2.39	2.58

Table 7 Comparative results based on unilateral paired t-test

H_0	H_1	t-value	p-value
	B_i =infinite		
EDA=HGA	EDA<HGA	-7.5116	0.000
EDA=HPSO	EDA<HPSO	-5.4095	0.000
	$B_i = 0$		
EDA=HGA	EDA<HGA	-9.9352	0.000
EDA=HPSO	EDA<HPSO	-10.0478	0.000
	$B_i = 1$		
EDA=HGA	EDA<HGA	-8.4649	0.000
EDA=HPSO	EDA<HPSO	-8.1492	0.000
EDA=PSOvns	EDA<PSOvns	-9.0326	0.000
	$B_i = 2$		
EDA=HGA	EDA<HGA	-7.7635	0.000
EDA=HPSO	EDA<HPSO	-5.4987	0.000
EDA=PSOvns	EDA<PSOvns	-7.4854	0.000
	$B_i = 3$		
EDA=HGA	EDA<HGA	-7.6886	0.000
EDA=HPSO	EDA<HPSO	-5.4696	0.000
EDA=PSOvns	EDA<PSOvns	-6.9382	0.000

6 Conclusion

This chapter presented an application of EDA to a FSP under limited buffer with respect to makespan criterion. Therefore, we developed a probabilistic model based on the main characteristics of the problem. After constructing new sequence of jobs through the EDA, a SVNS was added to improve the quality of the solution. The computational results show that our algorithm outperforms the approaches it was compared with in terms of the solution quality and gives good results in short CPU times. Moreover, the performance of our algorithm is proved by using the unilateral paired t-test procedure.

References

1. Baluja, S.: Populationbased incremental learning: A method for integrating genetic search based function optimization and competitive learning. Technical Report N°CMU-CS-94-163. Carnegie Mellon University, Pittsburgh (1994)
2. Baluja, S., Davies, S.: Using optimal dependencytrees for combinatorial optimization: Learning the structure of the search space. In: Proceedings of the 14th International Conference on Machine Learning, pp. 30–38. Morgan Kaufmann, San Francisco (1997)
3. Carlier, J.: Ordonnancements a contraintes disjonctives. RAIRO-Operations Research 12, 333–351 (1978)
4. De Bonet, J.S., Isbell, C.L., Viola, P.: MIMIC: Finding optima by estimating probability densities. In: Mozer, M.C., Jordan, M.I., Petsche, T. (eds.) Advances in Neural Information Processing Systems, vol. 9, pp. 424–431. The MIT Press, Cambridge (1997)

5. Dutta, S.K., Cunningham, A.A.: Sequencing two-machine flow-shops with finite intermediate storage. Management Science 21, 989–996 (1975)
6. Etxeberria, R., Larrañaga, P.: Global optimization with Bayesian networks. In: II Symposium on Artificial Intelligence. CIMAF 1999. Special Session on Distributions and Evolutionary Optimization, pp. 332–339 (1999)
7. Hansen, P., Mladenovič, N.: Variable neighborhood search: principles and applications. European Journal of Operational Research 130, 449–467 (2001)
8. Harik, G.: Linkage learning via probabilistic modeling in the ECGA. University of Illinois Genetic Algorithms Laboratory, Urbana (January 1999) IlliGAL Report No. 99010
9. Harik, G.R., Lobo, F.G., Goldberg, D.E.: The compact genetic algorithm. In: Proceedings of the IEEE Conference on Evolutionary Computation 1998 (ICEG 1998), pp. 523–528. IEEE Service Centre, Piscataway (1998)
10. Jarboui, B., Eddaly, M., Siarry, P.: An estimation of distribution algorithm for minimizing the total flowtime in permutation flowshop scheduling problems. Computers and Operations Research 36, 2638–2646 (2009)
11. Larrañaga, P., Etxeberria, R., Lozano, J.A., Peña, J.M.: Combinatorial optimization by learning and simulation of Bayesian networks. In: Proceedings of the Sixteenth Conference on Uncertainty in Artificial Intelligence, Stanford, pp. 343–352 (2000)
12. Larrañaga, P., Lozano, J.A. (eds.): Estimation of distribution algorithms: a new tool for evolutionary computation. Kluwer Academic Publishers, Boston (2002)
13. Leisten, R.: Flowshop sequencing problems with limited buffer storage. International Journal of Production Research 28, 2085–3100 (1990)
14. Liu, B., Wang, L., Jin, Y.H.: An effective hybrid PSO-based algorithm for flow shop scheduling with limited buffers. Computers and Operations Research 35, 2791–2806 (2008)
15. Lozano, J., Larraanaga, P., Inza, I., Bengoetxea, E.: Towards a New Evolutionary Computation: Advances in the Estimation of Distribution Algorithms. Springer, Heidelberg (2006)
16. Montgomery, D.C.: Design and analysis of experiments, 5th edn. Wiley, New York (2000)
17. Mühlenbein, H., Mahnig, T.: Convergence theory and applications of the factorized distribution algorithm. Journal of Computing and Information Technology 7, 19–32 (1999a)
18. Mühlenbein, H., Mahnig, T.: FDA - A scalable evolutionary algorithm for the optimization of additively decomposed functions. Evolutionary Computation 7, 353–376 (1999b)
19. Mühlenbein, H., Paaß, G.: From Recombination of Genes to the Estimation of Distributions I. Binary Parameters. In: Ebeling, W., Rechenberg, I., Voigt, H.-M., Schwefel, H.-P. (eds.) PPSN 1996. LNCS, vol. 1141, pp. 178–187. Springer, Heidelberg (1996)
20. Nowicki, E.: The permutation flow shop with buffers: a tabu search approach. European Journal of Operational Research 116, 205–219 (1999)
21. Papadimitriou, C.H., Kanellakis, P.C.: Flow shop scheduling with limited temporary storage. Journal of Association Computing Machine 27, 533–549 (1980)
22. Pelikan, M.: Hierarchical Bayesian Optimization Algorithm: Toward a New Generation of Evolutionary Algorithms. Springer, Heidelberg (2005)
23. Pelikan, M., Goldberg, D.E., Cantú-Paz, E.: BOA: The Bayesian optimization algorithm. In: Banzhaf, W., Daida, J., Eiben, A.E., Garzon, M.H., Honavar, V., Jakiela, M., Smith, R.E. (eds.) Proceedings of the Genetic and Evolutionary Computation Conference, GECCO 1999, vol. 1, pp. 525–532. Morgan Kaufmann Publishers, San Francisco (1999)
24. Pelikan, M., Mühlenbein, H.: The bivariate marginal distribution algorithm. In: Roy, R., Furuhashi, T., Chawdhry, P.K. (eds.) Advances in Soft Computing Engineering Design and Manufacturing, pp. 521–535. Springer, London (1999)

25. Qian, B., Wang, L., Huang, D.X., Wang, W.L., Wang, X.: An effective hybrid DE-based algorithm for multi-objective flow shop scheduling with limited buffers. Computers and Operations Research 36, 209–233 (2009)
26. Reddi, S.S.: Sequencing with finite intermediate storage. Management Science 23, 216–227 (1976)
27. Reeves, C.R.: A genetic algorithm for flowshop sequencing. Computers and Operations Research 22, 5–13 (1995)
28. Ronconi, D.P.: A branch-and-bound algorithm to minimize the makespan in a flowshop with blocking. Annals of Operations Research 138, 53–65 (2005)
29. Ruiz, R., Maroto, C., Alcaraz, J.: Two new robust genetic algorithms for the flowshop scheduling problem. Omega 34, 461–476 (2006)
30. Tasgetiren, M.F., Sevkli, M., Liang, Y.C., Gencyilmaz, G.: Particle swarm optimization algorithm for permutation flowshop sequencing problem. LNCS, pp. 382–389. Springer, Heidelberg (2004)
31. Wang, L., Zhang, L., Zheng, D.Z.: An effective hybrid genetic algorithm for flow shop scheduling with limited buffers. Computers and Operations Research 33, 2960–2971 (2006)

Solving Hierarchically Decomposable Problems with the Evolutionary Transition Algorithm

Tom Lenaerts and Anne Defaweux

Abstract. Capturing the metaphor of evolutionary transitions in biological complexity, the Evolutionary Transition Algorithm (ETA) evolves solutions of increasing structural and functional complexity from the symbiotic interaction of partial ones. From the definition it follows that this algorithm should be very well suited to solve hierarchically decomposable problems. In this chapter, we show that the ETA can indeed solve this kind of problems effectively. We analyze, in depth, its behavior on hierarchical problems of different size and modular complexity. These results are compared to the Symbiogenetic Model and it is shown that the ETA is more robust and efficient to tackle this kind of problems.

1 Introduction

In [3, 2, 4], a new evolutionary compositional search algorithm was proposed for tackling optimization problems. Different from other evolutionary approaches is its capacity to construct more complex solutions out of the interaction of simpler ones. The algorithm draws its inspiration from the biological theory of evolutionary transitions [16, 19] and is hence referred to as the Evolutionary Transition Algorithm (ETA). One of the observations made in previous simulations is that the compositional search technique seemed to be particularly efficient on problem instances which exhibit an inherent structure in the organization of the gene interactions. However, this conclusion could not be confirmed by the Binary Constraint Satisfaction problem (BINCSP) simulations alone as the instances were randomly generated and

Tom Lenaerts
MLG, Département d'Informatique, Université Libre de Bruxelles, Boulevard du Triomphe CP212, 1050 Brussels, Belgium
e-mail: tlenaert@ulb.ac.be

Anne Defaweux
COMO, Vrije Universiteit Brussel, Pleinlaan 2, 1050 Brussels, Belgium
e-mail: adefaweu@gmail.com

R. Chiong, S. Dhakal (Eds.): Nat. Intel. for Sched., Plan. and Pack. Prob., SCI 250, pp. 111–143.
springerlink.com

therefore could not guarantee the presence of structure in the network of constraints. In this chapter, we evaluate the correctness of this intuition by applying the ETA on a hierarchical problem, the hierarchical if-and-only-if (H-IFF) problem [29, 17], which is known to be difficult to solve by a classical Genetic Algorithm (GA).

From a real-world perspective, like planning and scheduling problems, examining these hierarchically structured problems is highly relevant since in that case, problems typically have some kind of inherent structure that allows one to first solve parts of the problems, which can be recombined later on while resolving the constraints between the different modules [25, 18, 26]. The choice for H-IFF is also motivated by existing work on this problem by other compositional approaches such as the symbiogenetic algorithm (SEAM) [30, 28] and the Hierarchical GA (HGA) [5]. Both SEAM and HGA are capable of resolving this problem. If ETA performs as well for structured problems as we think, then it should also have a good performance on the H-IFF. We show, here, the performance of the ETA on this particular problem and compare it to the performance of SEAM.

To apply the ETA to H-IFF in a direct and straightforward manner, the definition of H-IFF is re-expressed as a BINCSP. Apart from its straightforward application, the reason for performing the analysis on a BINCSP interpretation and not on the H-IFF fitness function itself is to explicitly validate the observations on the BINCSP performance of ETA mentioned earlier [3, 2, 4]. The transformation of the hierarchical fitness function of H-IFF into a set of weighted constraints provides such a structured problem and is equivalent to the original problem. We show here the behavior of a GA, the ETA and SEAM on this problem and provide an in depth analysis of the ETA and SEAM on different variations of H-IFF problem in terms of the size of the genotype, the domains of the variables and the modularity of the problem. By translating H-IFF to BINCSP and given the fact that real-world problems can easily be thought of in terms of satisfying a number of (conflicting) constraints, we also show that H-IFF is relevant for real-world applications.

In the next section, an overview of the related work is given. Afterwards, the H-IFF test case is introduced together with its transformation to a BINCSP. We then briefly describe the implementation details of ETA and SEAM on this BINCSP implementation of H-IFF. Finally, the simulations are shown and evaluated. These simulations are divided into two parts: the first one concerns the performance of ETA on H-IFF, and the second part, the comparison between the ETA and SEAM on this problem. This chapter then ends with a summary and some conclusions.

2 Related Work

There are two ideas behind the compositional search approach: the first idea concerns the use of symbiotic relations to identify good collaborations and the second idea is concerned with the aggregation of complex solutions from the interaction of partial ones. The related works listed here fall in either one or both of these categories. Note that certain publications might be missing. The goal here is to provide the main lines of research that are related to this topic.

The very first GA-related reference that can be found on the evolution of complete solutions from the combination of partial ones is the work on the messy GA [6, 7, 8]. The messy GA has the structure of a classical GA. When two partial solutions are selected for reproduction in the messy GA, their combined genetic material creates a bigger solution that is defined at a higher level of complexity. Messy GAs therefore have the idea of combining partial solutions. However, this approach differs from the one we propose here: fitness effects that may follow from the interactions of the partial solutions are not taken into account. Partial solutions are evaluated and selected on their own as in the classical GA. It is during the process of reproduction that combination occurs. Messy GAs therefore lack the idea of a transition that operates on the behavior of good symbionts.

Identifying good symbiotic relations is an essential issue in evolutionary algorithms (EAs). In the standard GA, which examines promising partial solutions implicitly, recombination and mutation operators often produce bad results when they are problem-independent. It has therefore always been important to design operators that can in some way identify the relevant partial solutions even when they are implicitly hidden in a complete solution representation. Otherwise, it has been shown that the required number of evaluations (and population size) will grow exponentially with the number of variables [27]. The methods that try to find the relevant partial solutions and exploit this in the recombination process have been called *linkeage learning* methods. In [21], a linkeage learning mechanism in combination with a powerful niching mechanism and effective representation of partial solutions has been applied to tackle hierarchical problems succesfully. Using decision graphs to model partial solutions, they are able to identify and reuse interesting solutions to solve more complex levels. As such, the goals of this algorithm and ours lie very close, even though they are addressed from a different angle.

Other approaches focus on the collaboration between partial solutions to construct a full solution for a particular problem. In the Parisian approach [23, 24] for example, the algorithm does not intend to evolve a full solution to a problem but rather a collection of partial solutions that together solve the entire problem. This approach takes an additional step into the direction of compositional evolutionary search. However, the Parisian approach lacks a transition step that merges the set of partial solutions into a full solution. In this respect, the Parisian approach is more related to previous work on multi-level selection [15] and the Random Mutation Hill-Climber proposed in [12, 13]. In this latter work, it was also shown that by the introduction of a particular multi-level selection model, a random mutation hill-climber is able to tackle hierarchically decomposable problems. A comparison of this method with the approach discussed here is left for future work.

There are also approaches that introduce an evolutionary "divide and conquer" mechanism, like the cooperative co-evolutionary GA [22, 31]. In this algorithm, the problem consists of a collection of sub-problems that can be solved in isolation. Afterwards, these partial solutions can be recombined into a full solution for the global problem. The major difference between this algorithm and the ETA resides in the explicit divide and conquer framework. In the cooperative co-evolutionary GA, the process divides the problem into collaborating sub-problems; for each sub-problem, the algorithm evolves solutions that collaborate with one another in order to address the entire problem. The way to divide the problem can become an issue when the problem structure to be learned is not trivial or cannot be easily identified.

A truly compositional approach that uses both the concept of symbiosis and transitions is the Symbiogenetic Model (SEAM) [29, 30, 28]. This model considers a population of partial solutions that are interacting with one another through the mechanism of symbiosis. In this population, the good symbiotic relations are identified and produce a new partial solution through a transition. This model initially appears very similar to the ETA. There is, however, a major difference between them: SEAM does not generate a succession of populations by reproducing the parent population through the mechanism of fitness proportional selection. It randomly selects pairs of individuals and places them into a symbiotic relation. It then replaces the symbiotic relation with a new individual when it performs better than the parents in isolation. The condition under which this transition occurs also differs from the ETA: It uses the concept of Pareto optimality with respect to an evolutionary context to define whether a symbiosis performs better or worse than the parents alone. The result of this approach is that it requires the model to include features about the structure of the problem that needs to be solved, like for instance the degree of modularity of the problem. Furthermore, in order for it to work, it requires the problem to be fully hierarchically decomposable; this covers only a subset of the structured problem class. The HGA [5] is an algorithm very similar to SEAM. The major difference resides in the fact that the HGA introduces, in its implementation, explicit information concerning the hierarchical nature of the problem that needs to be solved (by hard-coding the hierarchical search in the process and therefore disclosing the emergent aspect present in SEAM) and optimizes the sampling approach so that only the best samples are present for evaluation at all time, considerably speeding up the search process. Recently, a new symbiogenetic algorithm was proposed to overcome a number of limitations in SEAM [20].

3 Hierarchical IF and Only iF

In this section, we introduce the H-IFF problem. We first describe the IF and only iF (IFF) problem defined in terms of its original context and the BINCSP. Afterwards, this problem description is extended to the H-IFF problem.

3.1 The IF and Only iF (IFF) Problem

The IFF problem is defined as follows:

Definition 1. *The* IF and only iF (IFF) *problem consists of finding the right assignment of n variables each taking values in a given domain d in such a way that the following condition is fulfilled:*

Given $x_1, \ldots x_n \in V$ where V is the set of variables, we have :

$$x_1 \leftrightarrow x_2 \leftrightarrow \ldots \leftrightarrow x_n$$

where, if we denote by d_i and d_j the values taken by the variables x_i and x_j respectively, we have

$$(x_i \leftrightarrow x_j) \Leftrightarrow (d_i = d_j)$$

This means that we need to find an assignment for each of the n variables $x_1, \ldots x_n$ in such a way that each pair of variables is assigned the same value. For example, if we have 4 variables, each one taking a value in $\{0,1\}$, we have the compound labels $\langle x_1,1\rangle, \langle x_2,1\rangle, \langle x_3,1\rangle, \langle x_4,1\rangle$ and $\langle x_1,0\rangle, \langle x_2,0\rangle, \langle x_3,0\rangle, \langle x_4,0\rangle$ which are solutions for this IFF problem; any other compound label defined on these four variables will not satisfy any one of the $\leftrightarrow$ relations defined on these variables. For example, the solution $\langle x_1,1\rangle, \langle x_2,0\rangle, \langle x_3,0\rangle, \langle x_4,0\rangle$ does not respect the relation that binds x_1 with the three other variables as the following conditions are not met: $x_1 \leftrightarrow x_2$;$x_1 \leftrightarrow x_3$ and $x_1 \leftrightarrow x_4$.

The IFF problem is trivial to solve as the solutions consist of the assignment of the same value to each variable. If the intersection of all variable domains is a set of length m, this problem has exactly m solutions. Furthermore, any single mutation hill-climber will find the solution in at most n good mutations where n corresponds to the number of variables.

3.2 *Rewriting the IFF Problem as a BINCSP*

The re-expression of IFF as a BINCSP is straightforward and each instance will have the following properties:

- The variable set X corresponds to the set of variables described in IFF: $X = \{x_1, \ldots, x_n\}$.
- The set of domains D is simply the set of the domains of each variable in X: $D = \{D_{x_1}, \ldots, D_{x_n}\}$.
- The set of constraints C contains the $\binom{n}{2}$ binary constraints that can be defined on the variable set and that enforce the $\leftrightarrow$ relation to be satisfied:

$$\begin{array}{rlll} C = \{ & x_1 \leftrightarrow x_2, \; x_1 \leftrightarrow x_3, & \ldots & x_1 \leftrightarrow x_n \\ & \quad\quad\quad\;\; x_2 \leftrightarrow x_3, & \ldots & x_2 \leftrightarrow x_n \\ & & \ddots & \vdots \\ & & & x_{n-1} \leftrightarrow x_n \} \end{array}$$

3.3 *The Hierarchical IF and Only iF*

H-IFF adds a hierarchical partitioning or modular structure to the IFF problem. The parts in this partitioning are modules which are built out of variables from the variable set at the first level or, recursively, by combining other modules into a higher level. We begin this description by defining precisely what we mean by a module and a level:

Definition 2. *A* module *is a set of a given length l composed of other modules chosen from the previous level. At the first level, modules are built directly by assembling l variables. The modules defined at each level are disjunct from one another. The parameter l defines the* modularity *of the hierarchical problem.*

Definition 3. *The* level *of the hierarchy corresponds to a set of disjunct modules. The disjunct union of all modules at a given level contains the entire set of modules from the previous level.*

This means that, at each level, each variable of the variable set belongs to exactly one of the modules that is defined on that level.

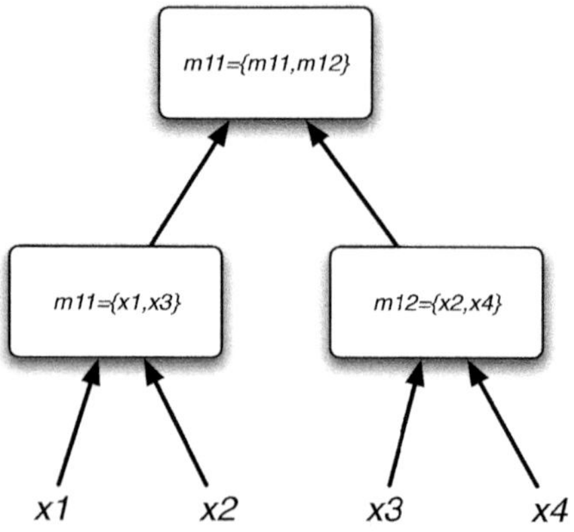

Fig. 1 A simple H-IFF example

H-IFF requires the system to evolve an optimal solution for the lowest level before considering the next level. It induces a hierarchy in the way the problem can be solved.

Definition 4. *The* H-IFF problem *consists of an IFF problem which is structured in multiple levels on which modules are defined hierarchically. The* Shuffled H-IFF!definition *problem is a H-IFF problem for which the structure in the modules at each level is randomly generated.*

Shuffled H-IFF problem instances, by the fact that the structure in the modules is randomly generated, have an a-priori unknown structure and prevent us from en coding the internal structure of H-IFF in the genotypes of the solutions. Therefore, (shuffled) H-IFF becomes very difficult to solve for single mutation hill-climbers or classical GAs as we illustrate in the following small example.

Let us consider the variable set of length 4:

$$V = \{x_1, x_2, x_3, x_4\}$$

where each variable is defined on the same domain:

$$D_{x_i} = \{0, 1\} \, \forall i.$$

The two bit-string 0000 and 1111, where each variable has the same values, are the perfect solutions for this problem.

Suppose now that the H-IFF instance consists of the randomly generated modules shown in Figure 1. Let us now consider the solutions $s_1 = 1010$ and $s_2 = 0101$. Both solutions satisfy the first level of the hierarchy as the variables in m_{11} and m_{12} are assigned the same values within each module. The second level, however, is not yet satisfied. For a single mutation hill-climber, any mutation of 1 position in the bit-string results in the violation of one of the modules at the first level. The hierarchical nature of the problems requires the first level to be satisfied (s) before considering the second level. If the solution fails (f) one module of the first level, the second level automatically fails as well. The problem is solved in the situation where a solution satisfies all levels. In the case of s_1 and s_2, a single bit mutation will always yield a solution which is less fit than the original. This is a direct consequence of the regression in the hierarchy as the previous level is no longer satisfied. We illustrate this situation in Table 1 where each position of s_1 is tried in turn for a mutation.

Table 1 Each mutation of one position of s_1 produces a solution which is less fit than s_1 as one of the modules in the first level is no longer satisfied. This means that to escape this local optimum, this algorithm needs to introduce specific heuristics so that, for example, 2 mutations may occur at this step instead of 1.

	level 1		level 2
	m_{11}	m_{12}	m_2
1010	s	s	f
0010	f	s	f
1110	s	f	f
1000	f	s	f
1011	s	f	f

Neither a GA with one-point crossover nor the two-point crossover on the two parents s_1 and s_2 is capable of producing a solution that will not corrupt at least one of the modules at the first level. In other words, the offspring will always be less fit than the parents. We illustrate this in Table 2.

The expected quality of the crossover operation is correlated to the tightness of the linkage within the genotype. Crossover performs best when closely related genes lie close to one another in the genotype. In the case of tight linkage, the chance that the crossover will disrupt a highly correlated sub-set of genes is reduced. However, one specification of shuffled H-IFF is that we do not know what the structure of the

Table 2 Here a one-point crossover is applied with the crossing in the middle of the genotype to both parents. Both offspring are less fit than their parents.

		level 1		level 2
		m_{11}	m_{12}	m_2
Parents	1010	s	s	f
	0101	s	s	f
Offspring	1001	f	f	f
	0110	f	f	f

genotype will be. Therefore, we cannot assume tight linkage. This assumption is one of the reasons classical GAs do not perform well on this kind of problems.

In our simulations, we will use shuffled H-IFF problem instances to illustrate that the ETA is able to learn the internal structure of the data.

We will now describe how H-IFF actually enforces lower level modules to be solved first. We place this description in the framework of BINCSP in order to obtain an implementation that is as close as it can be to the implementation we used in our previous simulations on the BINCSP [4, 2, 3].

In the BINCSP approach, for each constraint of the constraints set of IFF, the fitness will associate a positive reward if the constraint is met, but no reward otherwise. The H-IFF function will use the same approach but assigns a weighted reward to each constraint. The way in which weights are defined depends on the modules to which the variables in the constraint belong.

To define the H-IFF problem instance, we therefore need to answer two questions. The first question concerns the creation of the hierarchical structure of the modules. The second question concerns the choice of the right weights: the weights have to be selected in such a way that the values within the modules are optimized first before the optimization process looks for the best combination of modules. In the remainder of this section, these two questions are answered.

3.4 H-IFF as a BINCSP: Building Hierarchical Modules

A module is a set of variables: As the basic modules are defined over variables and all the other modules are defined recursively from these basic modules, modules are therefore defined by variables but may vary in size (with respect to the level of the hierarchy). It is supposed, therefore, that variables in a module are tightly linked to one another; they are closely related with one another and less related to variables which are present in other modules. The underlying idea is that modules may be constructed in a hierarchical way either bottom-up or top-down.

In the bottom-up approach, variables that should belong to the basic modules are grouped together (i.e. the variables that are very closely related). Then, the upper level considers the grouping of two or more such basic modules. Aggregation goes on until the next aggregation combines all variables back together.

For example, consider the IFF problem defined on 8 variables $x_1, \ldots x_8$. In the bottom-up approach, we start with 8 singletons at level 0. If we want to define modules of length 2, at the first level, pairs of these singletons are taken to form the 4 modules of the first level. For example, the first level may look as follows:

$$\begin{array}{l} \text{level } 0: \{x_1\}, \{x_2\}, \{x_3\}, \{x_4\}, \{x_5\}, \{x_6\}, \{x_7\}, \{x_8\} \\ \text{level } 1: \{x_1, x_4\}, \{x_2, x_8\}, \{x_3, x_7\}, \{x_5, x_6\} \end{array}$$

Going to the second level means aggregating pairs of modules defined at the first level which yields, this time, sets of 4 variables:

$$\text{level } 2: \{x_1, x_4, x_5, x_6\}, \{x_2, x_3, x_7, x_8\}$$

And finally, at the third level, we aggregate the two modules from level 2 and obtain the entire variable set:

$$\text{level } 3: \{x_1, x_2, x_3, x_4, x_5, x_6, x_7, x_8\}$$

This means that, for example, x_1 is closely related to x_4 and less related to x_5 because x_1 and x_4 are in the same module at level 1 while x_1 and x_5 are not in the same module at the same level. However, x_1 and x_5 are in the same module at level 2, which means that although x_1 is less related to x_5 than it was to x_4, it is still more related to x_5 than, for example, to x_2 which is defined in another module at level 2.

This hierarchical structure induces, therefore, a network of preferences between the variables. These preferences will be captured in the weights we define on the constraints in the evaluation function.

The top-down approach works on a similar basis but instead of aggregating modules, it splits the groups into smaller ones until singletons are found.

Linking this modularization feature with the weights of the H-IFF function means that the weight assigned to variables belonging to a lower module should be significantly greater than the weights attributed to the module one level higher, where in turn these weights will be bigger than the weights on a module in an even higher level and so on.

3.5 *H-IFF as a BINCSP: Defining Weights*

As stressed in [28], the weights should be chosen to enforce the close intra-module dependency in relation to the inter-module dependency. It is therefore recommended to give a significant advantage to intra-module weights in comparison to inter-module weights. However, the difference between the intra-module strength and inter-module strength should not become too large in order to prevent the inter-module dependency from becoming insignificant relative to the intra-module dependency.

The idea is that the weights are assigned in relation to the number of links between the modules. Figures 2 and 3 show how weights can be assigned for pairwise interactions and length-3 module interactions respectively (length-3 means that the modularity of the problem is 3).

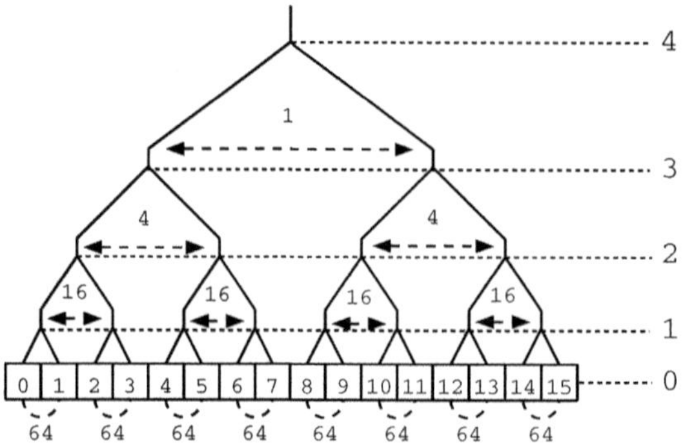

Fig. 2 A length-2 modularization example: In this size 16 genotype example, modules are defined on 4 levels of the hierarchy. To define the weights between a constraint on two variables, we have to observe the level of the hierarchy at which these two variables are linked. This level yields the resulting weight.

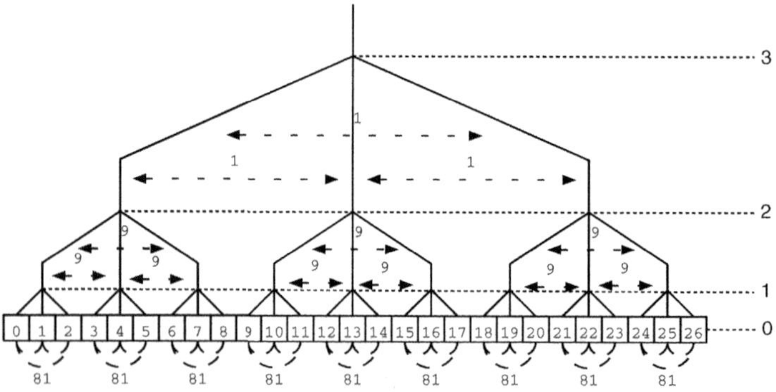

Fig. 3 A length-3 modularization example: This size 27 genotype example works in a similar way as the 16-bits. This time, however, variables and modules are joined three by three instead of pairwise.

Let us suppose that we have v layers of modules in the hierarchical structure, next to the layer that contains the variables themselves, layer 1 to the first basic modules and so on up to layer v which is the highest level for modules.

We assume that variables interacting in layer v are the variables which are not strongly related to one another and therefore should obtain the lowest possible weight, for example 1. Then, the variables interacting on one layer beneath $(v-1)$ should be bound with a greater strength and the weights increase as we proceed along the hierarchy all the way down to the lowest layer.

The way the weights are increased is important as the weights correspond to the forces that link together the elements within a module. If not strong enough, the module will fail to counter-balance the epistasis caused by the links between module elements. In [28], it was proposed to relate the weights with the number of links between the modules. We will now formally discuss how these weights can be assigned to modules of an arbitrary length l.

Let us start at the first level, i.e. layer 1. At this level, we have the very basic modules, that is, modules composed of exactly l variables. Each variable within the module is bound to each other with exactly one link, yielding a total of $\binom{l}{2}$ links within each module.

Now, going one level higher, i.e. layer 2, each module can on its turn be linked to other $(l-1)$ modules yielding $\binom{l}{2}$ inter-module links. When two modules are linked, it can actually occur through links between each variable of the first modules with every variable of the second module, i.e. $l \times l$ links. We therefore have a total of $\binom{l}{2} l^2$ links at layer 2.

Going to the next level, i.e. layer 3, we resume the description we made for layer 2; this time, however, each module is made of l^2 variables which results in $(l^2)^2$ links between each module-pair yielding a total of $\binom{l}{2} (l^2)^2$ links at that layer. As a generalization we suppose that at layer λ, we will have $\binom{l}{2} (l^2)^{\lambda-1}$ links between the modules.

The idea is to enforce the intra-module strength by assigning as weights, between the elements of the lowest layer, the number of links between modules generated at the highest layer, i.e. $(l^2)^{\nu}$. The weights between the elements on the next layer correspond to the number of links between modules generated at one layer beneath the highest, i.e. $(l^2)^{\nu-1}$ and so on up to the highest layer which receives the smallest weight i.e. 1. In general, if we assume ν layers (next to the independent variables themselves), we obtain the weight w_λ between the elements of a module at layer λ (≥ 1) as:

$$w_\lambda = (l^2)^{\nu-\lambda}$$

In Figure 2, we observe the weight assignments for a genotype of size 16 with a modularity of 2 ($l = 2$). We can see that the weights assigned to pairs of variables decrease by a factor of 4 as the link is placed one step higher in the hierarchy. More concretely, the constraints operating on variable x_6 and x_7 are bound at level 1. As there are 4 such levels, the corresponding weight is $(2^2)^{4-1} = 64$. Variables x_7 and x_8 however are bound at the highest level of the hierarchy, that is, level 4 and a constraint on these two variables will be assigned a weight of 1. Figure 3 gives the weights for a 27-size genotype with a modularity of 3. We observe, this time, that the weights decrease with a factor of 9 ($l^2 = 3^2 = 9$). Furthermore, where we had exactly one link between each component of the modules at each layer in the former example, we now have 3 such links which correspond to the $\binom{3}{2}$ inter-module links.

```
weight[n][n] := [0];
level := 1;
size := n/l;
subgroups := 1;
while size ≥ 1 do
  for subg := 0 to l * (subgroups − 1) step l do
    for g1 := 0 to l − 2 step 1 do
      for g2 := g1 + 1 to l − 1 step 1 do
        for n1 := (g1 + subg) * size to (g1 + subg + 1) * size − 1 step 1 do
          for n2 := (g2 + subg) * size to (g2 + subg + 1) * size − 1 step
          1 do
            weight[n1][n2] := level;
            [weight[n2][n1] := level;]
  level := level * l^2;
  subgroups := subgroups * l;
  size := size/l;
```

Fig. 4 Computing the weights for H-IFF

The pseudo-code for computing weights is given in Figure 4. In this program, n refers to the number of variables, and l to the module-length. It fills in a matrix *weight* with the resulting weights between each pair of variables. This matrix is initialized with 0 at each position, and as this is a symmetric matrix, only the upper or lower diagonal needs to be computed.

4 General Description of the Evolutionary Transition Algorithm

As can be seen in Figure 5, the transition model follows a classical evaluate-select-reproduce evolutionary loop. As a GA it works with a limited fixed size population which is evaluated and reproduced at each iteration. In the evaluation phase, individuals are paired up and a fitness score is assigned based on their own properties and the properties they obtain from the interaction with the other individual. The combination of both properties is referred to as the induced phenotype, which will be explained in more detail later. The fitness score will guide the selection process as in the GA and the selected individuals can reproduce in three different ways. First, they can just reproduce their own genetic information. Second, to maintain interesting links, we also provide the possibility that all the members of an interaction are reproduced. Third, good symbionts which solve the partial problem completely are reproduced as individuals at a higher level of complexity, meaning that their genetic information is combined into a new individual. Repeated iterations of this process produces individuals of increasing complexity that solve an increasing proportion of the problem.

The following sections explain each part of the ETA in detail.

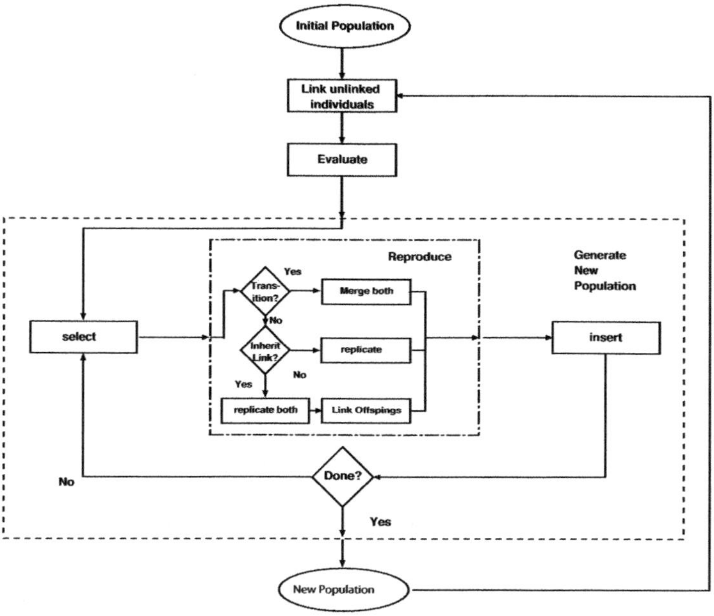

Fig. 5 Schematic overview of the ETA. The algorithm consists of three phases: 1) the pairing up of partial solutions, 2) the evaluations of the interacting solutions and 3) the reproduction of the solutions. Every iteration of these three steps produces a new population.

4.1 *Representation of Basic Elements and Partial Solutions*

A partial solution is a compound label which is not defined on all variables of the variable set X. A compound label on, for instance, variable x_1, x_3 and x_7 where each variable is instantiated with values d_1, d_3 and d_7 respectively is denoted by

$$(\langle 1, d_1 \rangle, \langle 3, d_3 \rangle, \langle 7, d_7 \rangle)$$

In the current description of ETA, we make a distinction between a *partial solution* and a *solution*: A partial solution is a compound label which does not consist of all variables listed in the variable set. When all variables are present, we obtain a solution to the problem. The underlying idea behind this distinction is that a solution, as opposed to a partial solution, is the final outcome of the compositional search algorithm. We therefore speak of partial solution as long as the algorithm is still in the process of evolving complexity, meaning it is still trying various combinations of partial solutions with the hope of obtaining a full solution to the problem. The representation of a (partial) solution is also called the *genotype*.

4.2 *Interactions and the Induced Phenotype*

In the ETA, pairs of (partial) solutions are bound by *symbiotic relations*. When bound, the two solutions exchange information. The result of this exchange is captured in the *induced phenotype* of the solution which can alter, positively or negatively, the fitness of the solution. We explicitly make the distinction between phenotype and induced phenotype to stress that the latter is phenotype produced from both the genetic information of an individual and the effect of external factors, like other population members and the environment, on the final phenotype.

Since the induced phenotype of an individual is constructed by combining the information contained in its own genotype with the information contained in the genotype of the other members of the symbiotic relation, conflicts in this information may occur. In case of a conflict (i.e. two different values are assigned by the partial solutions to the same variable), a conflict mediation strategy is applied that consists of randomly choosing one of the possible values in the conflicting set of values. Interactions are randomly defined in the population of solutions, i.e. the solutions can be linked to any other solution in the population. Alternative strategies, based on multi-level selection, are currently under investigation.

Let us briefly illustrate the way in which the induced phenotype is obtained. Suppose we have a partial solution s that interacts with another partial solution sp. Both partial solutions are represented through the following genotypes:

$$\begin{aligned} s &= \langle 1,\alpha_1\rangle, \langle 2,\alpha_2\rangle, \langle 9,\alpha_9\rangle \\ sp &= \langle 2,\beta_2\rangle, \langle 4,\beta_4\rangle, \langle 9,\alpha_9\rangle \end{aligned}$$

In these two solutions, we observe that s is well defined for the variables 1, 2 and 9 while sp for the variables 2, 4 and 9. The values assigned to variable 9 is the same for both partial solutions. However, the values assigned to variable 2 are not the same, meaning that the solutions have a conflict to solve for this variable: we need to choose between value α_2 or β_2. The conflict mediation strategy randomly chooses, before evaluation, between one of the conflicting values yielding two possible induced phenotypes in our specific example:

$$\varphi(s,sp) = \langle 1,\alpha_1\rangle, \left\langle 2, \begin{pmatrix} \alpha_2 \\ \beta_2 \end{pmatrix} \right\rangle, \langle 4,\beta_4\rangle, \langle 9,\alpha_9\rangle$$

where α_2 or β_2 is chosen randomly with equal probability.

When the conflict resolution is done, the resulting phenotype assigns exactly one value for each variable that was assigned in s or sp. This phenotype is then used by the fitness function to evaluate the partial solution. The details concerning the fitness function and the way partial solutions are evaluated, are provided in the following section. Note first that the induced phenotypes of the members of the symbiotic relation do not have to be the same since the process decides randomly which value to use for each member independently.

4.3 Evaluation Functions for H-IFF

In this section, we provide the fitness functions that will be used to guide the evolutionary and the transition process in the H-IFF context. Remember from Section 3.4

and 3.5 that we redefined the H-IFF problem as a BINCSP. As such, the evaluation functions we define here correspond strongly to the standard evaluation functions found in the BINCSP literature.

If we denote by w_c the weight associated with constraint c, we have the following fitness function for evaluating the (partial) solution s interacting with a symbiotic partner sp:

$$f_{weighted}(s) = \frac{1}{W} \sum_{c \in C} w_c eval(\varphi(s, sp), c) \tag{1}$$

where W is the sum of all the weights: $W = \sum_{c \in C} w_c$ and $eval(s, c)$ returns 1 if the constraint c is satisfied and 0 otherwise:

$$eval(\varphi, c) = \begin{cases} 1 \text{ if } (\varphi \ covers \ c) \text{ and } (satisfies(\varphi, c)) \\ 0 \text{ otherwise} \end{cases} \tag{2}$$

In this approach, the weights enforce the fact that the lower levels in the hierarchy are preferred over the higher levels. We will observe how this preference works on a concrete example. Let us consider a genotype that accomplishes no success at all at the lowest levels such as this size 4 genotype, non-shuffled H-IFF problem with a modularity of 2:

1010

In this situation, the lowest levels are defined on the first two and the last two variables respectively. For these two modules, the assigned values are 10 which are invalid assignments in the IFF context. However, at the length-4 level, we have 2 links that are satisfied between variables 1 - 3 and 2 - 4 ($1 \leftrightarrow 1$ and $0 \leftrightarrow 0$) respectively. These valid links therefore will be considered in the fitness evaluation although the lowest modules are not satisfied. The fitness is therefore[1]:

level	weight	value	part of the genotype processed
1	4	0	10**
		0	**10
2	1	1	1*1*
		0	1**0
		0	*01*
		1	*0*0

This yields a fitness of 2 (summation of $value \times weight$) which is counter-intuitive as the lower level modules are not satisfied at all. However, the choice of the fitness makes sure that such genotypes will always be assigned a fitness that will be lower than the fitness assigned to any genotype that satisfies at least one lower module: for instance, 11** and **11 will each receive a fitness of 4. This small example illustrates the importance of the right choice of weights to enforce the hierarchy.

As in previous work on the ETA and BINCSP [3, 2, 4], a second fitness function is introduced: the *covering fitness* function. The only difference with previous work

[1] We adopt the convention that the symbol $*$ corresponds to a position for which the value is not relevant.

is again the use of weighted constraints in the evaluation of the fitness function. The equation of $f_{weighted}$ provided by Equation 1 is easily adapted to a covering version of the same fitness:

Definition 5. *The* covering set *of constraints for a given genotype γ on the constraints set C is denoted by $C_{cov}(\gamma)$ and consists of all the constraints of C which are covered by γ:*

$$C_{cov}(\gamma) = \{c \in C \mid \gamma \, covers \, c\}$$

With this definition of the covering set, we define the covering fitness as follows:

$$f_{weighted,(cov)}(s) = \frac{1}{W_{C_{cov}(s)}} \sum_{c \in C_{cov}(s)} w_c eval(\varphi(s,sp),c) \tag{3}$$

where $C_{cov}(s)$ is the covering set of s and $W_{C_{cov}(s)}$ is the sum of all the weights of the constraints covered by the solution s: $W_{C_{cov}(s)} = \sum_{c \in C_{cov}(s)} w_c$.

This covering fitness is used to evaluate whether a transition should occur or not. If the covering-weighted fitness of s interacting with its symbiotic partner sp equals 1.0, this means that the sub-problem defined by the interaction is solved and therefore the symbiotic relation may be merged through the process of a transition.

4.4 Replication and Transitions

When selected for reproduction, a partial solution will have three possible options:

1. Simple replication: In this case, the solution is replicated, that is, its genotype is copied and passed on to its offspring (with possible mutations). It occurs when the solution scored sufficiently well to survive one more generation but did not perform well enough within its symbiotic relation to see this relation survive or fixed through a transition.
2. Inherit symbiotic link: In this situation, the solution will not only replicate itself but also replicate its symbiotic partner. The symbiotic link that binds the parent is also inherited by the offspring. This reproduction mode is a step toward a transition; it is, however, a reversible step since both individuals still exist independently. For the current problem, we allowed this mode to occur randomly at each generation.
3. Perform transitions: This case occurs when the induced phenotype of an individual actually solves the sub-part of the problem completely, i.e. when the covering fitness is equal to 1.0. In this case, the outcome of the symbiotic relation is replicated instead of the original genotype of the solution. This means that it is the induced phenotype (the combined information of both partners that solved the partial problem) that becomes the genotype of the offspring.

The second reproduction mode might look like an unnecessary step in the entire process as it occurs randomly without any relation to the transition itself. Nevertheless, this mode may be very important for various reasons:

1. Even though this event occurs randomly, it is performed on individuals selected according to some fitness proportionate selection scheme. In this way, only individuals which are potentially better get the chance to reproduce their symbiotic partner. Consequently, it gives more chance to the conflict mediation strategy to find the good combination.
2. It is a way to keep possibly good genetic material in the population as the symbiotic partner might possess the genetic information that actually causes this solution to be selected. In other words, it is because of the combination of both partners that the fitness is good; so we need to explore this relationship further.
3. Through the possible errors in the copy process, a nearly good combination may become the right combination.

Further evaluation is required for this part of the algorithm to obtain a deeper insight into the relevance of this step.

The third reproduction mode, the transition, creates solutions of higher complexity. This way, small genotypes aggregate into larger genotypes that in turn, through interactions with other genotypes, aggregate into better defined genotypes until a full solution evolves. Let us illustrate this on a small example. Consider the two partial solutions s and sp we defined earlier as follows:

$$s = \langle 1, \alpha_1 \rangle, \langle 2, \alpha_2 \rangle, \langle 9, \alpha_9 \rangle$$

$$sp = \langle 2, \beta_2 \rangle, \langle 4, \beta_4 \rangle, \langle 9, \alpha_9 \rangle$$

The induced phenotype of s interacting with sp is one of the two possibilities in the following expression:

$$\varphi(s, sp) = \langle 1, \alpha_1 \rangle, \left\langle 2, \begin{pmatrix} \alpha_2 \\ \beta_2 \end{pmatrix} \right\rangle, \langle 4, \beta_4 \rangle, \langle 9, \alpha_9 \rangle$$

Suppose that the current induced phenotype of s is the expression that resolves the conflict by choosing the value β_2 for variable 2. Then, if the induced phenotype perfectly solves the sub-problem defined by the sub-set of the constraints defined on variables 1, 2, 4 and 9, the offspring of s when a transition occurs will be:

$$offspring = \langle 1, \alpha_1 \rangle, \langle 2, \beta_2 \rangle, \langle 4, \beta_4 \rangle, \langle 9, \alpha_9 \rangle$$

This offspring genotype is a new solution that can in turn interact with another genotype in the new population and then, through a good symbiotic relation, become even better defined until a complete solution is reached.

5 Simulations

The simulations described in this section were conducted to achieve the following goals:

1. The very first goal is to evaluate the ETA on structured problem instances and observe whether our intuition on ETA regarding structure is correct.
2. We evaluate the impact of the parameters of shuffled H-IFF such as the problem size (the number of variables), the problem range (by observing different variable domain setup) or the modularity (the size of the modules) on the results. This way, we can guarantee that the results obtained to verify our first goal do not result from a lucky choice in the parameters setup.
3. Finally, our last goal is to study the behavior of ETA and compare it with the behavior of SEAM on the same problem instances. This should provide insight into the workings of both algorithms.

Before going into detail of the simulation setup and the results of each of these three goals, we first illustrate the fact that GAs do not perform well on H-IFF[2] as they soon get stuck in sub-optimal solutions and cannot easily find the good recombination of genes because of the high epistasis induced by H-IFF. This allows us to demonstrate that our compositional approach or any other approach that also uses a compositional approach such as HGA [5] or SEAM [28, 29, 30] is able to learn the internal structure of the genotype and therefore beat the poor results obtained from GAs. Another aspect that will briefly be addressed in this section concerns the comparison of both fitness functions, the weighted version of the BINCSP fitness function and the hierarchical fitness function proposed by [28], and show that they exhibit similar behaviors for GAs and are as such equivalent.

5.1 Classical Genetic Algorithm on H-IFF

5.1.1 Simulation Setup

For this simulation, we used a typical 1-point crossover GA with following parameters:

- Generational GA: at each generation, fitness proportional selection is used and the entire population is replaced by the same number of offspring.
- Crossover: one-point crossover.
- Mutation: offspring are mutated with a probability of 0.05. When mutating, one position is chosen randomly for mutation and flipped to a new domain value.
- Population size: 100.
- Number of generations: 10000.
- Problems are non-shuffled and shuffled H-IFF instances.

We generated 100 runs each with a different seed. All the runs were performed on problem instances of 128 bits with a modularity of 2.

[2] We perform this test on the two evaluation functions, i.e. the weighted constraints fitness function and the hierarchical fitness function.

5.1.2 Results

The results are shown in Figure 6. The figure shows the results on shuffled and unshuffled H-IFF instances for the weighted constraints fitness function $f_{weighted}$. A solution for this problem is found when either fitness value has reached 1.0. The fitness values (maximum, average and minimum) are averaged over the 100 runs.

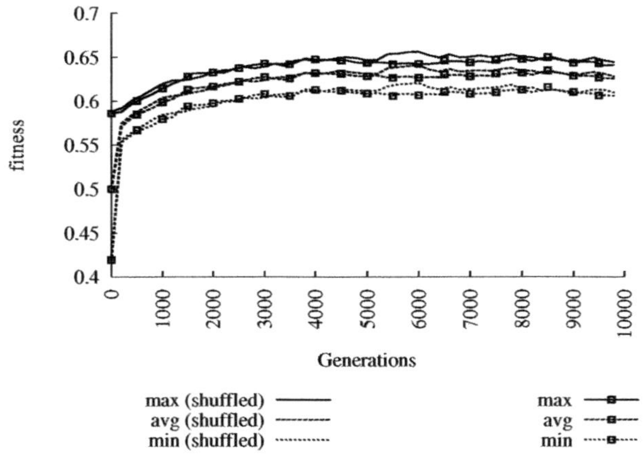

Fig. 6 Simple GA with 1 point crossover on H-IFF. The figure gives the minimum, average and maximum fitness of the weighted constraints fitness for both shuffled and not shuffled instances.

The GA fails to evolve a solution. Furthermore, the trends of the average and worst fitness indicate that the population converges toward a local optimum. This reduction in variety of the population lowers the chance to improve the current best fitness or to escape the local optimum. After a few hundred generations, the population of solutions in the GA converges toward its maximum solution. This solution is, however, far from being optimal as the fitness is far from the 1.0 value that defines the optimal solution.

5.2 *Applying the ETA on a Shuffled H-IFF Problem Instance*

In this section, we examine the behavior of the ETA when applied to shuffled H-IFF problem instances. In these simulations, we monitor the performance of the ETA for problem instances of increasing size. This provides us with some scaling information regarding the ETA.

The analysis is performed on problem instances of modularity 2 for which the values, that can be taken by the variables, belong to the domain $\{0,1\}$. Other setups for the instances (where we vary, for example, the domain size or the modularity) are considered later in this section.

5.2.1 Simulation Setup

For this preliminary simulation on structured problem instances, we considered the following setup: we evaluate how evolutionary process dynamics vary as the genotype size increases. As the genotype sizes increase, so does the number of hierarchical levels. It is more likely that the ETA will require more time to solve the extra levels in the hierarchy. In this simulation, we observe how this increase of time is related to the resulting increase in the number of levels in the hierarchy.

We summarize the setup in the following points:

1. *Algorithm setup*: All these simulations use a generic ETA which has the following parameter setup:
   - Initial population: sound population (i.e. seeded with length-1 partial solutions where all possible assignments for each variable are present in the initial population).
   - Variation operator: a small probability (0.001) mutation is possible during replication.
   - Transition condition: the transition is based on the covering fitness which has to be equal to 1.0. Depending on the problem we optimize, we use the covering weighted constraints fitness function $f_{weighted,(cov)}$ described by Equation 3.
2. *The problem instances*: The simulations are performed on shuffled H-IFF problem instances of modularity 2 defined on binary strings of length 128, 256 and 512.
3. *Run setup*: Each run is limited to 30000 generations. For all simulations, a set of 30 or 64 runs (depending on the simulations[3]) each with a different random seed is performed.

5.2.2 Simulation Results

Table 3, Figure 7 and Figure 8 summarize the results obtained for these preliminary simulations of the ETA on structured problem instances. All the values in the table and figures are averaged over all the runs of the simulations.

Our analysis focuses on problems of size 128, 256 and 512. However, for the sake of completeness, in Table 3, we also present the results for a smaller problem setup, i.e. a size 64 problem. We observe that there is not much difference between the time required to find a solution between a 64 shuffled H-IFF instance and a 128 shuffled H-IFF instance. However, as the size increases toward the 512 shuffled H-IFF instance, we observe that the time required to find a solution increases as well (see Figure 8).

[3] We first made a few tests with 64 size samples to compare the dynamics of the two fitness functions. For the remainder of the simulations, we adopted the setup provided in [28] that restricted the number of runs to 30 and therefore speeds up the simulation process.

Table 3 Results of the preliminary simulations of ETA on structured problem instances (success ratio, average number of generations to reach an optimum, standard deviation in these numbers of generations and corresponding number of evaluations for this number of generations)

		Success Ratio	Mean Generations	Standard Deviation	Mean Evaluations
Scaling	64	1.0	128.6	91.6	16512
	128	1.0	100	54	25600
	256	1.0	473	358	242176
	512	1.0	314	225	321536

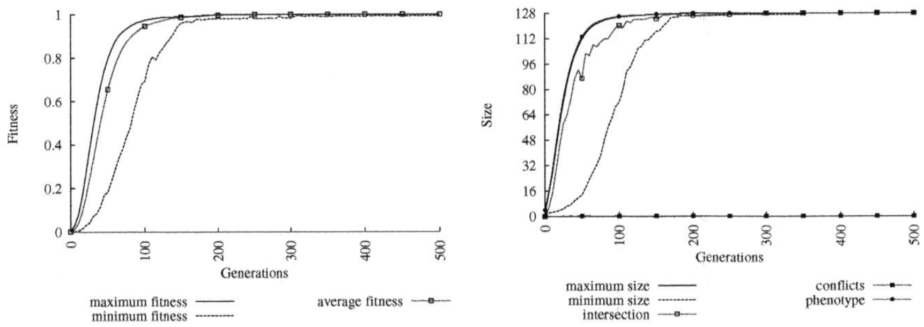

Fig. 7 Evolution of fitness and genotype size over evaluations. Left plot shows the fitness evolution. Right plot shows the evolution in genotype size.

From this, we can conclude that the size of the genotype influences the number of evaluations required by the process to find a solution. However, as the size of the genotype is directly related to the modularity of a problem, we cannot distinguish whether the modularity or the size is the most influencing factor. In the next section, we observe how the domain size of the variables or the modularity of the problem influences the efficiency of our algorithm. This eventually will allow us to answer whether the modularity or the size of the genotype is the most influencing in terms of the time required to find a solution.

5.3 *Impact of the Parameters of H-IFF on the Performance of ETA*

In this section, we examine how the ETA performs when applied to larger domain sizes. Indeed, in the current definition of H-IFF, the parameters n and m which are the number of variables and their respective domain influence the hardness of the problem. We have already varied n from 128 to 512 and observed that it took the

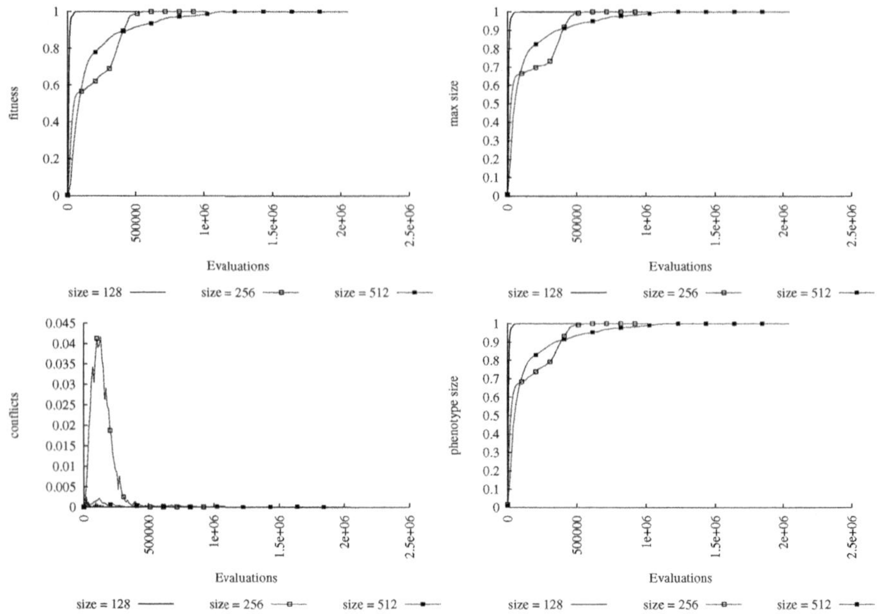

Fig. 8 Comparing the evolution of the fitness and the scaled maximum size, phenotype size and the number of conflicts for different problem sizes (128,256 and 512)

system a larger amount of time to reach a solution as the number of variables increased. Here, we will fix the number of variables to $n = 64$ and let m vary from 2 (binary variables) to 10. We wish to see how the ETA addresses the increased difficulty.

Another aspect that we observe in this section concerns the variation in the modularity of the H-IFF instance. So far, the modularity of all the instances was 2, meaning that the length of the modules was exactly 2 at each level. We will see how an increase in the modularity influences the performance of the ETA. To do so, we fix this time the depth of the hierarchy (i.e. the number of hierarchical levels) to 4 and observe the problems of the following different modularities: 2, 3 and 4. Of course, given a certain depth, the modularity has a direct influence on the genotype length. For example, a depth 4 H-IFF problem of modularity 2 requires genotype strings of length 16 to be solved while a depth 4 H-IFF problem of modularity 3 and 4 requires genotype strings of length 81 and 256 respectively. The results will therefore capture the impact of the modularity but also be influenced by the genotype size as we have illustrated before that this size influenced the performance of the ETA. However, we are able to compare the length 256 H-IFF problem of modularity which was performed in the preliminary simulation with the same length problem but of modularity 4 provided in this set of simulations. This way, we are able to decide which of the modularity lengths influences the outcome of the ETA most.

5.3.1 Simulation Setup

We use the same ETA as in the previous section, and perform each time 30 runs (each with a different random seed). We limit each run to a maximum of 30000 generations.

The only difference in the setup concerns the problem instances on which the ETA will run:

1. Evaluating the impact of the domain size on the performance of the ETA: For this simulation we generate BINCSP instances of length 64 for which the domain of the variable varies from 2 to 10. These instances are then evaluated with a weighted constraints fitness function $f_{weighted}$ defined in Equation 1.
2. Evaluating the impact of the modularity of the hierarchical tree on the performance of the ETA: For this simulation, we generate three BINCSP instances of binary variables of size 16, 81 and 256. For each instance, we use set of weights that correspond to a hierarchically structured instance of modularity 2, 3 and 4 respectively.

5.3.2 Simulation Results

We discuss the results in two parts. We first address the results where the domain size varies and then the results where the modularity of the problem varies.

Varying the domain size

For this simulation, the only variation that was introduced in each setup concerns the domain size of the variables of the problem. This domain size varies from 2 to 10. In Table 4, we summarize the results for the 9 different domain size setups. In general, the ETA easily finds the solution for the different sizes addressed in this simulation because it finds a solution relatively quickly in almost every run. However, there is

Table 4 Success ratio and the average number of generations as well as the corresponding number of evaluations to reach an optimum for various domain sizes on the 64-length shuffled H-IFF problem of modularity 2

Domain Size	success ratio	Mean Generations	standard deviation	Mean Evaluations
2	1.0	128.6	91.6	16512
3	1.0	110.7	42.6	21312
4	1.0	134.5	51.4	34432
5	1.0	156.4	39.8	50048
6	1.0	234.0	237.5	89856
7	0.93	358.1	717.5	160429
8	1.0	247.1	149.7	126515
9	1.0	234.9	77.5	135302
10	0.97	300.2	227.2	192128

an increasing trend in the required number of evaluations to reach the solution. In the following table, however, we illustrate that although the difference in speed (see page 136 for the details of how the speeds are computed) is significant in most of the cases (p-value of the Mann-Whithney ranking test is < 0.05 in most cases), we do not observe a clear pattern in the increase when we compute the ratio of the required number of evaluations between two consecutive domain sizes (see Table 5).

Table 5 Results for Mann-Whithney ranking test between consecutive domain sizes

	$2 \rightarrow 3$	$3 \rightarrow 4$	$4 \rightarrow 5$	$5 \rightarrow 6$	$6 \rightarrow 7$	$7 \rightarrow 8$	$8 \rightarrow 9$	$9 \rightarrow 10$
trend	0.77	0.62	0.69	0.56	0.56	1.27	0.93	0.07
p-value	0.07	< 0.0001	< 0.0001	< 0.0001	0.0016	0.1858	0.085	0.0087

In Figure 9, we observe the evolution of the fitnesses, the sizes (maximum and minimum) of the induced phenotypes as well as the size of the intersection between the best partial solution and its symbiotic partner and the number of conflicts within this intersection. These plots confirm that, on one hand, although there is not much difference between the different domain sizes, we observe that as the domain size increases, so does the necessary average time to find a solution. The general dynamics, on the other hand, are the same for all domain sizes. This can be seen by coupling the graphs of the intersection with the graphs of the conflicts. This reveals that the process, once it has found a solution, converges toward this solution; and this applies for all domain sizes.

We conclude from all these results that the domain size has a slight but significant impact on the number of evaluations required to find a solution as this number increases with respect to the domain size of each variable of the problem. However, increasing the domain size does not dramatically affect the ETA's chance for success in relation to this class of problem and the change in performance is proportional to the difference in the sizes.

Varying the modularity

In Table 6, we summarize the success ratio and the average number of steps required to reach a solution. In all these setups, we fix the depth of the hierarchical tree and let the modularity vary between 2 and 4 yielding genotype sizes of 16 to 256. The purpose, therefore, is to see whether the increase in the number of generations required to find an optimum corresponds to the increase in the modularity or the increase in the genotype size. To address this issue, we also discuss in this table the results we obtained for the length 256 genotype in a problem of modularity 2.

The first observation that can be made is that an increase in modularity does not influence the probability of success of the ETA on this kind of problem. This means that even when the modules we wish to discover are of length 3 or 4, pairwise interactions are sufficient to evolve a solution. This result is interesting because it means that although there is no reason to limit the ETA to pairwise interactions, pairwise interactions do not limit the performance of the ETA, even when it is addressed to a

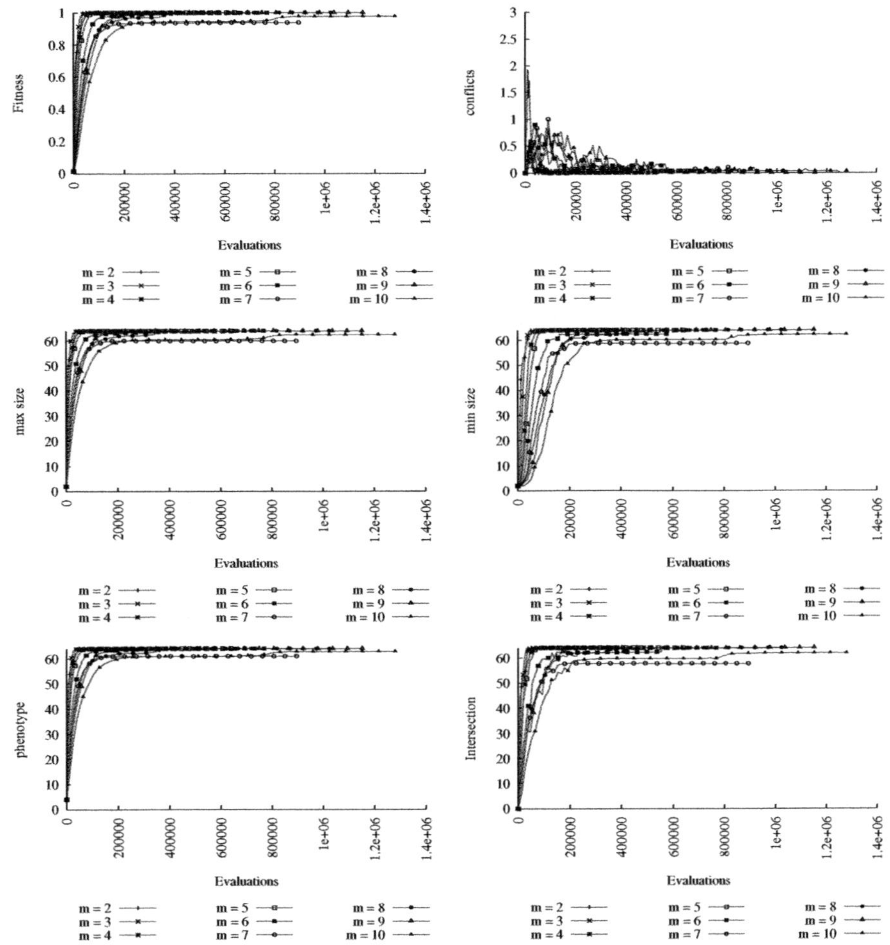

Fig. 9 Evolution of the fitness and sizes of the ETA for different domain size (m) on the 64 length shuffled H-IFF problem

Table 6 Success ratio and the average number of generations as well as the corresponding number of evaluations to reach an optimum for various modularities (and as a consequence, the sizes) of the shuffled H-IFF problem instances

modularity	size	success ratio	mean Generations	standard deviation	Mean Evaluations
2	16	1.0	48	38	1536
3	81	1.0	94	58	15228
4	256	1.0	285	226	145920
2	256	1.0	473	358	242176

problem for which the modularity would probably require another kind of interaction (group based interactions) to be solved by another compositional algorithm.

The second observation concerns the comparison between a 256 shuffled H-IFF problem of modularity 2 and the same problem addressed with a modularity of 4. In the first case, the depth of the hierarchical tree is 8 while in the second, it is 4. We observe that the speed seems to follow the ratio of the depth as the algorithm with modularity 4 seems to find a solution twice as quickly as the algorithm of modularity 2. Of course, other simulations should be performed to conclude that the speed of the algorithm is proportional to the depth of the hierarchical tree. However, this significant difference[4] allows us to conclude that the speed of the algorithm is undoubtedly related to the depth of the hierarchical tree.

With these sets of simulations however, we conclude that the ETA can quite easily solve hierarchically structured BINCSP instances. The small variations to the parameters of the problem instances show that the ETA is able to resolve increasingly difficult problem instances.

In the last section, we highlight how these results differ or are similar to the results obtained by SEAM on the same test cases.

5.4 Comparing Results from the ETA with Results from SEAM

In this section, we focus our discussion on the differences in behavior between the ETA and SEAM on the H-IFF problem. We first compare the fitness dynamics between the two algorithms on the 128 shuffled H-IFF problem. We then see how variation in the domain size influences both dynamics. Finally, we discuss the case of various modularities and the applicability of SEAM and the ETA on these problems.

5.5 How the Algorithms Are Compared

We will compare the results obtained by the SEAM described in [28] with those obtained by the ETA. In our implementation, we altered SEAM a little from its previous description so that the context set is no longer generated at each generation but every 100 generations. This modification is allowed since it does not influence the outcome of the algorithm but increases its overall speed. The implemented model is sketched in Figure 10.

To evaluate whether there are any significant differences between both algorithms, we will compare their speed. The speed at which a solution is found (a solution of maximum fitness = 1.0) is defined by:

$$speed(r) = \begin{cases} 0 & \text{if no solution found during } r \\ \frac{e_{max}-e}{e_{max}} & \text{if the optimum is found in } e \text{ evaluations} \end{cases} \quad (4)$$

[4] The Wilcoxon-Mann-Whitney test which compares the mean speed of these two algorithms yields a p-value of 0.0528 which means that the difference is not significant under a 5% error range. The significance of the difference is therefore quite important.

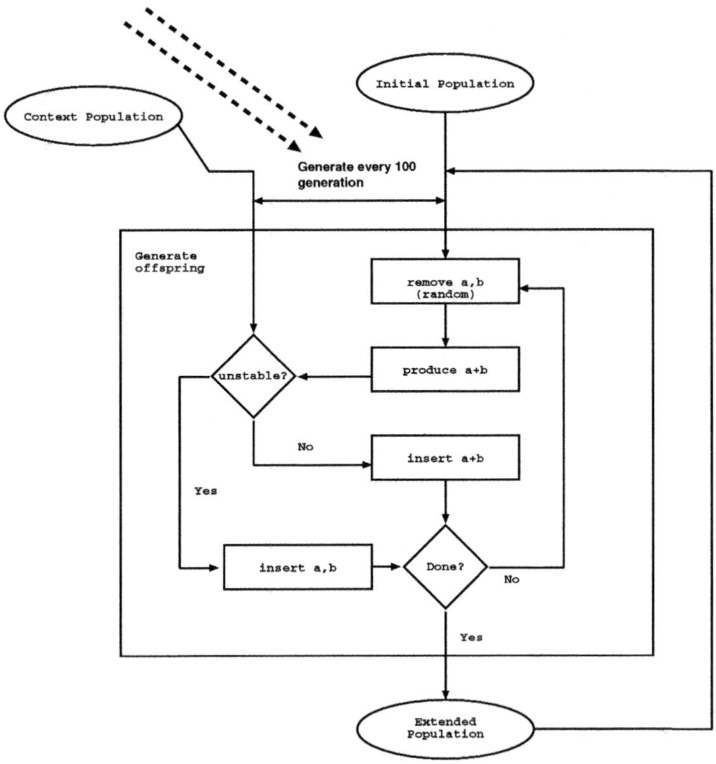

Fig. 10 Adaptation of SEAM from its basic description in [28] to a version where the context set is not built at each generation but every 100 generations

where e_{max} corresponds to the maximum number of evaluations compared in the simulation run.

The speed scales from 0 to 1 where 0 means that no solution is found and 1 means the solution is found from the very first run. The speed of each run is computed and yields for each simulation setup a set of values. Here, two sets of speeds for the ETA and SEAM are then compared with the Wilcoxon-Mann-Whitney rank test to evaluate whether the means are equivalent.

5.5.1 General Comparison of the Dynamics between SEAM and the ETA

In Figure 11, we observe on the left the best fitness evolution (averaged over 64 runs) of the ETA performing on a binary 128 shuffled H-IFF problem instances and on the right, the same observation but this time for the SEAM.

We observe that although both processes succeed in solving this problem instance, the ETA requires fewer evaluations to reach the result. Furthermore, the evolutionary behavior of both processes is different. The dynamics of the ETA are more comparable to the dynamics observed in typical EA where the fitness grows quickly

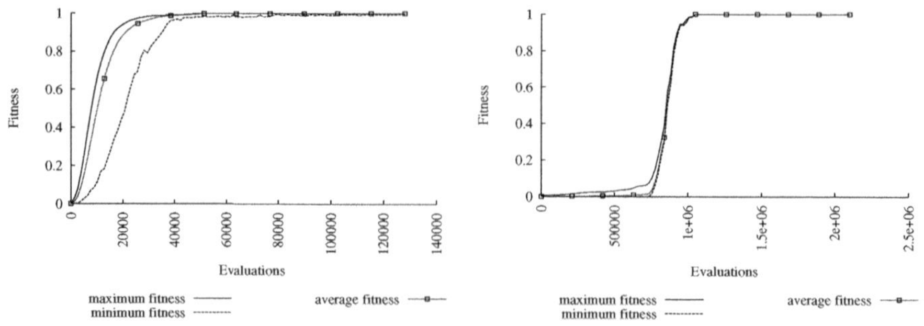

Fig. 11 Comparing the dynamics between the ETA (on the left) and SEAM (on the right)

to a nearly-optimal value in the beginning and converges slowly afterwards to the optimum. This is probably a direct consequence of the fitness proportional selection that favors, from the very beginning, good solutions to be selected and to spread into the population. The SEAM, on the other hand, was built to solve hierarchically decomposable problems. It therefore focuses, in the beginning, on the identification of the lowest level modules and requires these modules to be conquered first before addressing other levels. The resulting dynamics is a fitness function that slowly increases in the beginning (the lowest level of the module is conquered) and once the lower levels are conquered, the process speeds up exponentially as it conquers the higher levels a lot more easily. This difference in the evolutionary dynamics is one of the biggest differences that can be observed in the simulation of the SEAM and the ETA on shuffled H-IFF problem instances. Of course, as we just stressed, this difference is a direct consequence of the major differences between the two models: on one hand, a fitness proportional selection model including symbiosis and fitness specific transition mechanism (the ETA) and on the other hand, an ecosystem that does not use any selection mechanism and gradually increases the complexity of its individuals through symbiosis and a context based transition condition (SEAM).

5.5.2 Varying the Domain Size and the Modularity

In Table 7, we summarize the success ratio, average number of steps (and standard deviation) for the results of SEAM applied to a 64 shuffled H-IFF problem instance for which the domain size varies from 2 to 10. All values are averaged over the 30 runs of each simulation.

In this table, we observe that SEAM has no difficulty in solving problem instances with increasingly large domain sizes as it succeeds in creating a solution for each run. However, the required number of evaluations increases relative to the domain size. This is relatively different to the way ETA reacted to the size as, although we could observe an increase in the number of evaluations, we could not distinguish a clear trend in the required number of evaluations to find a solution. To observe this difference, in the following table, the ratio of the average number of steps at which a solution is found with a domain size m and the average number of

Table 7 SEAM: success ratio, average number of generations and corresponding number of evaluations to find a solution with standard deviation in this number of steps

$m =$	success ratio	Mean Generations	Standard Deviations	Approximated Mean Evaluations
2	1.0	101	6	212100
3	1.0	233	20	489300
4	1.0	410	27	861000
5	1.0	628	44	1318800
6	1.0	897	52	1883700
7	1.0	1236	54	2595600
8	1.0	1621	72	3404100
9	1.0	2072	92	4351200
10	1.0	2511	90	5273100

steps at which a solution is found with a domain size $m+1$ are computed for the ETA and SEAM. We can see that, on one hand, no clear relation appears between consecutive domain sizes in the ETA and, on the other hand, SEAM shows a clear relation between consecutive domain sizes as the ratios increase slowly toward 1.0 (see Table 8).

Table 8 Results for Mann-Whithney ranking test between consecutive domain sizes

	$2 \to 3$	$3 \to 4$	$4 \to 5$	$5 \to 6$	$6 \to 7$	$7 \to 8$	$8 \to 9$	$9 \to 10$
ETA	0.77	0.62	0.69	0.56	0.56	**1.27**	0.93	**0.07**
SEAM	0.43	0.57	0.65	0.7	0.72	0.76	0.78	0.82

As we have said, we observe that the required number of evaluations in the SEAM for a certain size is related to the required number of evaluations of the previous domain size and this ratio increases slowly to 1 (meaning that the difference between the two decreases slowly), yielding a trend in the way a domain size influences the expected success. For the ETA, however, there is no obvious relation from one size to another as we illustrated earlier in this simulation section (see page 133). This weaker relation between the domain size for the ETA with respect to SEAM is illustrated in Figure 12, where we compare the evolution of the best fitness for the different domain sizes for the ETA (on the left) and SEAM (on the right).

In this graph, we observe that for SEAM, the time to find the lowest level modules increases in proportion to the domain size. This can be seen as the fitness is shifted to the right with respect to the domain size. For the ETA, the trend of the fitness is relatively equivalent for all the domain sizes, making it very difficult to distinguish a difference in the fitness evolution in relation to the size. From this, we can conclude that the ETA is less influenced by the domain size than SEAM and can therefore efficiently address a larger search space.

The last interesting aspect that we observed in our ETA simulations on shuffled H-IFF concerned the variation in the modularity of the problem. Unfortunately,

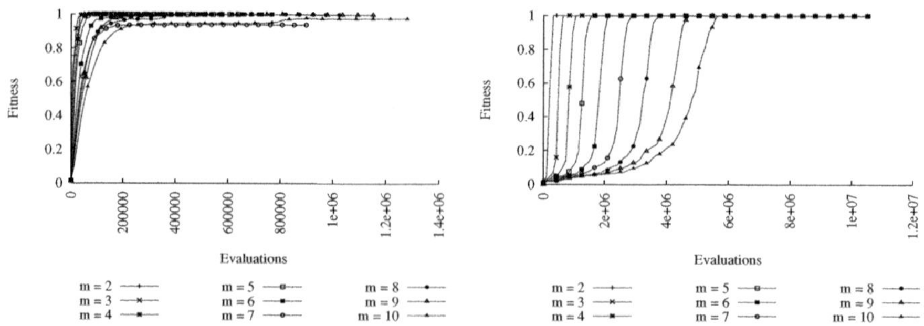

Fig. 12 Comparing the dynamics between the ETA (on the left) and SEAM (on the right)

SEAM could not be straightforwardly applied to higher modularity problems as pairwise interacting H-IFF can only address problems with a modularity of 2 [28]. To extend H-IFF to higher modularity, would require symbiotic groups that grow in size with respect to the modularity. For example, for a modularity of 3, interacting groups of 3 are required. The Pareto optimality of a group is therefore compared against each component of the group and a transition is performed when the resulting group is Pareto optimal with respect to a given context set. This adaptation is not too difficult to introduce but is beyond the scope of this work and is left for future studies. It is, however, trivial to understand that the required time to find an optimum will grow combinatorially with respect to the modularity.

Let us take the modularities 2, 3 and 4 as an example. When modularity is 2, each time a pair is selected, it has a probability of $\frac{1}{nm(nm-1)}$ being a good pair where n is the number of variables and m is the domain size. However, at the lowest level, $\frac{nm}{2}$ such pairs exist. This means that, at the lowest level, we have a probability of $\frac{1}{2(nm-1)}$ of selecting a good pair. This probability is therefore in a factor $\frac{1}{nm}$.

In the case of a modularity 3, we have this time $\frac{nm}{3}$ modules at the lowest level. Each module has a probability of $\frac{1}{nm(nm-1)(nm-2)}$ to be selected which yields a probability of $\frac{1}{3(nm-1)(nm-2)}$ of selecting a good module at the lowest level. As we can see, this probability now is in a factor of $\frac{1}{3n^2m^2}$.

In general, we will have that for a problem of modularity k, the probability of generating a good group at the lowest level is:

$$\frac{1}{k\frac{(nm-1)!}{(nm-k)!}}.$$

This probability is a factor of $\frac{1}{k(nm)^{k-1}}$ which decreases exponentially with respect to k. Therefore, increasing the value of the modularity results in a SEAM algorithm that will require exponentially more time to resolve the lowest levels of the hierarchy.

This increase of time was not observed in the case of the ETA where, on the contrary, we observed that the 256 shuffled H-IFF problem of modularity 4 was resolved twice as fast as the equivalent 256 shuffled H-IFF problem of modularity 2. This may be due to another difference between the ETA and SEAM. However, we did not test it further as this was beyond the scope of this work.

6 Discussion

In this chapter, we investigated the performance of the ETA on hierarchically structured problems defined by H-IFF. Our goal was to confirm the results obtained on random BINCSP instances that showed that the ETA performs better on structured problem instances. To do so, we generated a weighted constraints version of the BINCSP that corresponds to the IFF problem. These weights reflect the hierarchical structure induced by H-IFF. We described H-IFF and the results obtained by a simple GA, the ETA and SEAM on instances of varying difficulty. The major conclusion of all these simulations is that the ETA performs very well on the shuffled H-IFF problem. Furthermore, we have observed that the parameter tuning did not produce significant effects on the success ratio of the algorithm, although the time required to find a solution could vary in proportion to the genotype size or the domain size of the variables. The modularity of the problem was not an issue for the ETA.

We also compared the dynamics of the ETA and SEAM on shuffled H-IFF. We observed that the ETA requires fewer evaluations to reach its goal and that both processes have very different dynamics as the dynamics of the ETA are more closely related to the dynamics of a classical fitness proportional selection EA while the dynamics of SEAM enforce the resolution of the lowest-levels modules first and therefore observe a slow increase of the fitness in the beginning of the process and then an exponential growth of the fitness as the basic modules have been discovered. This difference in the dynamics is observable everywhere and SEAM seems a lot more sensitive to the parameter tuning in the shuffled H-IFF instance than the ETA does. For example, the time required to find a solution grows in proportion in a significant manner to the domain size while this is not really the case for the ETA. Furthermore, SEAM, limited to pairwise interactions, is not able to address an increase in modularity and requires adaptations in the implementation of the model in order to work with group-wise interactions whose sizes grow in proportion to the size of the module. In this respect, the ETA reveals itself to be more general as it did not require any adaptation to address higher modularity.

As in any comparison, certain limitations exist and the current positive results should be moderated with some critical remarks. It should, for instance, be clear that the ETA is not suited for every type of problems. Moreover, as observed for learning classifier systems (LCS) [14, 10], finding a suitable transition function can be very hard. This problem of finding a good mechanism to evaluate partial solution is known as the credit assignment problem [9]. It is clear that, given the problem on which one wants to apply the ETA, one will spend some time in determining how to decide on the quality of a partial solution. Different from the assignment problem in LCS is that we will not have a propagation problem since two partial solutions are merged as soon as they are considered to be adequate. Furthermore, multiple design choices of the ETA need to be evaluated further. It is very likely that certain choices can be revised that would make the approach more suitable for the problem domain.

The current comparison with SEAM should also be extended to the newly proposed methods, i.e. the Reciprocal Synergy Symbiosis Algorithm [20] and the Random Mutation Hill-Climber proposed in [12, 13]. This investigation is left as future work.

To conclude, we believe that for many real-world problems related to some form of constraint optimization, like in planning and scheduling, the ETA might have a bright future.

References

1. Defaweux, A., Lenaerts, T.: Evolutionary transitions in sequence complexity: a proof of concept. In: Proceedings of the Annual Machine Learning Conference of Belgium and The Nederlands, Vrije Universiteit Brussel, pp. 38–45 (2004)
2. Defaweux, A., Lenaerts, T., van Hemert, J.: Evolutionary transitions as a metaphor for evolutionary optimisation. In: Proceedings of The VIIIth European Conference on Artificial Life, Canterbury, UK, pp. 342–352 (2005)
3. Defaweux, A., Lenaerts, T., van Hemert, J., Parent, J.: Complexity transitions in evolutionary algorithms: Evaluating the impact of the initial population. In: Proceedings of the IEEE Congress on Evolutionary Computation, Edinburgh, UK, pp. 2174–2181 (2005)
4. Defaweux, A., Lenaerts, T., van Hemert, J., Parent, J.: Transition models as an incremental approach for problem solving in evolutionary algorithms. In: Proceedings of The Genetic and Evolutionary Computation Conference, Washington DC, USA, pp. 599–607 (2005)
5. de Jong, E.D., Watson, R.A.: On the complexity of hierarchical problem solving. In: Proceedings of The Genetic and Evolutionary Computation Conference, Washington DC, USA, pp. 1201–1208 (2005)
6. Goldberg, D., Korb, B., Deb, K.: Messy genetic algorithms: motivation, analysis, and first results. Complex Systems 3, 493–530 (1989)
7. Goldberg, D., Korb, B., Deb, K.: Messy genetic algorithms revisited: Studies in mixed size and scale. Complex Systems 4, 415–444 (1990)
8. Goldberg, D.E., Deb, K., Clark, J.H.: Genetic algorithms, noise, and the sizing of populations. Complex Systems 6, 333–362 (1992)
9. Greffenstette, J.: Credit assignment in rule discovery systems based on genetic algorithms. Machine Learning 3, 225–245 (1988)
10. Holmes, J., Lanzi, P., Stolzmann, W., Wilson, S.: Learning classifier systems: New models, successful applications. Information Processing Letters 82(1), 23–30 (2002)
11. Hoos, H.H., Stützle, T.: Stochastic Local Search: Foundations and Applications. Morgan Kaufmann, San Francisco (2005)
12. Khor, S.: Rethinking the adaptive capability of accretive evolution on hierarchically consistent problems. In: IEEE Symposium on Artificial Life, pp. 409–416 (2007)
13. Khor, S.: Hill climbing on discrete HIFF: exploring the role of DNA transposition in long-term artificial evolution. In: Proceedings of the 9th annual conference on Genetic and evolutionary computation, pp. 277–284 (2007)
14. Lanzi, P.L., Stolzmann, W., Wilson, S.W. (eds.): IWLCS 1999. LNCS (LNAI), vol. 1813. Springer, Heidelberg (2000)

15. Lenaerts, T.: Different Levels of Selection in Artificial Evolutionary Systems: Analysis and Simulation of Selection Dynamics. PhD thesis, Department of Computer Science, Vrije Universiteit Brussel, Pleinlaan 2, 1050 Brussel (2003)
16. Maynard-Smith, J., Szathmáry, E.: The Major Transitions in Evolution. W.H. Freeman, San Francisco (1995)
17. McPhee, N.F., Crane, E.F.: A theoretical analysis of the HIFF problem. In: Proceedings of the conference on Genetic and Evolutionary Computation, Washington DC, USA, pp. 1153–1160. Morgan Kauffman, San Francisco (2005)
18. McKenzie, J.F., Castillo, L., Borrajo, D., Salido, M.A., Oddi, A.: Planning, scheduling and constraint satisfaction. IOS Press, Amsterdam (2005)
19. Michod, R.: Darwinian Dynamics: Evolutionary transitions in Fitness and Individuality. Princeton University Press, Princeton (1999)
20. Mills, R., Watson, R.A.: Symbiosis, synergy and modularity: introducing the reciprocal synergy symbiosis algorithm. In: Almeida e Costa, F., Rocha, L.M., Costa, E., Harvey, I., Coutinho, A. (eds.) ECAL 2007. LNCS (LNAI), vol. 4648, pp. 1192–1201. Springer, Heidelberg (2007)
21. Pelikan, M., Goldberg, D.E.: Escaping hierarchical traps with competent genetic algorithms. In: Proceedings of the 2001 conference on Genetic and Evolutionary Computation, pp. 511–518. Morgan Kaufmann, San Francisco (2001)
22. Potter, M.: The Design and Analysis of a Computational Model of Cooperative Coevolution. PhD thesis, Department of Computer Science, George Mason University (1997)
23. Raynal, F., Collet, P., Lutton, E., Schoenauer, M.: Individual gp: an alternative viewpoint for the resolution of complex problems. In: Proceeding of the Genetic and Evolutionary Computation Conference (GECCO), pp. 974–981. Morgan Kaufmann, San Francisco (1999)
24. Raynal, F., Collet, P., Lutton, E., Schoenauer, M.: Polar ifs + parisian genetic programming = efficient ifs inverse problem solving. Genetic Programming and Evolvable Machines Journal 1(4), 339–361 (2000)
25. Russell, S., Norvig, P.: Artificial Intelligence: A Modern Approach. Prentice-Hall, Englewood Cliffs (1995)
26. Salido, M.A., Garrido, A., Barták, R.: Introduction: special issue on constraint satisfaction techniques for planning and scheduling problems. Engineering Applications of Artificial Intelligence 21(5), 679–682 (2008)
27. Thierens, D., Goldberg, D.E.: Mixing in genetic algorithms. In: Proceedings of the fifth international conference on genetic algorithms, pp. 38–45. Morgan Kaufmann, San Mateo (1993)
28. Watson, R.: Compositional Evolution: Interdisciplinary Investigations in Evolvability, Modularity, and Symbiosis. PhD thesis, Brandeis University (2002)
29. Watson, R.A., Pollack, J.B.: Symbiotic combination as an alternative to sexual recombination in genetic algorithms. In: Deb, K., Rudolph, G., Lutton, E., Merelo, J.J., Schoenauer, M., Schwefel, H.-P., Yao, X. (eds.) PPSN 2000. LNCS, vol. 1917, pp. 425–434. Springer, Heidelberg (2000)
30. Watson, R.A., Pollack, J.B.: A computational model of symbiotic composition in evolutionary transitions. Biosystems Special Issue on Evolvability 69(2-3), 187–209 (2002)
31. Wiegand, R.P.: An Analysis of Cooperative Coevolutionary Algorithms. PhD thesis, George Mason University, Fairfax, VA (2004)

Electrical Load Forecasting Using a Neural-Fuzzy Approach

Ajoy K. Palit, Walter Anheier, and Dobrivoje Popovic

Abstract. Electrical load forecasting plays a vital role in the operation, scheduling, maintenance and planning of electrical power generation and distribution companies. A novel hybrid system based on the neural implementation of fuzzy logic models of the nonlinear and chaotic electrical load time series is presented in this chapter. Being a universal function approximator, the hybrid neural-fuzzy system can very accurately capture the nonlinearity and chaotic behavior of electrical load series. A hybrid network's performance is further largely improved from the application of moving average filter and fuzzy clustering on electrical load data along with least square error estimation of rules' consequents. This, in turn, makes the short-term and mid-term forecasting of electrical load feasible. The results presented at the end confirm the efficacy of the neural-fuzzy system as a good electrical load forecaster.

1 Introduction

A precise forecasting of the electrical demand of its customers is very important and essential for the day-to-day operation, scheduling, routine maintenance and planning of electrical power generation and distribution companies [3]. For instance, an accurate short-term electrical load forecasting, which includes prediction of quarterly to hourly loads (demand) for the next hour to several days out, helps in avoiding risky operational conditions of power systems, especially in a deregulated and competitive market where many systems are being pushed to function in an overstressed situation close to their security margin [8]. Therefore, it is necessary to take operational planning seriously based on the accurate and effective short-term forecasts of electrical load to enforce system security. Apart from that, the quality of short-term hourly load forecasts has a large impact on the economic operation of the electric utility [8]. This is due to the fact that many decisions such as economic scheduling of the generating capacity, scheduling of fuel

Ajoy K. Palit, Walter Anheier, and Dobrivoje Popovic
Faculty of Electrical Engineering, University of Bremen, Otto-Hahn-Allee-1, D-28359 Bremen, Germany
e-mails: palit@uni-bremen.de, anheier@item.uni-bremen.de, popovic@iat.uni-bremen.de

R. Chiong, S. Dhakal (Eds.): Nat. Intel. for Sched., Plan. and Pack. Prob., SCI 250, pp. 145–173.
springerlink.com

purchases, system security assessment, and planning of energy transactions, which have significant economic consequences, are based on these forecasts. This indirectly means that proper information about the short-term electrical load or demand forecasts can avoid the unnecessary higher costs of power generation and distribution, allowing utilities to commit their production resources to optimize the energy prices and exchange with vendors and clients. On the other hand, the mid-term and long-term electrical load forecasts, which include the prediction of weekly or monthly electrical load for the former and yearly forecasts for the latter, are applied to maintenance planning, expansion planning, inter-tie tariff setting and long term capital investment return problems, etc [3]. Spatial forecasting is mainly about forecasting future load distribution in a special region, such as a state, or the whole country. Thus, there is a need for accurate load forecasting techniques, in general, and the importance of the same will further increase in the future because of the dramatic changes occurring in the structure of the utility industry.

In this chapter, a fuzzy model of a given electrical load time series is presented as a multilayer feed-forward network, whose parameters are thereafter tuned with a classical training algorithm of neural networks. The hybrid network's performance is further largely enhanced by the application of moving average filter and fuzzy clustering on electrical load data along with least square error (LSE) estimation of rules' consequents. Once the parameters of the fuzzy model have been fine-tuned, the adaptive fuzzy model can be used as an efficient load forecaster.

The organization of the rest of the chapter is as follows. Section 2 focuses on the related works. Section 3 summarizes the basic principle of fuzzy logic systems. In Section 4, the structure of neural-fuzzy systems is presented. The training algorithm for the neural-fuzzy network, the classical backpropagation algorithm and its modification for faster training convergence, are described in Section 5. Section 6 presents the preprocessing of electrical load data and its subsequent application in short-term and mid-term electrical load forecasting by neural-fuzzy networks. Section 7 presents the rearrangement of normalized electrical load data for mid-term forecasting, fuzzy clustering techniques and moving average filtering applied to electrical load for further improvement of the network's training performance. The simulation experiments for mid-term forecasting and the performance measure of neural-fuzzy system are also described in Section 7. Finally, the conclusions and future work are given in Section 8.

2 Related Works

A large variety of techniques have been investigated in the literature [3, 8] for the load forecasting application, namely the traditional time series analysis (Box-Jenkin's ARMA or ARIMA method) or regression approach [1], group methods of data handling (GMDH), feed-forward neural networks based approaches [8], recurrent neural networks based approaches, and evolutionary techniques, etc. The load has complex and nonlinear relationship with climatic factors, past usage patterns, the day of the week, and the time of the day, etc. Modeling such complex nonlinearities with conventional techniques such as time series analysis and regression analysis will always lack the desired accuracy [3]. Moreover, most of the conventional models are not portable; implying that a model developed for one site cannot be easily modified for use at another utility. On the other hand, the nature of the load forecasting problem lends itself well to the artificial neural networks (ANNs) technology since ANNs can model complex and nonlinear relationships through a training process involving historical data patterns. Consequently, several studies over the past few years have considered the application of ANN to the load forecasting problem.

Recently, hybrid approaches based on neural-fuzzy systems have received considerable attention from researchers for various nonlinear modeling applications [2, 7, 11]. The main advantage of such a hybrid approach is that it exploits the merits of both neural networks and fuzzy logic based modeling techniques. For instance, the fuzzy models are considered, to some extent, as "white-box models", since they are based on fuzzy "If-Then" rules and are, to a certain degree, transparent to human interpretation and analysis. On the other hand, the neural networks based models have a unique learning ability, although such models are usually considered as black-box models. This is the main motivation for us to develop an electrical load forecasting system based on the neuro implementation of a fuzzy logic model of nonlinear, chaotic electrical load time series. The hybrid forecasting system is first trained using past load data to capture the nonlinearity and chaotic behavior of electrical load time series. Thereafter, the same is applied for short-term and mid-term forecasting of electrical load for several weeks.

3 Fuzzy Logic Systems

Fuzzy logic systems are based on fuzzy concepts, such as fuzzy sets, linguistic variables, and fuzzy logic. Fuzzy systems are unique in the sense that they can simultaneously process numerical data and linguistic knowledge provided by a human expert. From the mathematical point of view, a fuzzy logic system is a nonlinear mapping of an input feature (data) vector into a scalar output.

The block diagram of a fuzzy logic system is shown in Figure 1. From the figure, it can be seen that the fuzzy logic system takes a crisp input value ($x_{in} = a$) and this is then fuzzified (converted into corresponding membership grades in the

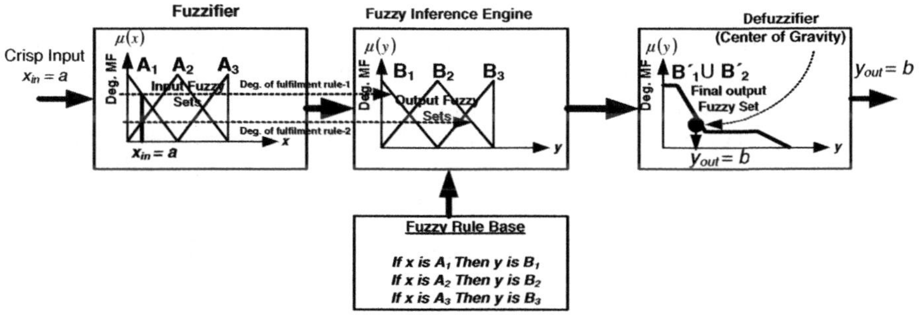

Fig. 1 Fuzzy sets and block diagram of fuzzy logic system

input fuzzy sets A_1 and A_2, also called degrees of fulfillment for rule-1 and rule-2), and thereafter, it is applied to the fuzzy inference engine. Using the stored "If-Then" fuzzy rules from the rule base and the degrees of fulfillment, the inference engine produces a fuzzy output $\left(B_1^{'} \cup B_2^{'}\right)$, that undergoes further (center of gravity) defuzzification (opposite of fuzzification) resulting in a crisp output (y_{out}) (see [7] for details).

Fuzzy sets can be involved in a fuzzy logic system in different ways:

- in the system description
- in the representation of input, output and system states, etc.

When involved in the system description, fuzzy sets appear as linguistic terms or labels representing the state of the linguistic variable in the fuzzy rule. An illustration of the first case can be presented considering Boyle's law for a given mass of gas. Using fuzzy linguistic rules, Boyle's law can be stated as:

Rule-1: If Pressure is **high** Then Volume is **low**
Rule-2: If Pressure is **medium** Then Volume is **medium**
Rule-3: If Pressure is **low** Then Volume is **high**

In the above case, Boyle's law has been described by three "If-Then" fuzzy rules. In fact, any system can be described by a collection of such "If-Then" linguistic rules. The fuzzy logic systems are actually a rule-based system and usually defined using the If-Then rules. The general form of such a rule is: If "antecedent propositions" Then "consequent propositions". The examples of such antecedent propositions can be "Pressure is high" or even "x is $\boldsymbol{A}$". Here, the term "High" is a linguistic term or label, also called a fuzzy term, represented by a fuzzy set (membership function) on the universe of discourse (UD) of the linguistic variable "pressure". Similarly, a fuzzy set $\boldsymbol{A}$ is a representation of a linguistic label/term. Sometimes linguistic hedges (modifiers) are used to modify the linguistic label/fuzzy set without redefining the fuzzy set completely. Examples of this can be "very $\boldsymbol{A}$" or "more or less $\boldsymbol{A}$," etc.

Also, fuzzy sets may appear as the only means of expressing human perceptions or even noisy or uncertain data or information that have to be used as a system's input, output variables, and system's state. As an illustration, consider the input to a system that can be noisy data (reading from unreliable sensors), or even human perceptions such as hot, warm, comfortable, uncomfortable, beautiful and tasty, etc. Fuzzy logic based systems can process such types of information by defining their suitable ranges and criteria with fuzzy sets.

Fuzzy inference is the core part of a fuzzy logic system. In practice, the following fuzzy inference systems have been employed most frequently and have been the subject of most of the theoretical and practical studies:

- Mamdani type fuzzy inference systems
- Relational (Pedrycz) fuzzy logic systems
- Takagi-Sugeno type fuzzy inference systems

Mamdani (1977) proposed the first fuzzy inference system with the objective to control a combination of a steam engine and a boiler [4], based on a set of linguistic control rules built from the extracted knowledge of a human expert. As an example of Mamdani type of fuzzy rules, consider a single-input single-output system that describes the relationship between the heater-current and temperature trend as follows:

Rule-1: If Heater-current is **high** Then Temperature-rise is **fast**
Rule-2: If Heater-current is **medium** Then Temperature-rise is **moderate**
Rule-3: If Heater-current is **low** Then Temperature-rise is **slow**

These types of rules are known as **Mamdani-type fuzzy rules**. The main features of such rules are that both the If (antecedents) parts and the Then (consequents) parts of the rules are fuzzy (imprecise) in nature. That is, fuzzy sets are used here in order to describe both the input and output variables and the system's behavior.

In Relational (Pedrycz) fuzzy logic systems, similar to Mamdani-type fuzzy logic system, both the If (antecedent) parts as well as the Then (consequent) parts are fuzzy. However, there is a slight difference in the rule's representation: in this case, an antecedent proposition is allowed to be associated with several different consequent propositions via a fuzzy relation [9]. This can be illustrated, again, based on the above single-input single-output system, which is now described by the following rules:

Rule-1: If Heater-current is **high** Then Temperature-rise is **slow (0.0), moderate (0.1), fast (0.9)**
Rule-2: If Heater-current is **medium** Then Temperature-rise is **slow (0.1), moderate (0.95), fast (0.0)**
Rule-3: If Heater-current is **low** Then Temperature-rise is **slow (0.85), moderate (0.1), fast (0.0)**

In the first relational fuzzy rule, the consequent fuzzy set **fast** (0.9) implies that the output variable (Temperature-rise) mostly belongs to the fuzzy set **fast** with a degree of affiliation (also called degree of membership) equal to 0.9. Similarly, slow (0.0) and moderate (0.1) imply that the same output variable (here, Temperature-rise) does not belong to the fuzzy set slow at all (as degree of membership in slow is 0.0), whereas the same output variable belongs a little to the fuzzy set moderate (as the degree of membership in moderate is 0.1). It is to be noted from the above three relational fuzzy rules that the summation of the degrees of membership of the output variable in the consequent (output) fuzzy sets may or may not be 1.0. From the above three rules, it is easy to understand that relational fuzzy rules can be regarded as a generalization of Mamdani-type fuzzy rules.

In contrast, in Takagi-Sugeno (TS) type of fuzzy logic systems, the If (antecedent) part of the fuzzy rule is fuzzy in nature, whereas the Then (consequent) part is a crisp function of an antecedent variable (as a rule, a linear equation) rather than a fuzzy proposition. The example presented above for a single-input single-output system can be written for Takagi-Sugeno fuzzy system correspondingly as:

Rule-1: If Heater-current is **high** Then Temperature-rise = a_H(Heater-current) + b_H
Rule-2: If Heater-current is **medium** Then Temperature-rise = a_M(Heater-current) + b_M
Rule-3: If Heater-current is **low** Then Temperature-rise = a_L(Heater-current) + b_L

In the above rules, the consequent parameters (a_H, b_H), (a_M, b_M), and (a_L, b_L) are constants and they are determined/tuned by the input-output information of the system. Note that, in the above rules, if the first constant parameters are all set to zero (i.e. $a_H = 0$, $a_M = 0$, and $a_L = 0$), then the rule consequents are singleton fuzzy sets. Similarly, with Mamdani-type fuzzy rules, if the consequent (output) fuzzy sets are singleton type (a real value) then they are identical to Takagi-Sugeno type fuzzy rules with singleton consequents (i.e. when $a_H = 0$, $a_M = 0$, and $a_L = 0$).

In this work, we will use multi-input single-output (MISO) and multi-input multi-output (MIMO) Takagi-Sugeno type of fuzzy logic models and their inference-systems for developing the electrical load forecasting system. The main reason for using the Takagi-Sugeno type of fuzzy logic system is that fuzzy inference of TS type system is very straight forward and easy in comparison with the other two types of fuzzy logic systems as the output obtained from the inference system is a crisp output, which is directly compatible to the real world signals.

4 Neural Implementation of Fuzzy Logic Model

The fuzzy logic model (1) considered in this chapter for constructing the hybrid load forecaster is based on multi-input multi-output (MIMO) Takagi-Sugeno (TS) type fuzzy rules with Gaussian membership functions, product inference rule, and a weighted average defuzzifier.

$$f_j(\mathbf{x}) = \sum_{l=1}^{M} \left(\beta^l \cdot y_{TS\_j}^l\right) \Big/ \sum_{l=1}^{M} \beta^l \tag{1}$$

where $j = 1, 2, 3, \ldots, m$; $l = 1, 2, 3, \ldots, M$; and

$$y_{TS\_j}^l = \left(\theta_{0j}^l + \sum_{i=1}^{n} \theta_{ij}^l x_i\right), \quad \text{with } (i = 1, 2, 3, \ldots, n) \text{ and} \tag{2}$$

$$\beta^l = \prod_{i=1}^{n} \mu_{G_i^l}(x_i); \quad \mu_{G_i^l}(x_i) = \exp\left\{-\left(x_i - c_i^l\right)^2 / \left(\sigma_i^l\right)^2\right\}. \tag{3}$$

Here, we assume the mean parameters i.e. $c_i^l \in U_i$, and the variance parameters of Gaussian membership functions i.e. $\sigma_i^l > 0$, and the l th rule (Takagi-Sugeno) consequent for j th output i.e. $y_{TS\_j}^l \in V_j$, where U_i, and V_j are the universes of discourse for the i th input and j th output respectively and the input vector $\mathbf{x} = (x_1, x_2, \ldots, x_n)$, $x_i \in U_i$. The corresponding l th rule (Takagi-Sugeno) from the above fuzzy logic system can be written as follows:

$$R^l : \text{If } x_1 \text{ is } G_1^l \text{ and } x_2 \text{ is } G_2^l \text{ and} \cdots \text{and } x_n \text{ is } G_n^l$$

$$\text{Then } \; y_{TS\_j}^l = (\theta_{0j}^l + \sum_{i=1}^{n} \theta_{ij}^l x_i). \tag{4}$$

In the above fuzzy logic model and also in the fuzzy rule, x_i with $i = 1, 2, 3, \ldots, n$; are the n inputs to the fuzzy logic system and f_j with $j = 1, 2, 3, \ldots, m$; are the m outputs from the fuzzy logic system and G_i^l, with $i = 1, 2, 3, \ldots, n$; and with $l = 1, 2, 3, \ldots, M$; are the $(M \times n)$ Gaussian membership functions (GMFs) with corresponding mean and variance parameters as (c_i^l, σ_i^l) respectively and $y_{TS\_j}^l$ is output consequent of the l th fuzzy rule.

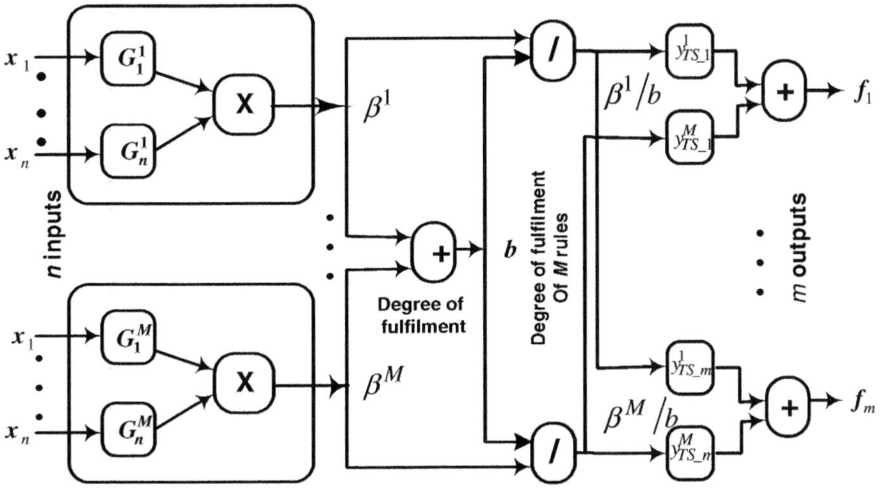

Fig. 2 Feed-forward network (neural) implementation of TS type multi-input multi-output fuzzy logic system

Note that the Gaussian membership functions shown in the rule antecedent parts actually represent the **linguistic terms** such as **low**, **medium** and **high**, etc. It is clear that because of the crisp output function (linear equation) the rule stated in (4) is a Takagi-Sugeno type of fuzzy rule. In the fuzzy logic system (1), the Gaussian membership function is deliberately chosen as this membership function is continuously differentiable at all points, which is an essential criterion for applying gradient based training algorithm. Furthermore, it is to be noted that the fuzzy logic system (1) is a universal function approximator, which implies that it is capable of approximating any nonlinear function to any degree of accuracy over a universe of discourse $U \subset \Re^n$, provided that sufficient numbers of model (tuned) parameters are available [11].

Now, by carefully observing the functional forms of (1), it can be seen that the above fuzzy logic system can be represented as a three layer multi-input multi-output feed-forward network as shown in Figure 2. Because of this feed-forward network (neural) implementation of the Takagi-Sugeno type of fuzzy logic system, Figure 2 actually represents a Takagi-Sugeno type of multi-input multi-output neural-fuzzy network. Instead of connecting weights and biases as in a neural network – we have the mean c_i^l and the variance σ_i^l parameters of the Gaussian membership functions, along with $\left(\theta_{0j}^l, \theta_{ij}^l\right)$ i.e. $y_{TS\_j}^l$ from the TS rules consequent, as the equivalent adjustable parameters. If the adjustable parameters of the neural-fuzzy network are suitably selected, then the above fuzzy logic system can correctly approximate any nonlinear system based on the given input-output data pairs.

5 Training Algorithms for the Neural-Fuzzy Network

The TS type of fuzzy logic system, once represented as an equivalent multi-input feed-forward network (Figure 2), can generally be trained using any classical training algorithm of neural network, such as the standard backpropagation algorithm [5-7, 11]. However, because of its relatively slow speed of convergence, this algorithm needs to be further improved.

5.1 Backpropagation Training of TS type Neural-Fuzzy Network

Let a set of N input-output data pairs $\left(\mathbf{x}^p, d_j^p\right)$, with p = 1, 2, 3, …, N; and $\mathbf{x}^p = \left(x_1^p, x_2^p, \cdots, x_n^p\right) \in U \subset \Re^n$, and $d_j^p \in V_j \subset \Re^m$ be given. The objective is to determine a fuzzy logic system $f_j\left(\mathbf{x}^p\right)$ in the form of (1), such that the performance function S (sum squared error), defined as:

$$S_j = 0.5 \cdot \sum_{p=1}^{N} \left(e_j^p\right)^2 = 0.5 \cdot E_j^T \cdot E_j \tag{5}$$

$$S = \sum_{j=1}^{m} S_j = \left(S_1 + S_2 + \cdots + S_m\right), \tag{6}$$

is minimized, where E_j is the column vector of errors $e_j^p = \left(f_j\left(\mathbf{x}^p\right) - d_j^p\right)$, and p = 1, 2, 3, …, N; for the j th output from the fuzzy logic system. In addition, we also assume that the number of fuzzy rules and the number of membership functions (to be implemented) M are known a priori. In this way the problem is reduced to the adjustment of $y_{TS\_j}^{l}$, i.e. the parameters $\left(\theta_{0j}^{l}, \theta_{ij}^{l}\right)$ from the rules consequent and the mean $\left(c_i^l\right)$ and the variance $\left(\sigma_i^l\right)$ parameters of the Gaussian membership functions, so that the performance function (6) is minimized. For convenience, we replace $f_j\left(\mathbf{x}^p\right)$, d_j^p, and e_j^p in the above definition of error by f_j, d_j, and e_j respectively, so that the individual error becomes $e_j = \left(f_j - d_j\right)$. Recall that the steepest descent rule used for training of neural-fuzzy networks is based on the recursive expressions:

$$\theta_{0j}^{l}(k+1) = \theta_{0j}^{l}(k) - \eta \cdot \left(\partial S / \partial \theta_{0j}^{l}\right) \tag{7}$$

$$\theta_{ij}^{l}(k+1) = \theta_{ij}^{l}(k) - \eta \cdot \left(\partial S / \partial \theta_{ij}^{l}\right) \tag{8}$$

$$c_i^{l}(k+1) = c_i^{l}(k) - \eta \cdot \left(\partial S / \partial c_i^{l}\right) \tag{9}$$

$$\sigma_i^{l}(k+1) = \sigma_i^{l}(k) - \eta \cdot \left(\partial S / \partial \sigma_i^{l}\right) \tag{10}$$

where S is the performance function (6) at the kth iteration step and $\theta_{0j}^{l}(k)$, $\theta_{ij}^{l}(k)$, $c_i^l(k)$, and $\sigma_i^l(k)$ are the free parameters of the network at the same iteration step, whose starting values are, in general, randomly selected. In addition, η is the constant step size or learning rate (usually, $\eta \ll 1$), $i = 1, 2, 3, \ldots, n$ (with n as the number of inputs to the neural-fuzzy network); $j = 1, 2, 3, \ldots, m$ (with m as the number of outputs from the neural-fuzzy network); and $l = 1, 2, 3, \ldots, M$ (with M as the number of Gaussian membership functions selected, as well as the number of fuzzy rules to be implemented).

From Figure 2, it is evident that the network output f_j and hence the performance function S_j and therefore, S depend on θ_{0j}^{l} and θ_{ij}^{l} only through $y_{TS\_j}^{l}$. Similarly, the network output f_j and, thereby, the performance functions S_j and S depend on c_i^l and σ_i^l only through β^l, where f_j, $y_{TS\_j}^{l}$, b, and β^l are represented by

$$f_j = \sum_{l=1}^{M} \gamma^l y_{TS\_j}^{l} \tag{11}$$

$$y_{TS\_j}^{l} = \left(\theta_{0j}^{l} + \theta_{1j}^{l} x_1 + \theta_{2j}^{l} x_2 + \cdots + \theta_{nj}^{l} x_n\right), \tag{12}$$

$$\gamma^l = \left(\beta^l / b\right), \quad \text{and} \quad b = \sum_{l=1}^{M} \beta^l \tag{13}$$

$$\beta^l = \prod_{i=1}^{n} \exp\left(-\left(\frac{x_i - c_i^l}{\sigma_i^l} \right)^2 \right). \tag{14}$$

Therefore, the corresponding chain rules are as follows:

$$\left(\partial S/\partial \theta_{0j}^l\right) = \left(\partial S_j/\partial f_j\right)\cdot\left(\partial f_j/\partial y_{TS\_j}^l\right)\cdot\left(\partial y_{TS\_j}^l/\partial \theta_{0j}^l\right) \tag{15}$$

$$\left(\partial S/\partial \theta_{ij}^l\right) = \left(\partial S_j/\partial f_j\right)\cdot\left(\partial f_j/\partial y_{TS\_j}^l\right)\cdot\left(\partial y_{TS\_j}^l/\partial \theta_{ij}^l\right) \tag{16}$$

$$\begin{aligned}\left(\partial S/\partial c_i^l\right) &= \left(\partial\left(\sum_{j=1}^{m} S_j \right) \Big/ \partial \beta^l \right)\cdot\left(\partial \beta^l/\partial c_i^l\right) \\ &= \sum_{j=1}^{m} \left(\partial S_j/\partial f_j\right)\cdot\left(\partial f_j/\partial \beta^l\right)\cdot\left(\partial \beta^l/\partial c_i^l\right)\end{aligned} \tag{17}$$

$$\begin{aligned}\left(\partial S/\partial \sigma_i^l\right) &= \left(\partial\left(\sum_{j=1}^{m} S_j \right) \Big/ \partial \beta^l \right)\cdot\left(\partial \beta^l/\partial \sigma_i^l\right) \\ &= \sum_{j=1}^{m} \left(\partial S_j/\partial f_j\right)\cdot\left(\partial f_j/\partial \beta^l\right)\cdot\left(\partial \beta^l/\partial \sigma_i^l\right)\end{aligned} \tag{18}$$

They can finally be written as:

$$\left(\partial S/\partial \theta_{0j}^l\right) = \left(f_j - d_j\right)\cdot\left(\beta^l/b\right) \tag{19}$$

$$\left(\partial S/\partial \theta_{ij}^l\right) = \left(f_j - d_j\right)\cdot\left(\beta^l/b\right)\cdot x_i \tag{20}$$

$$\left(\partial S/\partial c_i^l\right) = \left(\sum_{j=1}^{m} \left(y_{TS\_j}^l - f_j\right)\left(f_j - d_j\right) \right)\cdot\left(2\cdot\left(\beta^l/b\right)\cdot\left(x_i - c_i^l\right)/\left(\sigma_i^l\right)^2 \right) \tag{21}$$

$$\left(\partial S/\partial \sigma_i^l\right) = \left(\sum_{j=1}^{m} \left(y_{TS\_j}^l - f_j\right)\left(f_j - d_j\right) \right)\cdot\left(2\cdot\left(\beta^l/b\right)\cdot\left(x_i - c_i^l\right)^2/\left(\sigma_i^l\right)^3 \right) \tag{22}$$

Using the above results, the final update rules for the network's free parameters can be written as:

$$\theta_{0j}^l(k+1)=\theta_{0j}^l(k)-\eta\cdot(f_j-d_j)\cdot(\beta^l/b) \tag{23}$$

$$\theta_{ij}^l(k+1)=\theta_{ij}^l(k)-\eta\cdot(f_j-d_j)\cdot(\beta^l/b)\cdot x_i \tag{24}$$

$$c_i^l(k+1)=c_i^l(k)-\eta\cdot\left(\sum_{j=1}^{m}(y_{TS\_j}^l-f_j)(e_j)\right)\cdot\left(2\cdot(\gamma^l)\cdot(x_i-c_i^l)/(\sigma_i^l)^2\right) \tag{25}$$

$$\sigma_i^l(k+1)=\sigma_i^l(k)-\eta\cdot\left(\sum_{j=1}^{m}(y_{TS\_j}^l-f_j)(e_j)\right)\cdot\left(2\cdot(\gamma^l)\cdot(x_i-c_i^l)^2/(\sigma_i^l)^3\right) \tag{26}$$

where $e_j=(f_j-d_j)$, $\gamma^l=(\beta^l/b)$ with $b=\sum_{l=1}^{M}\beta^l$ and γ^l is the normalized degree of fulfillment (firing strength) of the l th rule.

Equations (19) – (26) represent the backpropagation training algorithm (BPA) for Takagi-Sugeno type multi-input multi-output neural-fuzzy network or the equivalent fuzzy logic system of the form (1) with linear fuzzy rules consequent part as:

$$y_{TS\_j}^l=(\theta_{0j}^l+\theta_{1j}^l x_1+\theta_{2j}^l x_2+\cdots+\theta_{nj}^l x_n).$$

In the above TS type of fuzzy rules' (linear) consequents, if the co-efficients $\theta_{ij}^l=0$, for i = 1, 2, 3, …, n; l = 1, 2, 3, …, M; and m = 1, then the equivalent neural-fuzzy network is identical with the multi-input single-output neural-fuzzy network described by [11]. The resulting fuzzy logic system can be seen as a special case of both the Mamdani and Takagi-Sugeno type systems, where the rule consequents are singleton (constant numbers) fuzzy sets. However, if the coefficients $\theta_{ij}^l\neq 0$, for i = 1, 2, 3, …,n; l = 1, 2, 3, …, M; and for m = 1, then the resulting fuzzy logic system is identical to the multi-input single-output neural-fuzzy network, as described by [5].

5.2 Improvement of Backpropagation Training Convergence

It is generally known that the backpropagation algorithm based on steepest descent rule, in order to avoid the possible oscillations in the final phase of training, uses a relatively low learning rate $\eta \ll 1$. Therefore, backpropagation training usually requires a large number of recursive steps or epochs. The acceleration of the training process with classical backpropagation is, however, achievable if the adaptive version of the learning rate, or the momentum version of the steepest descent rule is used, i.e.

$$w(k+1) = w(k) - \eta \cdot (1 - m_0) \cdot (\partial S / \partial w) + m_0 \cdot \Delta w(k-1) \tag{27}$$

where w represents the network's free parameters vector in general and $\Delta w(k-1) = w(k) - w(k-1)$. The momentum constant is usually less than one. Therefore, we can write $m_0 < 1$.

In order to further improve the training performance of the proposed neural-fuzzy network, the momentum version of the backpropagation algorithm was modified by adding the modified error index term $S_m(w)$ to the original sum squared error function $S(w)$, as proposed by [12].

The modified error index term has been defined as follows:

$$\begin{aligned} S_m(w) &= \frac{1}{2} \cdot \gamma \sum_{p=1}^{N} \left(e_j^{\,p}(w) - e_{avg} \right)^2 \\ &= \frac{1}{2} \cdot \gamma \left(\left(e_1(w) - e_{avg} \right)^2 + \left(e_2(w) - e_{avg} \right)^2 + \cdots + \left(e_N(w) - e_{avg} \right)^2 \right) \end{aligned} \tag{28}$$

Note that in the above equation, γ is a (positive) constant, and $e_1(w), e_2(w), \cdots, e_N(w)$ have been written, for convenience, to represent respectively the terms $e_j^{\,1}(w), e_j^{\,2}(w), \cdots, e_j^{\,N}(w)$ and e_{avg} is the average error defined as

$$e_{avg}(w) = \frac{1}{N} \cdot \left(e_1(w) + e_2(w) + \cdots + e_N(w) \right) \tag{29}$$

Therefore, the derivative of the modified error index term with respect to the network's free parameter w can be written as follows:

$$\frac{\partial S_m(w)}{\partial w} = \frac{\gamma}{2} \cdot \left(\begin{aligned} & 2 \cdot \left(e_1(w) - e_{avg} \right) \cdot \frac{\partial \left(e_1(w) - e_{avg} \right)}{\partial w} + \cdots \\ & \quad + \; 2 \cdot \left(e_N(w) - e_{avg} \right) \cdot \frac{\partial \left(e_N(w) - e_{avg} \right)}{\partial w} \end{aligned} \right) \tag{30}$$

Or,

$$\frac{\partial S_m(w)}{\partial w} = \gamma \cdot \left(\begin{aligned} & \left(e_1(w) - e_{avg} \right) \cdot \left(\frac{\partial e_1(w)}{\partial w} - \frac{\partial e_{avg}(w)}{\partial w} \right) + \cdots \\ & \quad + \; \left(e_N(w) - e_{avg} \right) \cdot \left(\frac{\partial e_N(w)}{\partial w} - \frac{\partial e_{avg}(w)}{\partial w} \right) \end{aligned} \right) \tag{31}$$

Or,

$$\frac{\partial S_m(w)}{\partial w}=\gamma\cdot\left(\begin{array}{l}\left\{\left(e_1(w)-e_{avg}\right)\cdot\frac{\partial e_1(w)}{\partial w}+\cdots+\left(e_N(w)-e_{avg}\right)\frac{\partial e_N(w)}{\partial w}\right\}\\-\left\{\left(e_1(w)-e_{avg}\right)\frac{\partial e_{avg}(w)}{\partial w}+\cdots+\left(e_N(w)-e_{avg}\right)\frac{\partial e_{avg}(w)}{\partial w}\right\}\end{array}\right) \tag{32}$$

It can be shown that the second part in the parentheses will lead to zero, i.e.

$$\begin{aligned}&\left\{\left(e_1(w)-e_{avg}\right)\frac{\partial e_{avg}(w)}{\partial w}+\cdots+\left(e_N(w)-e_{avg}\right)\frac{\partial e_{avg}(w)}{\partial w}\right\}\\&=\frac{\partial e_{avg}(w)}{\partial w}\left\{\left(e_1(w)-e_{avg}\right)+\left(e_2(w)-e_{avg}\right)+\cdots+\left(e_N(w)-e_{avg}\right)\right\}\\&=\frac{\partial e_{avg}(w)}{\partial w}\cdot\left\{\left(e_1(w)+e_2(w)+\cdots+e_N(w)\right)-\left(N.e_{avg}\right)\right\}=0\end{aligned} \tag{33}$$

Therefore,

$$\frac{\partial S_m(w)}{\partial w}=\gamma\cdot\left(\left(e_1(w)-e_{avg}\right)\cdot\frac{\partial e_1(w)}{\partial w}+\cdots+\left(e_N(w)-e_{avg}\right)\frac{\partial e_N(w)}{\partial w}\right) \tag{34}$$

Now, rewriting the original sum squared error function for the *j*th output as:

$$S_j(w)=\frac{1}{2}\cdot\sum_{p=1}^{N}\left(e_j{}^p(w)\right)^2=\frac{1}{2}\cdot\left(\left(e_1(w)\right)^2+\left(e_2(w)\right)^2+\cdots+\left(e_N(w)\right)^2\right) \tag{35}$$

We have once again,

$$\frac{\partial S_j(w)}{\partial w}=\frac{1}{2}\cdot\left(2\cdot\left(e_1(w)\right)\cdot\left(\frac{\partial e_1(w)}{\partial w}\right)+\cdots+\quad 2\cdot\left(e_N(w)\right)\cdot\left(\frac{\partial e_N(w)}{\partial w}\right)\right) \tag{36}$$

Therefore, $S_{new}(w)=S_j(w)+S_m(w)$ and $\frac{\partial S_{new}(w)}{\partial w}=\frac{\partial S_j(w)}{\partial w}+\frac{\partial S_m(w)}{\partial w}$ (37)

Or,

$$\begin{aligned}\frac{\partial S_{new}(w)}{\partial w}=&\left(\left(e_1(w)\right)\cdot\left(\frac{\partial e_1(w)}{\partial w}\right)+\cdots+\quad\left(e_N(w)\right)\cdot\left(\frac{\partial e_N(w)}{\partial w}\right)\right)+\\&\gamma\cdot\left(\left(e_1(w)-e_{avg}\right)\cdot\left(\frac{\partial e_1(w)}{\partial w}\right)+\cdots+\quad\left(e_N(w)--e_{avg}\right)\cdot\left(\frac{\partial e_N(w)}{\partial w}\right)\right)\end{aligned}$$

Thereby,
$$\frac{\partial S_{new}(w)}{\partial w} = \begin{pmatrix} \left(e_1(w) + \gamma \cdot \left(e_1(w) - e_{avg}\right)\right) \cdot \left(\frac{\partial e_1(w)}{\partial w}\right) + \cdots \\ + \left(e_N(w) + \gamma\left(e_N(w) - -e_{avg}\right)\right) \cdot \left(\frac{\partial e_N(w)}{\partial w}\right) \end{pmatrix} \quad (38)$$

It can be seen from the above equations that with the modified error index extension we only need to add a new vector term $\gamma \cdot \left(e(w) - e_{avg}\right)$ to the original error vector $e(w) = \left(e_1(w), e_2(w), \cdots, e_N(w)\right)^T$. As per [12], training of neural networks with the above extended error index term can fulfill the requirement that large network errors receive more attention than small errors in a better way, which is usually not guaranteed by the conventional backpropagation algorithm. The extended error index also ensures a smoother error surface than the standard Sum Square Error (SSE) function and thereby, the former can resist the oscillation during the training phase which finally leads to faster training convergence.

5.3 Adaptive Learning Rate and Oscillation Control

The proposed backpropagation algorithm described in section 5.1 with modified error index extension of the original SSE function as the new performance function, and also with a small momentum term, in general, is very efficient, and much faster than pure classical backpropagation algorithm in training the MIMO/MISO neural-fuzzy network. However, even with this improved backpropagation algorithm, the performance function (SSE) is not always guaranteed to decrease towards the desired error goal in every epoch of training. This means that training can sometimes proceed in a negative direction, giving rise to a continuous increase in the SSE, or even oscillation, resulting in long training time or undesirable training performance. In order to guarantee a satisfactory training performance, three sets of adjustable parameters are recommended to be stored. If the two new consecutive sets of adjustable parameters reduce the network's performance function in the following epochs, then those sets are used and the learning rate is increased by a factor of 1.1 times the earlier learning rate for further training. Otherwise, if the performance function tends to increase beyond some factor say, *WF* times the current SSE with the new sets of parameters, then the new sets of adjustable parameters are discarded and training is continued with the old sets of adjustable parameters. Thereafter, a new direction of training is sought with the old sets of adjustable parameters and with lower values of learning rate, e.g. 0.8 or 0.9 times the old learning rate. By this way, in every epoch the SSE is either decreased steadily, or at least maintained within the certain range of current performance value.

6 Neural-Fuzzy Networks in Short-Term Load Forecasting

The idea behind the use of fuzzy/neural-fuzzy models in load forecasting is simple: it is assumed that future load is dependent on past load data and external factors (such as temperature of the day, relative humidity, cloud coverage, intensity of day-light, wind speed, etc.), and the neural-fuzzy network is used to approximate this dependency. The inputs to the network consist of those external factors (e.g. day temperature) and past load values, and the output is the target load values (e.g. a load value of a certain hour, load values of many future hours, the peak load of a day, the total load of a day, etc.). Therefore, building of a neural-fuzzy model for load forecasting can be seen as a nonlinear system identification problem. The determination of the model's structure consists of selecting the input variables and deciding the network's structure accordingly. The fuzzy model parameters' estimation is carried out by training the network on load data from history. This requires choices concerning the training algorithm (here, improved backpropagation algorithm) and appropriate training data. The model's validation is carried out by testing on load data, which have not been used for training.

6.1 Preprocessing of Electrical Load Data

Although the daily load profile depends on various external factors such as the weather variables mentioned above, they have not been considered in this work due to their unavailability. The training data and validation data for neural-fuzzy model are both based on available electrical load time series only. As the peak values of daily load data are very large in magnitude and unsuitable for direct application to neural-fuzzy network, the load data have been first scaled/normalized within the desired range of say, 0 to 1, using the following formula.

Given a load time series represented as a row vector $\boldsymbol{u} = [u_1, u_2, u_3, \ldots., u_q]$, with q data points recorded at the time instants $t = 1, 2, 3, \ldots, q$ respectively, the scaled and normalized load time series in the row vector form $\boldsymbol{u}_{nsc}$ can be written as

$$\boldsymbol{u}_{nsc} = K_0[(u_1 - u_{min}), (u_2 - u_{min}), (u_3 - u_{min}), \ldots.., (u_q - u_{min})] + u_{lo}. \tag{39}$$

$$K_0 = (u_{hi} - u_{lo})/(u_{\max} - u_{\min}) \tag{40}$$

In the above equations, $u_{\max}$ and $u_{\min}$ are the maximum and minimum values respectively of the load time series ($\boldsymbol{u}$ vector), and $u_{hi} = 1$, $u_{\mathrm{lo}} = 0$ are the desired highest and lowest values respectively of the scaled or normalized load time series ($\boldsymbol{u}_{nsc}$ vector). Once the scaling or normalization of load time series has been performed, the scaled or normalized time series data are represented as either multi-input multi-output, or multi-input single-output form as required for the training and validation data of neural-fuzzy network.

6.2 Short-Term Forecasting of Electrical Load

In this subsection, we will discuss the fuzzy modeling of electrical load time series using the neural-fuzzy network and thereafter, short-term forecasting of electrical load demand will be performed, based on the scaled or normalized electrical load time series data. In this work, the load curve to be forecasted consists of load values recorded at every quarter of an hour interval, implying that a total of 4×24 = 96 observations are available per day. Alternatively, hourly load values can further be obtained from the same load curve, which can also be seen as a time series of real numbers, thereby restricting the number of observations to 24 per day. Again, the peak load time series consisting of the maximum load value of each day can be constructed from the above load curve, where the number of observations is restricted to 365 per year.

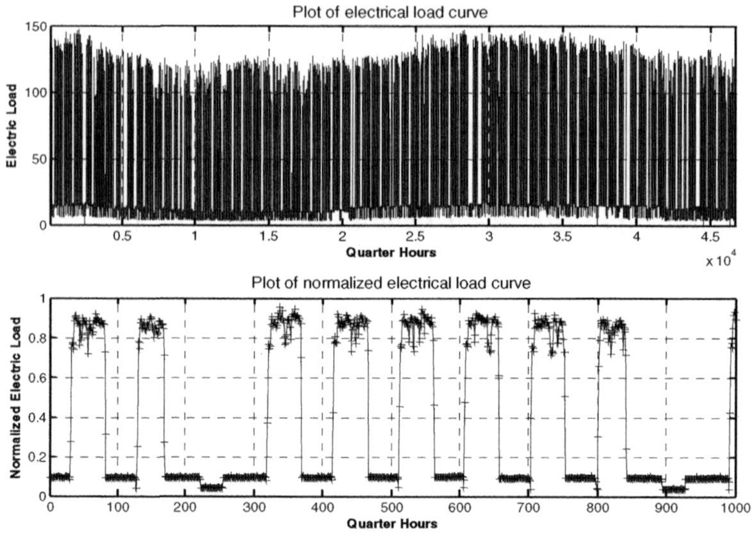

Fig. 3 Electrical load curve over the period March 2000 to June 2001 (top) and normalized electrical load curve (bottom)

The electrical load time series shown in Figure 3 (top) contains the electrical consumption values recorded at the interval of 15 minutes for 487 days. There are all together 487×96 = 46752 measurements recorded from March 2000 till June 2001. The maximum and minimum load values are shown to be 146.80 MW and 0 MW respectively. The normalized load curve is also shown in Figure 3 (bottom). For better clarity and presentation, only 1000 data points are shown in the normalized load curve. It can be seen from Figure 3 that the normalized load curve exhibits some sort of systematic load patterns (between 30 and 171 time points, and also between 318 and 843 time points), which are, however, not always observed regularly due to holidays in between, thereby giving rise to a chaotic behavior of load series besides the usual nonlinearity.

As mentioned earlier, short-term forecasting refers to the prediction of electrical demand (load) with a lead time of forecast (L) varying from several minutes to several hours or even to several days. This application concerns the short-term forecasting of electrical load demand, based on a load time series that predicts the load values at time ($t + L$) using the available observation data up to the time point t, where L is the lead time of prediction. For such modeling purposes, the given (normalized) load time series data $\mathbf{X} = \{X_1, X_2, X_3, \ldots., X_q\}$, for instance, have

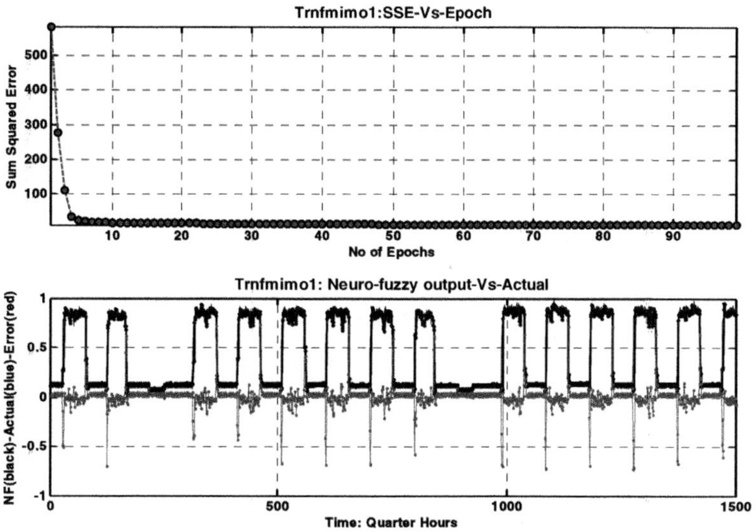

Fig. 4 Training performance of a neural-fuzzy network with the improved backpropagation algorithm for short-term forecasting of electrical load (normalized data). SSE versus epoch plot is shown in the top figure. Approximation error (red color line) of neural-fuzzy network, normalized load data (blue color line) and neural-fuzzy prediction (black color line) of normalized electrical load curve are shown in the bottom figure.

been rearranged in an **I**nput-**O**utput form **XIO**. The neural-fuzzy predictor to be developed for time series modeling and forecasting is supposed to operate with four inputs (i.e. $n = 4$) and three outputs (i.e. $m = 3$). Taking both the sampling interval and lead time of forecast to be one time unit, then for each $t \geq 4$ the input data have to be represented as a four-dimensional vector and the output data as a three-dimensional vector

$$\boldsymbol{XI} = [X(t - 3), X(t - 2), X(t - 1), X(t)],$$
$$\boldsymbol{XO} = [X(t + 1), X(t + 2), X(t + 3)].$$

Furthermore, in order to have sequential output in each row i.e. $\{X_5, X_6, X_7, X_8, X_9, X_{10}, \ldots, X_q\}$, the values of t should run as 4, 7, 10, 13, …, (q-3). The corresponding $\mathbf{XIO}_{\mathbf{STLF}}$ matrix will then look like (41), in which the first four columns represent the four inputs of the network and the last three columns represent its outputs.

$$\mathbf{XIO}_{STLF} = \begin{bmatrix} X_1, & X_2, & X_3, & X_4 & \Rightarrow & X_5, & X_6, & X_7 \\ X_4, & X_5, & X_6, & X_7 & \Rightarrow & X_8, & X_9, & X_{10} \\ \vdots & \vdots & \vdots & \vdots & & \vdots & \vdots & \vdots \\ X_{q-6}, & X_{q-5}, & X_{q-4}, & X_{q-3} & \Rightarrow & X_{q-2}, & X_{q-1}, & X_q \end{bmatrix} \tag{41}$$

In the first forecasting example, 15582 input-output data were generated, from which only the first 500 input-output data sets, i.e. the first 500 rows from $\mathbf{XIO_{STLF}}$ matrix were used for multi-input multi-output neural-fuzzy network training. The remaining 15082 rows of the $\mathbf{XIO_{STLF}}$ matrix were available for verification/testing of the forecasting results. The short-term training and forecasting performances achieved with the neural-fuzzy network are illustrated in Figures 4 – 5 respectively and the parameter settings of the improved backpropagation training algorithm is shown in Table 1.

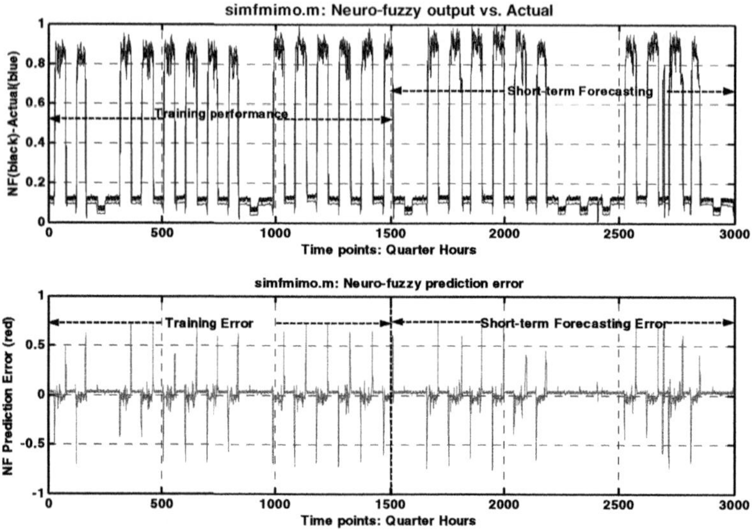

Fig. 5 Short-term forecasting performance of the neural-fuzzy network with electrical load (normalized data). Training performance and forecasting performance have been shown in the top figure. Approximation (training) error and short-term forecasting error (red color line) of neural-fuzzy network have been shown with normalized electrical load data (bottom figure).

For the purpose of short-term forecasting of electrical load, the neural-fuzzy network was first trained using the first 500 rows from $\mathbf{XIO_{STLF}}$ matrix and therefore, the neural-fuzzy network was operated with n = 4 inputs, m = 3 outputs and M = 15 fuzzy rules/Gaussian membership functions. Furthermore, the starting values of the network's adjustable parameters such as the mean and the variance parameters of the Gaussian membership functions and also the rules consequent parameters were all generated randomly. The improved backpropagartion algorithm was applied for training with a very small learning rate $\eta = 0.005$, modified error index coefficient $\gamma = 0.6$ and momentum constant m_o = 0.3. Note that by setting $\gamma = 0$ and m_o = 0, the modified algorithm will behave like a pure backpropagation algorithm. The backpropagation training was allowed to have a maximum 99 epochs. Figure 4 illustrates the training performance of the neural-fuzzy network with the proposed backpropagation algorithm. The top part of Figure 4 demonstrates that the improved backpropagation algorithm can bring the SSE down smoothly from its large initial value of 581.8213 to 11.5402 within only 99 epochs and the training does not exhibit any oscillation as can be seen in the SSE -versus-epoch plot. Note that the final SSE of 11.5402 is contributed due to the training error jointly by the first, second and third output of the neural-fuzzy network, whose respective values are 1.5632, 3.5615 and 6.4155. The bottom part of Figure 4 shows the corresponding approximation error (red color line), neural-fuzzy output (black line) and normalized actual load values (blue line) respectively after 99 epochs of network training.

Table 1 Parameter settings of the improved backpropagation algorithm

Backpropagation algorithm	Learning rate	Modified error index coefficient	Momentum constant	Epochs
parameters	0.005	0.6	0.3	99

Once the training has been performed, the final updated values of the network's adjustable parameters are stored; thereafter the same parameters are used for short-term forecasting. For short-term load forecasting, input data sets are only presented to the trained network and the corresponding output predictions are automatically performed by the trained network (in response to the applied input data), which are later compared with the desired load output data for verification of forecasting accuracy. The top part of Figure 5 demonstrates the short-term forecasting performance of electrical load by the neural-fuzzy network, where the first 1500 data points correspond to the training performance and remaining 1500 data points correspond to the short-term forecasting performance. It is important to note that although the data between the time points 2182 (quarter hour) and 2522 (quarter hour) are different from the training data set (i.e. first 1500 data points), the TS type multi-input multi-output neural-fuzzy network can still predict this data region with reasonably high accuracy. Also note that once the forecasting is performed for normalized electrical load series, the forecasted data have to be converted back to the original range using the formula mentioned below:

$$\boldsymbol{u} = [u_1, u_2, u_3, \ldots., u_q] = (\boldsymbol{u}_{nsc} - u_{lo}.)/K_0 + u_{min} \quad (42)$$

$$\text{with} \quad K_0 = (u_{hi} - u_{lo})/(u_{\max} - u_{\min}).$$

The forecasting accuracy achieved with the trained network is demonstrated by the bottom part of Figure 5. In the same Figure 5, the error curve (red line) corresponds to the total SSE = 21.5363, in which 11.5402 corresponds to the training error (final SSE of training) and the remaining 9.9961 is due to the forecasting error. Table 1 summarizes the important parameter settings of the improved backpropagation algorithm, which were used during network training.

7 Mid-Term Forecasting of Electrical Load Data

As mentioned earlier, mid-term forecasting refers to the prediction of electrical load demand with the lead time varying from several days to several weeks. For mid-term forecasting of electrical demand (load), the given (normalized) time series data have to be rearranged in a multi-input single-output form.

7.1 Data Rearrangement for Mid-Term Load Forecasting

The neural-fuzzy network considered for mid-term forecasting consists of seven input data (i.e. $n = 7$) and one output (i.e. $m = 1$) data. Therefore, the corresponding $\mathbf{XIO}_{\mathbf{MTLF}}$ matrix will look like (43).

$$\mathbf{XIO}_{MTLF} = \begin{bmatrix} Day_1, & Day_2, & Day_3, & Day_4, & Day_5, & Day_6, & Day_7 & \Rightarrow \hat{Day}_8 \\ Day_2, & Day_3, & Day_4, & Day_5, & Day_6, & Day_7, & \hat{Day}_8 & \Rightarrow \hat{Day}_9 \\ \vdots & \vdots & \vdots & \vdots & \vdots & \vdots & \vdots & \vdots \\ \hat{Day}_8, & \hat{Day}_9, & \hat{Day}_{10}, & \hat{Day}_{11}, & \hat{Day}_{12}, & \hat{Day}_{13}, & \hat{Day}_{14} & \Rightarrow \hat{Day}_{15} \\ \vdots & \vdots & \vdots & \vdots & \vdots & \vdots & \vdots & \vdots \end{bmatrix} \tag{43}$$

As can be seen from the first row of $\mathbf{XIO}_{\mathbf{MTLF}}$ matrix in (43), seven days of electrical load data are applied as the neural-fuzzy network's input to produce the eighth day data as the predicted output (indicated by hat symbol on Day_8) of the same network. Likewise, in the second row of $\mathbf{XIO}_{\mathbf{MTLF}}$ matrix input data set consisting of Day_2 till Day_7 data and also the previous prediction of Day_8 have been applied to the network to further predict the Day_9 data. Therefore, the input data set on the eighth row will consist of the previous predictions of Day_8 till Day_{14} data to further predict the Day_{15} (output) data. In a similar fashion, subsequent rows of the $\mathbf{XIO}_{\mathbf{MTLF}}$ matrix can be built up. Therefore, mid-term forecasting of electrical load data practically starts from the eighth row onwards, where each input data set consists of previously predicted data by the neural-fuzzy network. It is to be noted that in short-term load forecasting, only the desired output data (i.e. the eighth column of the $\mathbf{XIO}_{\mathbf{MTLF}}$ matrix) are replaced by predicted data, whereas all input data sets consist of original data values obtained from the normalized electrical load time series. As already mentioned, here, each day (24 hours) actually represents 96 data points, as four (load) measurements or observations were recorded within an hour at the sampling interval of every 15 minutes. Again, it is to be noted that, for mid-term forecasting also, the network training will be performed initially presenting only the actual (normalized) load data values to the network. However, after the satisfactory training of the network, for mid-term forecasting purpose, the method presented earlier will be followed. Usually, the first few hundreds of rows of the $\mathbf{XIO}_{\mathbf{MTLF}}$ matrix are needed for the network's training, whereas the remaining data (also a few hundreds of rows of the $\mathbf{XIO}_{\mathbf{MTLF}}$ matrix) are used for short-term and mid-term forecasting of electrical load.

7.2 *Fuzzy Clustering of Electrical Load Data*

In reality the electrical load data are not very smooth and they exhibit a lot of fluctuations (see the load curve in Figure 3). So, applying these (normalized) raw input data to the neural-fuzzy network will eventually produce output which does not properly follow the desired real data pattern (i.e. the network is extremely poorly trained). This, in turn, will certainly make the prediction and forecasting performances of the neural-fuzzy network unsatisfactory. Furthermore, unlike short-term forecasting, in the case of mid-term forecasting much higher training performance and thereby, better prediction or forecasting accuracy are required from the network. Therefore, in order to achieve sufficiently good training accuracy (i.e. low final SSE), one should usually provide a large number of randomly selected fuzzy rules and fuzzy membership functions, which may, in turn, make the network's training computationally expensive and very slow. On the other hand, the neural-fuzzy network can benefit from a fuzzy clustering algorithm which can provide an optimum number of fuzzy rules, fuzzy membership functions/parameters, and their good starting values for fast and very efficient training of the network.

7.2.1 Gustafson-Kessel Clustering Algorithm

The fuzzy clustering is actually performed on the given training data set even before the neural-fuzzy network's training. The purpose of applying a fuzzy clustering algorithm on the (normalized) electrical load data is, basically, to find the optimum or near optimal number of fuzzy (Gaussian) membership functions and/or their initial parameters (mean and variance), which will be used as the good starting parameters of the neural-fuzzy networks during its training. For detecting the clusters of different geometrical shapes in a given data set such as electrical data, the Gustafson-Kessel (GK) clustering algorithm is often proposed [7].

Given the data set:

$$Z = \{Z_1, Z_2, Z_3, ..., Z_N\}, \tag{44}$$

where Z is represented by a $n \times N$ pattern matrix, select the number of clusters $1 < c < N$, the weighting exponent or fuzziness exponent parameter $m_e > 1$, the termination tolerance $\varepsilon > 0$ and the cluster volume ρ_g.

Furthermore, the GK clustering algorithm proceeds with randomly generated initial partition matrix $U^{l=0}$ where $U = [\mu_{gs}]_{cXN}$.

Repeat for iterations $l = 1, 2, 3, ...$

Step 1: compute the cluster centers (means)

$$v_g^{(l)} = \frac{\sum_{s=1}^{N} \left(\mu_{gs}^{(l-1)}\right)^{m_e} \cdot Z_s}{\sum_{s=1}^{N} \left(\mu_{gs}^{(l-1)}\right)^{m_e}}; 1 \le g \le c. \tag{45}$$

Step 2: determine the cluster covariance matrices

$$F_g = \frac{\sum_{s=1}^{N} \left(\mu_{gs}^{(l-1)}\right)^{m_e} \cdot \left(Z_s - v_g^{(l)}\right)\left(Z_s - v_g^{(l)}\right)^T}{\sum_{s=1}^{N} \left(\mu_{gs}^{(l-1)}\right)^{m_e}}; 1 \le g \le c. \quad (46)$$

Step 3: calculate the distance

$$D_{gsA_g}^2 = \left(Z_s - v_g^{(l)}\right)^T \left[\rho_g \det\left(F_g\right)^{1/n} \cdot F_g^{-1}\right]\left(Z_s - v_g^{(l)}\right); \quad (47)$$
$$1 \le g \le c; \quad 1 \le s \le N.$$

Step 4: update the partition matrix
For $1 \le s \le N$
If $D_{gsAg} > 0$ for all $g = 1, 2, 3, \ldots, c$;

$$\mu_{gs}^{(l)} = \frac{1}{\sum_{h=1}^{c} \left(D_{gsAg}/D_{hsAg}\right)^{2/(m_e-1)}}; 1 \le g \le c; 1 \le s \le N. \quad (48)$$

else $\mu_{gs}^{(l)} = 0$ and $\mu_{gs}^{(l)} \in [0,1]$, with $\sum_{g=1}^{c} \mu_{gs}^{(l)} = 1$,

$$\text{until} \quad \left\| U^{(l)} - U^{(l-1)} \right\| < \varepsilon. \quad (49)$$

To apply the GK clustering algorithm for determining the initial parameters of TS-type MISO neural-fuzzy network for modeling and forecasting of electrical load data, the following steps are carried out:

1. Create $\mathbf{XIO}_{\mathbf{MTLF}}$ matrix from normalized electrical load data.
2. Generate the initial fuzzy partition matrix randomly.
3. Calculate cluster centers, cluster covariance, distances, and norm inducing matrices and update the partition matrix repeatedly till the tolerance limit is reached.
4. Compute the mean and the variance parameters of the fuzzy (Gaussian) membership functions from the cluster centers and the final fuzzy partition matrix using projection.
5. Calculate the TS rule's consequent parameters by applying the LSE estimation technique (see [7]).
6. Calculate SSE for the given training data set using the TS fuzzy rules' antecedents and consequent parameters. Save all fuzzy rules' parameters.
7. Perform the TS-type MISO neural-fuzzy network training using the stored fuzzy rules' parameters from step-6 as the initial parameters of the neural-fuzzy network. Note that the initial performance index, i.e. the initial SSE value, from the neural-fuzzy network should be the same as the final SSE result from the GK Clustering and LSE.

7.3 Moving Average Window Filter for Smoothing of Load Data

The idea of smoothing the real load data using the moving average filter is in fact trying to find good starting values of the network's parameters, which will reduce the performance function (SSE) down to the desired value much faster. After having reached the desired error goal, the original data can be reconstructed by gradually reducing the window size down to one, which is the same as the original data.

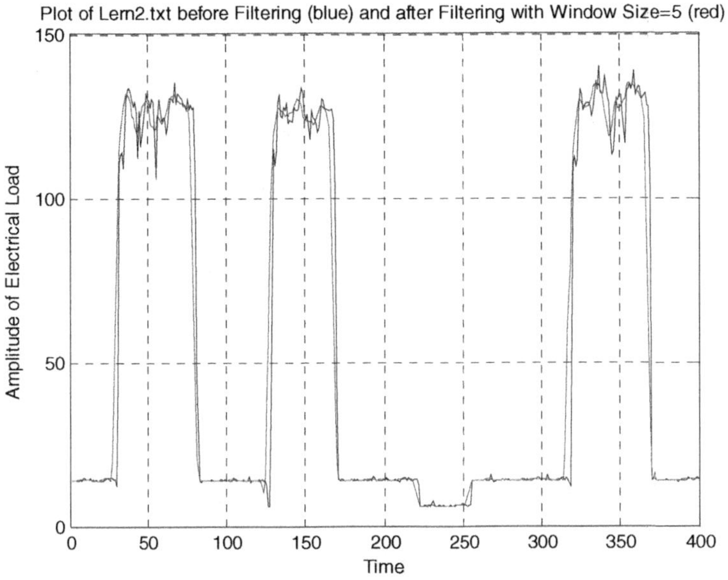

Fig. 6 The effect of using moving average filter with window size $ws = 5$ on raw electrical load data

If electrical load data consisting of q data points can be expressed as

$$\mathbf{X}_Q = \left[x_1, x_2, x_3, x_4, x_5, \cdots, x_q\right] \tag{50}$$

then, the filtered electrical load data with q data points can be written as

$$\mathbf{X}_{QF} = \left[x_{1F}, x_{2F}, x_{3F}, x_{4F}, x_{5F}, \cdots, x_{qF}\right], \tag{51}$$

where,

$$\begin{aligned} x_{1F} &= \left(x_1 + x_2 + x_3 + \cdots + x_{ws}\right)/ws, \\ x_{2F} &= \left(x_2 + x_3 + \cdots + x_{(ws+1)}\right)/ws, \cdots, \text{so on,} \end{aligned} \tag{52}$$

where the window size (ws) can be chosen properly by some experiments.

The data filtering process with moving average filter mainly involves three steps. With randomly generated initial fuzzy parameters or the initial fuzzy parameters obtained through the GK clustering, the neural-fuzzy network is first trained with moving average filtered data, where the window size of five is used in the beginning. Thereafter, the same neural-fuzzy network, with the last stored parameter values (after the first phase of training), is further trained with moving average filtered data but with window size equal to (*Initial WindowSize - StepWindowSize*). The neural-fuzzy network's training is repeated again and again with moving average filtered data and increasing *StepWindowSize* until the window size finally reaches one, *i.e.* practically with the original load data. The idea behind using this varying window size filtering technique is to reduce the performance function (SSE) of the network as low as possible by gradually finding the best parameters while training the neural-fuzzy networks successively with very smooth data to completely raw data.

7.4 Simulation Experiments for Mid-Term Load Forecasting

The fuzzy clustering of (normalized) electrical load data and LSE estimation of TS rules' consequents parameters were performed and the results of the same are depicted in Figure 7. The TS rules' parameters (both antecedents and consequents) which are determined by clustering and LSE techniques, result in the *SSE* = 0.5026, *MSE* = 0.0021 and *RMSE* = 0.046.

These performances are used as the initial performance functions of the neural-fuzzy network. Figures 9 and 10 show the graph of *SSE* versus training epochs and the neural-fuzzy network's training performance with 500 data points respectively. The training algorithm has brought the SSE down to 0.3431 from its initial SSE of 0.5026. The corresponding final *MSE* = 0.0014 and *RMSE* = 0.037.

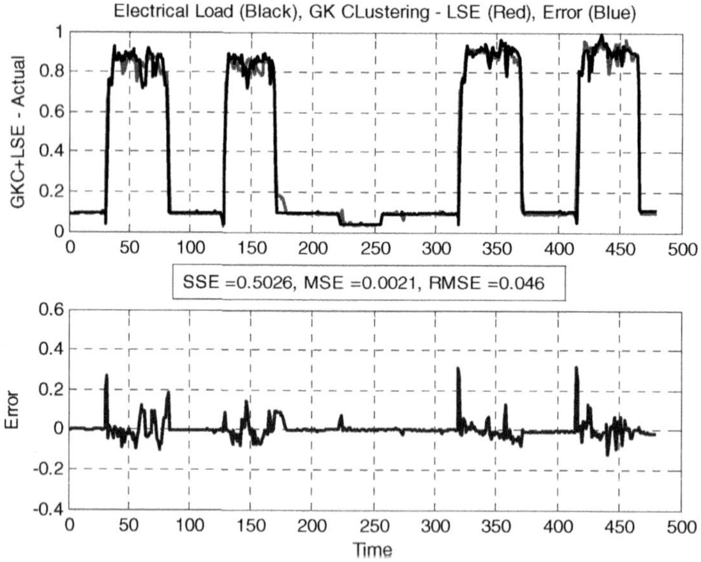

Fig. 7 Gustaffson-Kessel's fuzzy clustering and LSE estimation of TS rules consequents' parameters with normalized electrical load data, n = 7, m = 1, lead time of forecast L = 96, training data = 480, number of clusters c = 5, window size ws = 5, Fuzziness exponent $m_e = 2$

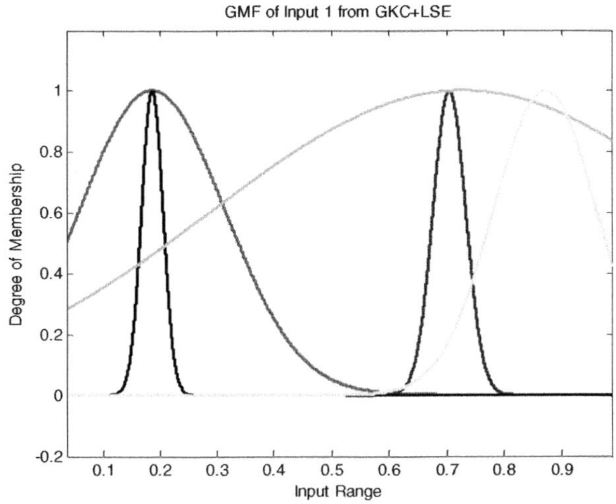

Fig. 8 Five GMFs tuned for first input of the network, produced by GK clustering and LSE estimation, training data = 500, $c = 5$, input = 7, output = 1, ws =5

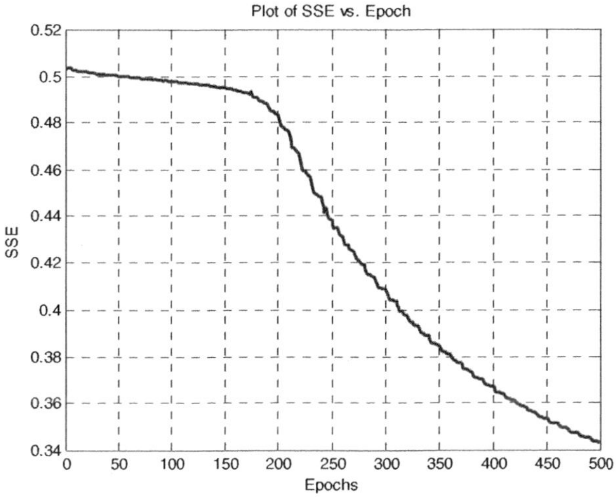

Fig. 9 Graph of SSE versus Epochs of TS-type MISO neural-fuzzy network training, n= 7, m = 1, Epochs= 499, d=96, training data=500, and c = 5

This means that the neural-fuzzy network's training has reduced the SSE by 32% compared to the initial SSE in 499 epochs. Figures 8 and 11 show the Gaussian membership functions obtained through the fuzzy clustering and from the neural-fuzzy network's training respectively.

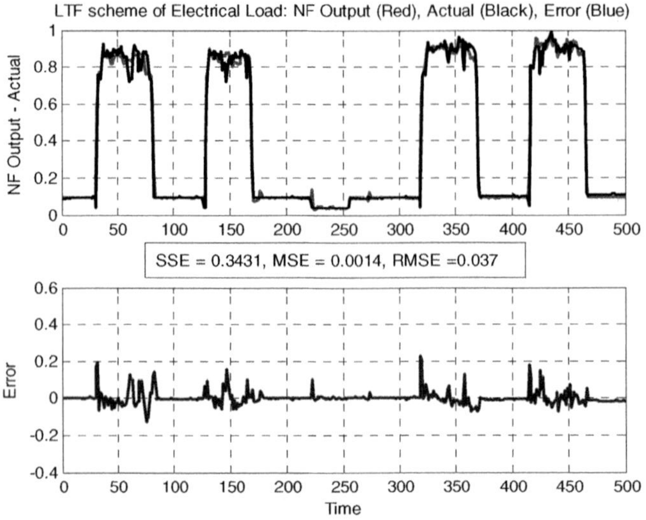

Fig. 10 Training performance of the TS-type MISO neural-fuzzy network with improved back-propagation, $n = 7$, $m = 1$, Epochs = 499, lead time of forecast $L = 96$, training data = 500, $c = 5$, $ws = 5$

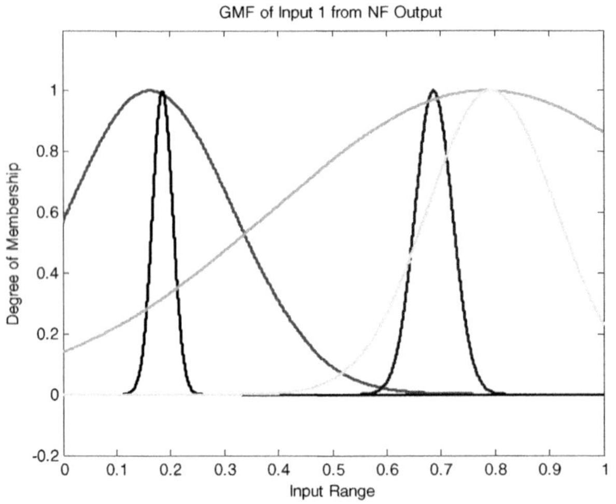

Fig. 11 Gaussian membership functions tuned for the first input universe of discourse, produced by the neural-fuzzy network, training data = 500, $c = 5$, $n = 7$, $m = 1$, lead time of forecast $L = 96$

Furthermore, for forecasting purposes, 2800 electrical load data were used as illustrated in Figure 12. The corresponding results of forecasting are: SSE (forecasting) = 77.8518, MSE (forecasting) = 0.0104 and RMSE (forecasting) = 0.102. Full

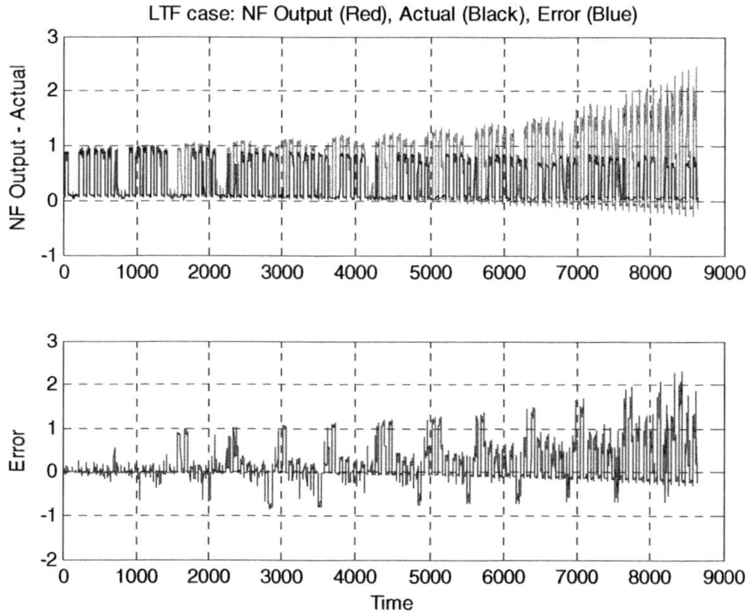

Fig. 12 Forecasting performance of the TS type MISO neural-fuzzy network, $n = 7$, $m = 1$, training epochs = 499, lead time of forecast $L = 96$. Sufficiently accurate forecasted data = 1500

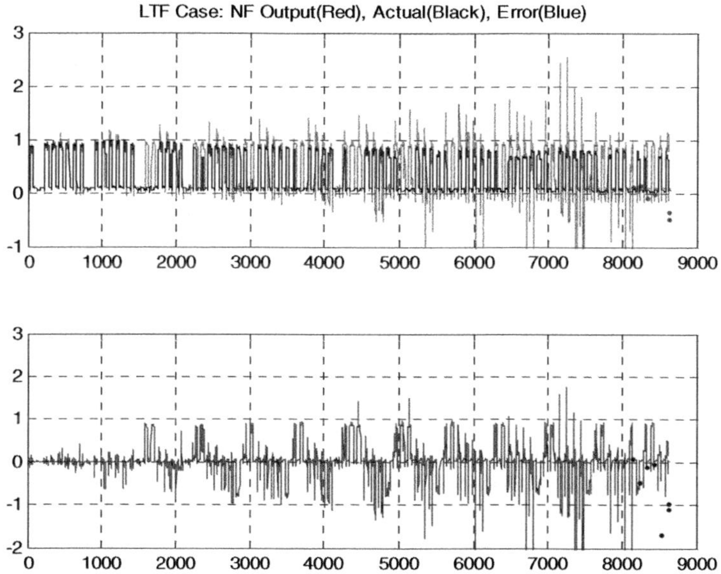

Fig. 13 Forecasting performance of the TS type MISO neural-fuzzy network, $n = 7$, $m = 1$, epochs = 499, lead time of forecast $L = 96$. Sufficiently accurate forecasted data = 1500

forecasting of electrical load starts from the 763rd data point (i.e. 7 days*96 data + 1) and mid-term forecasting of electrical load (*i.e.* neural-fuzzy network output) can follow until the next 1500 normalized actual data points (approximately 15 days) with reasonably good accuracy, as the actual error values within this data range are small (see blue curve in the bottom part of Figure 12). By using the fuzzy clustering, another performance is shown in Figure 13 that produces training *SSE* = 0.2397, *MSE* = 0.00096 and *RMSE* = 0.022. Figure 13 also shows that full mid-term forecasting of neural-fuzzy network output can follow until the next 1500 data points (15 days) with reasonably good accuracy.

8 Conclusions and Future Work

In this chapter, a neural-fuzzy approach has been presented for short-term and mid-term forecasting of electrical load. It has been demonstrated that neural-fuzzy networks, trained with improved backpropagation algorithm, provide sufficient accuracy in modeling and forecasting of nonlinear electrical load time series. Furthermore, the hybrid network's initial parameters were selected by applying fuzzy clustering and LSE estimation technique on electrical load data, which greatly improved both the network's training and forecasting performances. Apart from this, the neural-fuzzy approach is preferred to both neural networks and fuzzy logic technology as this approach exploits the merits of both technologies. Several issues need to be addressed in the future which mainly include verification of model transparency or interpretability of generated fuzzy rules obtained through adaptive training. For the latter issue, set theoretic similarity measures [10] should be computed for each pair of fuzzy sets and the fuzzy sets which are highly similar, should be merged together into a single one. Similarly, after model transparency testing, if any irrelevant fuzzy rule is found in the fuzzy rule base, that should also be discarded from the rule list to make the rule base more compact. Furthermore, in order to improve the forecasting accuracy further, the weather variables should be taken into account as model input.

References

1. Box, G.E.P., Jenkins, J.M.: Time Series Analysis: Forecasting and Control, Revised edn. Holden-Day, San Francisco (1976)
2. Jang, J.S.R.: ANFIS: Adaptive network based fuzzy inference system. IEEE Trans. System, Man and Cybernetics 23(3), 665–685 (1993)
3. Khotanzad, A., Afkhami-Rohani, R., Lu, T.L., Abaye, A., Davis, M., Maratukulam, D.J.: ANNSTLF – A neural-network based electric load forecasting system. IEEE Trans. on Neural Networks 8(4), 835–845 (1997)
4. Mamdani, E.H.: Application of fuzzy logic to approximate reasoning using linguistic systems. Fuzzy Sets and Systems 26, 1182–1191 (1977)
5. Palit, A.K., Babuška, R.: Efficient training algorithm for Takagi-Sugeno type neuro-fuzzy network. In: Proc. of FUZZ-IEEE, Melbourne, Australia, vol. 3, pp. 1538–1543 (2001)

6. Palit, A.K., Doeding, G., Anheier, W., Popovic, D.: Backpropagation based training algorithm for Takagi-Sugeno network to forecast electrical load time series. In: Proc. of FUZZ-IEEE, Honolulu, Hawaii, USA, vol. 1, pp. 86–91 (2002)
7. Palit, A.K., Popovic, D.: Computational Intelligence in Time Series Forecasting, Theory and Engineering Applications. Springer, London (2005)
8. Park, D.C., El-Sharkawi, M.A., Marks II, R.J., Atlas, L.E., Damborg, M.J.: Electric load forecasting using an artificial neural network. IEEE Trans. on Power Systems 6(2), 442–449 (1991)
9. Pedrycz, W.: An identification algorithm in fuzzy relational systems. Fuzzy Sets and Systems 13, 153–167 (1984)
10. Setnes, M., Babuška, R., Kaymark, U.: Similarity measures in fuzzy rule base simplification. IEEE Trans. System, Man and Cybernetics 28, 771–778 (1998)
11. Wang, L.X., Mendel, J.M.: Back-propagation fuzzy system as nonlinear dynamic system identifiers. Proc. of FUZZ-IEEE 2, 1409–1418 (1992)
12. Xiaosong, D., Popovic, D., Schulz-Ekloff, G.: Oscillation resisting in the learning of backpropagation neural networks. In: Proc. of 3rd IFAC/IFIP workshop on algorithm and architectures for real-time control (AARTC), Ostend, Belgium, pp. 21–25 (1995)

Quantised Problem Spaces and the Particle Swarm Algorithm

Tim Hendtlass

Abstract. The particle swarm algorithm has shown ability to optimise in continuous problem spaces, although it can struggle in problem spaces containing multiple optima. A variant, called Waves of Swarm Particles (WoSP), has been shown to be able to handle problem spaces containing multiple optima by sequentially exploring these optima. In this chapter, the WoSP algorithm is adapted to suit complex quantised problem spaces and applied to a highly constrained problem with many constraint-violating solutions but few constraint-satisfying solutions. The performance obtained is remarkably good and reflects the power of the WoSP algorithm, which combines the search abilities of particle swarm with that of evolution.

1 Introduction

Many algorithms are the result of biological inspiration and Particle Swarm Optimisation (PSO) is no exception. Originally introduced by Kennedy and Eberhart in 1995 [12] it was inspired by the flocking of birds and the schooling of fish. However, the PSO algorithm has slightly different end goals to the biological behaviour that provides its inspiration and so needs to differ from its biological inspiration in some, perhaps un-biological, ways. In the real world birds and fish group for the purposes of protection and for the efficient searching for food. The swarm needs to be compact for protection and once food is found the flock should settle to feed. In artificial particle swarm optimisation the aim of the algorithm is to find an optimum solution to some problem, rather than the protection or food sought in the natural

Tim Hendtlass
Faculty of Information and Communication Technology,
Swinburne University of Technology, Victoria 3122, Australia
e-mail: thendtlass@swin.edu.au

R. Chiong, S. Dhakal (Eds.): Nat. Intel. for Sched., Plan. and Pack. Prob., SCI 250, pp. 175–193.
springerlink.com

environment. Once an optimum is (accurately) found our interest in it is exhausted and the artificial swarm should go and search elsewhere in the hope of finding better optima. Since it does not require the whole swarm in order to explore the found optima, it is not necessary for all the particles to visit it; many could just set out on a new search.

The environment in which real birds and fish move has a continuous fitness function, which formally means that the derivative of the fitness exists at all points. But in our artificial world the fitness function may include instantaneous step functions, in which case the fitness function is said to be quantised (that is it can only take one of a finite number of discrete values and changes instantaneously between these discrete values). Assignment problems, in which a number of objects have to be distributed between various containers, are examples of problems with a quantised fitness function - an object is either in a container or not and the fitness differs accordingly (depending on whether it is in or not).

Of particular interest to readers of this book are a common and practical subset of assignment problems, scheduling and timetabling. Scheduling and timetabling are not normal activities for birds or fish and so it should come as no surprise that the basic particle swarm optimisation algorithm, in addition to modifications to allow it to sequentially discover a number of solutions of interest, will also require non-biological modification in order to be able to solve this class of problem. The modifications will suit PSO to the general class of assignment problems, and so the title refers to 'quantised problem spaces' even though the example given is a simple timetabling problem as befits a chapter in this book.

The rest of this chapter is organised as follows. After a description of the basic particle swarm optimisation algorithm the modification needed to allow it to perform sequential search is introduced. These two descriptions are in the context of a continuous problem space. Next the modification required to adapt either of these algorithms for use in a quantised problem space are introduced.

A simple timetabling problem is then introduced, adequate to show the difficulties inherent in such a class of problems and enable the performance of the revised swarm algorithm to be established. Results for this problem are presented and discussed and the chapter finishes with a discussion of the effect the problem scale is likely to have on the performance of algorithms of this type.

2 The Basic Particle Swarm Optimisation Algorithm

In all particle swarm algorithms the position of a particular particle in some way encodes a possible solution to the problem for which we seek, ideally an optimal, but at least a very good solution. Particles move under the combined influence of a number of factors in such a way that they tend to converge towards an optimum. A number of slight variants of the particle swarm algorithm have been proposed since the original algorithm was introduced by [12]. All of these try to balance several aspects of the behaviour. See, for example, [2, 3, 4, 5, 13, 6, 7, 8, 10, 11]. For an excellent reference on classical particle swarm optimisation, see 'Fundamentals of Swarm Intelligence' by A.P. Englebrecht [6].

For an optimisation algorithm to be more efficient than a random search, the choice of the next position to be searched must be guided by the positions previously tested and how good a solution those positions represent. Some way must be found that allows this previous knowledge to be exploited so that, on average, the next positions explored represent better solutions than those that would have been found if the next positions had just been picked randomly. Obviously storing all positions previously explored would rapidly produce an unsupportable demand for memory and manipulating them is, any way, an unsupportable computational load. So in the PSO only a few, often only two, kinds of good positions are kept by each particle. Probably some, and possibly all, the information about a particular particle's stored positions is shared with some or all of the other members of the swarm. There are a number of candidates for the types of position to keep.

All particles try to exploit (are influenced to move towards) at least one good position already found by some particles in the swarm. Often the position that is exploited is the best position yet found by any member of the swarm, a position known as the global best or *gbest* position. This obviously requires communication between the members of the swarm sharing the best position each has found and forms a sort of social collective memory of the current global best (*gbest*) position.

A second position used by a particle uses the local best position *lbest* found by some subset of the swarm, the subset to which this particle belongs. Often this is defined as the N closest particles (in problem space) to this particle but can also be described as the N closest in index (assuming that the particles are held in an indexed array). Over time because of the interaction of particles in the subset this second definition becomes remarkably similar to the first (but simpler to calculate). Either the best current position or the best position ever found by all the particles that make up the subset can be used. In this case information is only shared among members of the subset, but as all the particles are in general moving, the exact composition of any particular particle's subset may change as time progresses and so information could, in time, spread across more of the swarm.

The last position commonly used by a particle is the best position yet found by this particle, the personal best *pbest*. Since each particle is attracted to its own personal best position this does not depend on any inter-particle communication at all (although of course this is the same information that is shared as part of the calculation of whichever of *gbest* or *lbest* is also being used).

All particles are assumed to have mass so that they cannot change direction instantaneously and each time their velocity is updated it must contain a component that is in the same direction as their previously calculated velocity. This effectively provides a low pass filter to any change in their velocity and smooths the track the particles follow through the problem space. The fraction of the previous component that is retained is called the momentum and is a number greater than zero but less than one.

With only the use of momentum and *gbest*, particles would engage in a headlong rush toward the first reasonable position found by any particle, only changing this if some particle happens upon a better position during this rush. Eventually all particles would reach the same best-known position and exploration would stop. The particles would in time come to rest as, once a particle over-shot a position, it would be attracted back to it and, with the momentum term being less than one, the velocity would drop with each reversal. This behaviour would mimic real life birds rushing to and settling at the best-known food source, but little exploration would occur during the headlong convergence and this would represent the balance between exploration and exploitation being tipped firmly in favour of exploitation.

With only the use of momentum and *lbest*, the behaviour would depend on whether each particle is sharing the current best or the best ever position it has found. If the best ever is being kept the swarm could tend to divide and small groups of particles converge on a number of different positions. Particularly in the early stages the exact membership of the subsets would change, quite possibly changing the position towards which the current subgroup's particles are being attracted. This is less social than using *gbest* and more exploration will occur as the sub swarms move towards their (generally) individual convergence positions. The balance between exploration and exploitation is still tipped strongly towards exploitation, but not as strongly as using *gbest*.

If using momentum and *lbest* and the position recorded is the current best, then the position that the particles in the sub-swarm are converging on is likely to change often, but can as easily change to a lesser fitness as to a greater fitness. With no collective memory of good positions found, the final position where the swarm members end up is largely a matter of chance. The balance between exploration and exploitation is now tipped quite strongly towards exploration, but some small exploitation occurs as the particles that make up a sub swarm share their current but not their historic information.

With only the use of momentum and *pbest*, each particle will end up on the best position it has happened across during its travels. While this obviously means that there will be some, probably many, positions explored by the swarm with poorer fitness than those positions on which the swarm members finally end up, there is no guarantee that the final positions will be good in a global sense. This is pure exploration with no attempt at exploitation at all.

It is when momentum is combined with an attraction to two of these positions that the performance of the swarm improves sufficiently to make it attractive as a practical optimisation algorithm. Which two points are chosen determines how greedy the algorithm would be.[1] A very greedy algorithm will converge fast, but not necessarily to a good position: a less greedy algorithm will converge more slowly, possibly much more slowly, but the probability that it will converge to a good position is enhanced.

One form of the equation that governs the updating of each particle's velocity is given as Equation 1. This form differs from the more common one in that the magnitude of the attraction to each of the two points is made independent of the actual distance. In the common form this attraction is a function of distance and the whole right hand side of the formula has to be multiplied by a constriction factor [3]. Without either of these approaches it is possible that the swarm may become unstable as particles accelerate without limit.

$$\begin{aligned} \overline{V}_{T+t} = {} & M * \overline{V}_T + \\ & (1-M) * \left(\left[\frac{G*R1*(\overline{gbest - X_T})}{t*\left|\overline{gbest - X_T}\right|} \right] + \left[\frac{(1-G)*R2*(\overline{lbest - X_T})}{t*\left|\overline{lbest - X_T}\right|} \right] \right) \end{aligned} \tag{1}$$

The parameter $M \in [0,1]$ is the momentum of the particle and controls how fast the particle is able to respond to the two positions of attraction. A high momentum will make the particle slow to respond, which encourages exploration (and may enable the particle to move through small local optima entirely), conversely a low momentum makes the particle very quick to respond. Each of the two points of attraction has a constant that sets its maximum importance but the two random numbers $\in [0,1]$ introduce a stochastic element into the actual attraction at any time. The introduction of the $1/t$ term is to ensure that the dimensions of each part of the equation are the same, namely the dimensions of velocity. The units of time are usually arbitrary, allowing t to be set to one and so saving it from having to be written, but this time (whatever its value) is of great significance and has been included in Equation 1 deliberately so it is not overlooked. It must be remembered that, unlike real life, the algorithm assumes that a particle's velocity remains unchanged between fitness evaluations. A particle may travel a considerable distance, and even pass through several possible optima, between evaluations if t is large. Further implications of using a finite value for t and this constant velocity assumption will be discussed in Section 3.

Commonly the attraction to each position is also made dependent on the distance the particle is from the point of attraction: the further the particle is away the higher the attraction. While this can work well in simple problem

[1] A very greedy algorithm is one which attempts to optimise each step without regard to the final result, a less greedy algorithm is prepared to make one or more less optimal steps for now in the hope that these would set it up to make a 'killer' move later on.

spaces it may become unhelpful in more complex problem spaces. When a particle is far from a point of attraction, the acceleration of the particle towards that point of attraction would be high. By the time the particle reaches the point of attraction the particle's speed would be very high and so the distance it travels between evaluations is also large. An alternative approach (as expressed in Equation 1) is to make the attraction to a point independent of its distance from that point. Although a particle will still accelerate as a result of the continued attraction to a point, this approach has the effect of limiting the maximum velocity a particle can have and thus the maximum distance it can travel between fitness evaluations.

Alternatively, the magnitude terms in the denominator of each ratio in Equation 1 can be omitted, and instead a limiting speed introduced. Then no particle is permitted to go faster than this: if the magnitude of the right hand side of Equation 1 is greater than this limiting speed, the new speed of the particle is set to this limiting value but with the direction calculated from the equation. However, the initial acceleration will be faster than if Equation 1 as stated above is used and more of the journey will tend to occur at a high velocity, reducing the number of fitness evaluations along the journey.

To complete the description of Equation 1, it only remains to discuss which pair of attraction points to use. Originally, *gbest* and *pbest* were used, the former providing the social exploitation and the latter the personal exploration. Replacing *pbest* with *lbest* is a viable alternative as long as the size of the neighbourhood remains a modest fraction (say 10%) of the total swarm. If this is done the algorithm is made slightly greedier but a significant amount of exploration is still undertaken. Using *lbest* and *pbest* is the least greedy version of all and provides the least exploitation and the most exploration. The swarm may finally converge owing to the overlap between the sub swarms of each of the particles, but it will, on average, be the slowest of the three combinations.

Having calculated the new velocity for each particle, the only other step in the algorithm is to move each particle ready for the next iteration. As mentioned above, it is assumed that the velocity of the particle does not change during the time t between updates (and evaluations). The position update equation is given as Equation 2, again with the inter-evaluation time t explicitly shown so as to make the equation dimensionally consistent.

$$\overline{X}_{T+t} = \overline{X}_T + t * \overline{V}_{T+t} \tag{2}$$

The stopping condition could be the achieving of acceptable performance *or* the swarm having converged without achieving adequate performance *or* some maximum number of fitness evaluations having been made without either of the first two conditions being met. A swarm has converged when all the particles are (eventually) at the same position and the velocity of each particle is approaching zero.

Algorithm 1. The basic PSO algorithm

Randomly assign each particle to some point in problem space with some (small) random velocity and evaluate the fitness of each particle at its current position
while no stopping condition is met **do**
  Update *gbest* for the swarm, *pbest* for each particle if these are to be used
  If *lbest* is to be used, find the other particles that make up the sub swarm of each particle and update *lbest* for each particle from either the current best or the personal best fitness of it and its neighbours
  Calculate the new velocity of each particle using Equation 1 (if not using either *gbest* or *lbest* replace them with the position(s) used)
  Move each particle using Equation 2 and evaluate its fitness at this new position
  IF best fitness is better than some user defined threshold
  OR IF the user specified maximum number of iterations reached
  THEN stopping condition met
end while
end

3 WoSP: A PSO Algorithm for Sequentially Exploring Many Optima

In the Waves of Swarm Particles (WoSP) algorithm an extra term is added to the basic velocity update equation of the classical swarm algorithm, as shown in Equation 3. The terms $\widehat{b_T}$ and $\widehat{l_T}$ are the unit vectors towards the global and local best positions and are a short form of the full expression as written in Equation 1.

$$\overline{V}_{T+t} = M * \overline{V}_T + (1 - M) * \left(R1 * G * \widehat{b_T} + R2 * L * \widehat{l_T} + \overline{V_{SRF}}\right) \tag{3}$$

The extra term is a short range force of attraction between a particle and all other particles. This force becomes enormous when two particles are close but drops off very rapidly with distance. The ith component of the short range force exerted on particle x by particle y is given by:

$$SRF_{xyi} = SRF_{factor} * \frac{\overline{V}_{xyi}}{\overline{D}_{ij}^{SRF_{power}}} \tag{4}$$

where:

V_{xyi} is the ith component of velocity of particle x with respect to particle y,
D_{xy} is the distance from particle x to particle y,
SRF_{factor} is the short range force magnitude at unit distance, and
SRF_{power} sets how fast this force decreases with distance.

Figure 1 shows how sharply this varies with distance for a range of values of SRF_{power}.

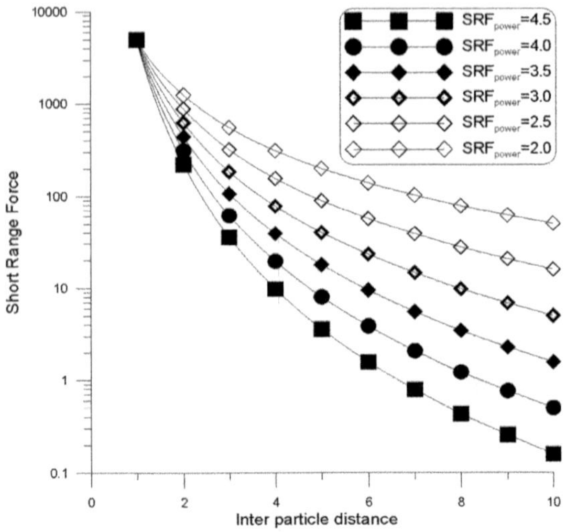

Fig. 1 How the short range force varies with the value used for the parameter SRF_{factor}

Equation 3 defines the new velocity of a particle at time $T+t$. The symbols on the right hand side of Equation 3 mostly have the same meanings as in Equation 1. $\overline{V_T}$ is the velocity this particle had at time T, t is the time between updates, M is the momentum of the particle and $R1$ and $R2$ are random numbers in the range $[0,1]$. G and L set the relative attention to be placed on the best global and local position known to this swarm at time T. The unit vector to the global best position is $\widehat{b_T}$ while the unit vector to the global best position is $\widehat{l_T}$. M, G and L are all bounded to the range $[0,1]$. The last term on the right hand side $\overline{V_{SRF}}$ represents a velocity produced by a short-range force of attraction between the particles. It would be of little use if it were not for the way the swarm is updated. The velocity is calculated using Equation 3 at a time T and assumed to be unchanged until the next update at time $T+t$. This can introduce an aliasing effect when two particles are so close that the magnitude of this attractive component is large and results in a high velocity for the particles. By the time of the next evaluation, the particles have crossed and are far apart and the magnitude of this now reversed attractive component is small. As a result the particles keep much if not all of the velocity that they acquired at the previous evaluation (for a fuller discussion of this aliasing effect see [8]). This is shown diagrammatically in Figure 2 for three successive iterations, the large arrows show the magnitude and direction of the short range force and the thin arrows the velocity. Note that the short range forces are very large in the center iteration which produces high velocities, and, because of the non-biological condition that velocity does not alter between iterations, this velocity is substantially kept in the third iteration when the short range forces are too weak to significantly attenuate it.

Each particle belongs to a wave (identified by a number) and only communicates with other particles with the same wave number. When the ratio of

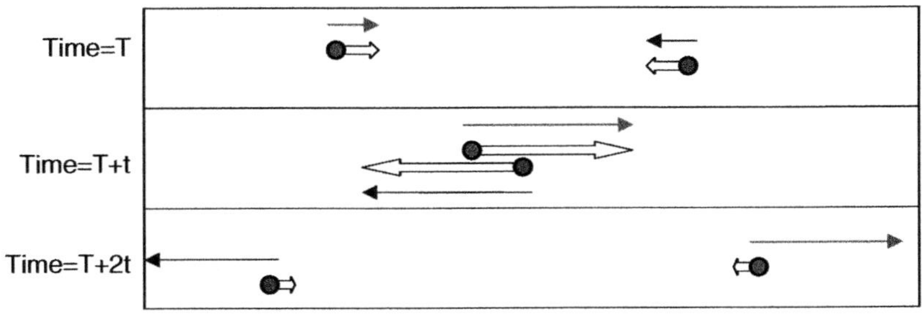

Fig. 2 The velocities and short range forces acting on two particles at three discrete sequential time intervals

the magnitude of the attractive velocity component $\overline{V}_T$ compared to the net magnitude of the three other velocity components exceeds a threshold (normally because particles have come close while converging on an optimum), the particles concerned join the most recently created (highest numbered) wave or, if they already belong to the most recently created wave, they create an even newer one. This process is referred to as promotion and the position in problem space where it occurred as a promotion point. Particles keep individual records of their promotion points.

Like the other forces, the short-range force is assumed to apply unaltered for the basic time interval t since the last velocity update was performed. The sum of all the components defined by Equation 4 produces a change in velocity $\overline{V}_{SRF}$ (the constant of proportionality between the force and the velocity is assumed to have been absorbed into SRF_{factor}).

Identical to the conventional particle swarm algorithm, the position of each particle is calculated at each new time using Equation 2.

Algorithm 2. The WoSP algorithm

Randomly assign each particle to some point in problem space with some (small) random velocity and evaluate the fitness of each particle at its current position
while no stopping condition is met **do**
  Update *gbest* for the swarm if necessary
  Find the other particles that make up the sub swarm of each particle and update *lbest* for each particle from the current best fitness of it and its neighbours
  Calculate the short range force acting on each particle using Equation 4 (practically this can be restricted to just the contributions from the close particles that form the neighbourhood).
  Calculate the new velocity of each particle using Equation 3
  Move each particle using Equation 2 and evaluate its fitness at this new position
  If the short range force is large enough, promote this particle
  IF best fitness is better than some user defined threshold
  OR IF the user specified maximum number of iterations reached
  THEN stopping condition met
end while
end

Each wave keeps track of the best fitness it has found and, should a later wave be outperforming an earlier wave, the particles in the earlier wave are forcibly recruited to the later, higher performing wave (which need not be the highest numbered wave). Recruited means that a particle acquires the new wave number and only responds to other particles in this wave. Importantly, since the wave has its own common *gbest*, the recruited particle tends to go and join the rest of its new wave. This 'survival of the fittest' introduces a type of evolutionary pressure which makes a significant contribution to the performance of the algorithm as a whole.

It is possible, although unusual, for only one of a pair of particles to meet the promotion criterion and for a particular wave to eventually decay to one particle. If this should happen the sole remaining particle is recruited to the highest numbered wave. Recruitment is not promotion and the positions of recruitment events as described in the previous paragraph are not recorded by the particles involved. Once a wave has no particles left it is considered 'dead' and plays no further role in the search.

Particles are actively repulsed from their promotion points if within a user specified parameter (the search scale) of them. This effectively sets the coarseness of the search. The repulsion has two components, $C1$ directly away from the promotion point in question and $C2$ in the direction of the smallest component of the first. Without this latter component, the search by a particular particle tends to be confined to the hyper plane defined by the positions of the particle's first few promotion points (there are undoubtedly many ways in which preventing this could be achieved, this one is used for its simplicity).

Like all forms of Swarm, WoSP benefits from the addition of a local search algorithm. Only invoking this when a wave dies is computationally efficient as a wave commonly dies owing to its particles being promoted away as they converge to a local optimum. The exact nature of the best local search is problem specific.

4 Mapping to a Quantised Problem Domain

PSO inherently operates in a continuous problem space in which each possible position represents a potential solution to the problem and in which the slightest movement of a particle changes the solution it currently describes. Timetabling does not fit this description as room and time allocation can only take on certain specific and discrete values: for example there are a finite number of rooms in which some activity may take place and the activity must be confined to one room, not spread across parts of two (or more) rooms. Similarly, activities are discrete, as must be the allocation of people to activities.

For PSO to work with timetabling (or other quantised) problems it is necessary to map this continuous space to the quantised problem space before the fitness (number of constraint violations) of a particular solution is calculated. Each axis corresponds to a unique quantised property of the problem (for example the room a class will take place in, or the time at which the class will commence). The length of each axis is chosen as some arbitrary value L (100 was used in the experiments described below) and the axis is considered to be circular so that the point L and 0 are in fact the same point. This length L is then divided into N equal parts (cells), where N is the number of quantum values of the relevant property in the problem. It is the position of the particle, decoded into quantum values, that is passed to the fitness evaluation routine so that the number of constraint violations of this possible solution can be returned. A simple instance of this is shown in Figure 3 in which the axis is divided into six cells. For this simple example all cells are the same size, although there is no requirement that they be so. Every time that the particle is moved along the axis its position as given by Equation 2 is replaced by this position modulus L.

There is obviously a length of L/N on an axis that will correspond to the same quantised value. The fitness function is required to be continuous in the sense that adjacent positions should in general return different fitness values so as to provide guidance to the swarm as it converges. For this reason the distance of a particle from the closest cell center is also recorded (expressed as a fraction of L/N). The fitness function is comprised from the sum of two parts, the number of constraint violations plus the average of these fractional distances across all axes. The use of the average ensures that the number of constraint violations remains the most important component of the fitness and it is not possible for a solution with more constraint violations to get a better (lower) score than a solution with less constraint violations just because its particle happens, on average, to be closer to the centre of the cell on every axis.

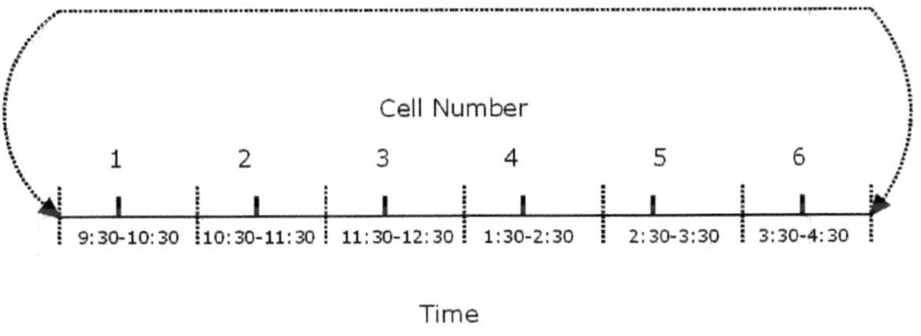

Fig. 3 A simple axis with six cells, each corresponding to the time a class can be held. The small vertical marks show the cell centre. The dotted line shows that the axis is continuous and that the start and end position are in fact the same.

5 The Particular Quantised Problem Discussed in This Chapter

The sample quantised problem space discussed in this chapter is a simple timetabling problem involving scheduling 19 classes for four groups of students in three rooms for one day of six time periods . While all classes can occur in any of the six available time periods, there are various constraints as to the rooms each class may occur in and the group(s) of students that will be involved. The aim is to timetable the classes so that these constraints are met and no student is required to undertake two classes at once and no room is required to contain more than one class at a time. The constraint details are shown in Table 1, with classes, room and groups identified by numbers.

Table 1 Timetable problem constraints for a simple timetable problem scheduling 19 classes into 4 rooms. Any class can run at any time.

Class	Possible rooms	Groups involved
1, 2, 3	1, 2, 3	1
4	4	1
5, 6, 7	1, 2, 3	2
8	4	2
9, 10, 11	1, 2, 3	3
12	4	3
13, 14, 15	1, 2, 3	4
16	4	4
17	1, 2, 3	1, 2
18	1, 2, 3	3, 3, 4
19	1	1, 2, 3, 4

There are approximately $1.7 * 10^{26}$ ways in which these classes can be arranged but only slightly more than 2500 that do not violate any constraint. While this is a trivial problem as far as timetabling is concerned, it is more than adequate to explore the behaviour of swarm particles in quantised problem spaces as it is easy to comprehend and has the advantage of fast fitness evaluation. The aim of the PSO is to minimise the number of constraint violations (classes that share rooms or times when they are not supposed to), ideally finding solutions with no constraint violations at all.

Two quantised variables were associated with each class, the time it was to be scheduled and the room it was to occur in. Each of these was mapped to a different axis in problem space. In principle a total of 38 axes were therefore required to schedule these 19 classes and the number of quantised values on these axes varied from 1 to 6. In practice it is not necessary to provide an axis when there is no choice in the value of a parameter although the algorithm will work fine if you do.

6 Experimental Details

Details of the parameter values used are shown in Table 2. It is not claimed that the values used were optimum but follow the general guidelines given in [4].

Table 2 The parameter values used

Number of particles	30
Momentum (M)	0.9
Global attraction coefficient (B)	0.3
Local attraction coefficient (L)	0.7
Maximum fitness evaluations per repeat	1,500,000
Short Range Force parameters	
$SRFfactor$	500
$SRFpower$	3.5
Promotion coefficient (P)	2
Repulsion coefficient (R)	5000
Search scale (S)	2
Local heuristic parameters (see 6.2)	
$LH1$	50
$LH2$	50

6.1 The Repulsive Force C2 Used

Providing the contribution of each axis to the overall fitness function can be quantified, an estimate may be made of the error contribution to each axis. For the timetabling problem used in this chapter, each class requires two axes in problem space: one to define the class time and a second to define the classroom. The fact that a group is being required to do more than one class at a given time suggests that the times of all the classes involved should be changed and so a weighting reflecting the number of clashing classes is added to the velocity of the particle along each of the axes that defined the times of these classes. The fact that two or more classes are scheduled in one room suggests that the rooms used for each of these classes should be changed and a weighting reflecting the number of classes in the room at this time was added to the velocity of the particle along each of the axes defining the rooms of these classes. The rejection component C2 along each axis was made proportional to the net weighting on that axis. In this way it is possible to focus change on the axes on which it is most needed.

6.2 The Local Heuristic Used

A simple problem specific local heuristic was used that took a solution with one or more constraint violations and repeated the following set of steps until the number of constraint violations was reduced to zero or a user specified number of attempts had been made.[2] A list of all classes that were involved in these violations was made. One class was chosen at random from this list and a second list made of all the other room time combinations to which it could be moved. One possible move was chosen from this second list and the change this possible move would make in the constraint violation count was calculated. Only if this was positive was the move actually made. A new list of clashing classes was then made and the process repeated until either no clash occurred or a total of $LH1$ tries had been made.

The algorithm described in the above paragraph is very greedy and can easily result in a local optimum being reached that does not meet all constraints. As a result the original quantised position passed to the local heuristic was saved and the algorithm described above was tried LH2 times, with the original quantised position being restored at the start of each time. The best result found in any of these $LH2$ tries was the result actually used.

The extra computational load was insignificant with the values of $LH1$ and $LH2$ used as the local heuristic was only run when a wave died – typically a few hundred times per run.[3]

7 Results

Table 3 shows that, in terms of the best performing repeat out of each group of 100, there is a steady improvement from left to right as progressive enhancements are made to the classical particle swarm algorithm. Interestingly, adding waves alone produces a greater positive effect than adding the local heuristic alone but the best results are obtained when both of these are used. While Table 3 shows only the best result for each repeat, Tables 4 and 5 show statistics derived from all the repeats for all combinations.

Comparing the columns in Table 4 that do not involve the local heuristic, it is clear that the addition of waves consistently and substantially improves the performance. Again it can be observed that the performance with waves alone is better than the performance with the local heuristic alone.

Inspection of Table 5 shows that the improvement made by the use of the local heuristic was essentially independent of the use of waves.

---

[2] This local heuristic is most certainly not the only one that could be used, it is in all probability not the best local heuristic designed for timetabling. Since this chapter talks about 'quantised problem spaces' of which timetabling is just one (common) example, the use of an over customised local heuristic might obscure the main message of the chapter.

[3] Since the results quoted in this chapter were obtained, further experimentation has showed that the values used for these two parameters were not critical and could be reduced by an order of magnitude without significantly affecting the results for this problem.

Table 3 An overview of the performance of 100 independent repeat runs for each of the four possible combinations of waves and local heuristic

Using waves	No	No	Yes	Yes
Local heuristic	No	Yes	No	Yes
Constraint violations				
Number of times one or more solutions found with none	0	0	8	80
Average number per run	5.72	3.81	2.25	0.2
Number in best solution found	1	1	0	0
Number in worst solution found	8	8	3	1

Table 4 The average number of constraint violations for 100 repeats of each of all combinations of waves and local heuristic

Waves	No	No	Yes	Yes
Local heuristic	No	Yes	No	Yes
Number of constraint violations in best solution				
0			8	80
1		6	26	20
2	1	10	32	
3	1	32	17	
4	15	22	10	
5	24	13	5	
6	29	14		
7	27	2	2	
8	3	1		

Table 5 The reduction in constraint violations obtained using the local heuristic for 100 repeats with and without waves

	Without waves	With waves
Average	1.91	2.05
Minimum	5	7
Maximum	0	0

The best performing combination by far is when both waves and the local heuristic are used and these results have been examined in more detail. During the 100 independent repeats, each run reported on average 653 (max 671, min 634) candidate solutions that were passed to the local heuristic. A candidate solution corresponds to the best position found by a wave during its existence. The range of the number of constraint violations in all these candidate solutions is from 0 to 12.

During the runs a grand total of 1242 solutions that satisfied all constraints (absolute solutions) were found. Twenty runs produced no absolute solutions;

the other 80 runs produced on average 15.7 each (the maximum for any wave was 51, the minimum was 1).

However, because each particle maintains an individual list of promotion points, some optima were explored by more than one wave during a single run. On average, the runs that found absolute solutions found 5.8 different absolute solutions each (the maximum for any wave being 21, the minimum 1). Overall, the 100 repeats found a grand total of 446 unique absolute solutions out of the approximately 2500 that exist.

8 Discussion

The method of mapping the particle's position in continuous space to its position in quantised space described in section 4 of this chapter permits any particle swarm algorithm to work on quantised problem spaces. For problems with many local optima, few of which meet all the problem constraints, the WoSP algorithm has two distinct advantages over the conventional swarm.

The first advantage comes from the use of waves that allow the sequential exploration of optima. Inherent within the wave mechanism is the integration of evolution into the swarm process, which produces substantial performance gains.

The second advantage lies in the ability to actively repulse particles from previous promotion points (almost always previously explored optima), in particular, to be able to adjust the magnitude of each rejection component so that it is proportional to the number of problems introduced by the quantised value this axis position represents. Change is thus directed where it is most needed.

The performance of the full WoSP algorithm has previously been shown to be strongly encouraging for continuous domain problems [7, 8]. The work described in this chapter shows that its performance on at least one class of quantised domain problems is also strongly encouraging.

Considering the illustrative problem used in this paper, a mere 1,500,000 fitness evaluations gives an 80% chance of finding one of the 2,500 solutions that meet all constraints out of the total of $1.7 * 10^{26}$ possible solutions. 1,500,000 fitness evaluations using a systematic sequential search would give a $2.2 * 10^{-15}$% chance of finding one of these absolute solutions. The value that might be expected from a random search with the same number of evaluations is even (substantially) lower as re-exploration of solutions would inevitably occur. The combination of collective intelligence and evolutionary pressure has dramatically improved the efficiency of the search.

The problem described, which was only chosen as an easily understood example of a quantised problem space, required the particles to move in a 38 dimensional space: as the complexity of the quantised problem increases (for example, timetabling a more practically realistic number of classes, time periods and rooms) so will the dimensionality of the space the particles move in. Little information on how the performance of the WoSP algorithm scales beyond 100 dimensions is as yet available.

Although not yet attempted, there is no known reason why the WoSP algorithm should not show similar performance when run on hybrid optimisation problems as it does on both continuous and quantised problems. Hybrid optimisation problems involve some continuous and some quantised parameters, a class of problem into which many real life industrial problems fall. Overall, WoSP, with suitable problem mapping, has the ability to extend the range of applicability of particle swarm far beyond the ability to be successfully associated with the conventional swarm algorithm.

9 Problem Scaling

The problem described here is a 'toy problem' and it should be noted that as the number of events to be timetabled is incremented the dimensionality of the problem increases by two. While there is no theoretical limit to the number of dimensions that a PSO can optimise, practicality implies otherwise. As the number of dimensions grows, the time taken to perform fitness evaluations grows. This is no different for PSO than for any other algorithm. But PSO as a general algorithm makes more evaluations on its way to a solution than a more specialist algorithm explicitly tailored to this class of problem.

As such, it may be felt that PSO could never compete. However, PSO has one interesting feature: it is not necessary to perform a true fitness evaluation at each step, the performance does not degrade if many, maybe up to 90%, of evaluations are actually quick estimates. A brief explanation of this follows, the interested reader will find more information in [16, 9, 15].

The estimation process has two additional requirements: that it is possible to estimate the distance a particle has moved in problem space and that each fitness value used has a reliability associated with it. A new fitness value f_{child} is estimated from the fitness f_1 and reliability r_1 of this particle's last iteration and the fitness f_2 and reliability r_2 of the closest particle's *last iteration* to the position of this particle in this iteration using Equation 5. Simultaneously with the fitness being estimated the reliability of this fitness estimation is calculated using Equation 6. If the reliability $[0, 1]$ is above a user chosen threshold, then this fitness estimate is used and no true fitness calculation is done. If it is below the threshold then the fitness estimation is discarded and a true fitness calculation performed. True fitness calculations always have a reliability of unity.

$$f_{child} = \frac{w_1 * f_1 * r_1 + w_2 * f_2 * r_2}{w_1 * r_1 + w_2 * r_2} \tag{5}$$

$$r_{child} = \frac{(w_1 + r_1)^2 + (w_2 + r_2)^2}{w_1 * r_1 + w_2 * r_2} \tag{6}$$

where

$w_1 = d_1/(d_1 + d_2)$ and $w_2 = d_2/(d_1 + d_2)$
d_1 is the distance moved by the particle between last iteration and this, and
d_2 is the distance from the particle's current position to the position of the closest particle in the last iteration.

The basic method described above works well, achieving the same level of performance as that obtained when always using true fitness calculation but achieving it much faster. Using an adaptive threshold and performing a simple error correction every time both a true and an estimated fitness value becomes available can further enhance performance – whether this will make a PSO approach competitive with customised timetabling problems remains to be seen.

10 Concluding Comment

As a general observation, the WoSP variant of PSO shows considerable promise as an optimiser for quantised problem spaces (such as timetabling) even if it may at times be outperformed by highly optimised algorithms developed specifically for certain problems. It certainly has promise for other quantised problems for which specialist customised software is not available, a not insignificant fact given that most real world problems have at least some element of quantisation about them. Further work is needed to establish just how much of that promise may be realised in practice on real life problems.

References

1. Blackwell, T., Branke, J.: Multi Swarms, Exclusion and Anti-Convergence in Dynamic Environments. IEEE Transaction on Evolutionary Computing 10(4), 459–472 (2006)
2. Brits, R., Engelbrecht, A.P., van den Bergh, F.: A Niching Particle Swarm Optimizer. In: Proceedings of the 4th Asia-Pacific Conference on Simulated Evolution and Learning 2002 (SEAL 2002), Singapore, pp. 692–696 (2002)
3. Clerc, M., Kennedy, J.: The Particle Swarm-explosion, Stability and Convergence in a Multi Dimensional Complex Space. IEEE Transactions on Evolutionary Computing 6(1), 58–73 (2002)
4. Eberhart, R.C., Dobbins, P., Simpson, P.: Computational Intelligence PC Tools. Academic Press, Boston (1996)
5. Eberhart, R.C., Shi, Y.: Comparing Inertia Weights and Constriction Factors in Particle Swarm Optimisation. In: Proceedings of the IEEE Congress on Evolutionary Computation, vol. 1, pp. 84–88 (2000)
6. Engelbrecht, A.P.: Fundamentals of Computational Swarm Intelligence. Wiley, West Sussex (2006)

7. Hendtlass, T.: A Particle Swarm Algorithm for High Dimensional, Problem Spaces. In: Proceedings of the IEEE Swarm Workshop, pp. 149–154 (2005)
8. Hendtlass, T.: WoSP: A Multi-Optima Particle Swarm Algorithm. In: Proceedings of the IEEE Congress on Evolutionary Computation, vol. 1, pp. 727–734 (2005)
9. Hendtlass, T.: Fitness Estimation and the Particle Swarm Algorithm. In: Proceedings of the IEEE Congress on Evolutionary Computation, pp. 4266–4272 (2007)
10. Janson, S., Middendorf, M.: A Hierarchical Particle Swarm Optimizer. In: Proceedings of the IEEE Congress on Evolutionary Computation, pp. 1666–1670 (2003)
11. Janson, S., Middendorf, M.: A Hierarchical Particle Swarm Optimizer for Dynamic Optimization Problems. In: Raidl, G.R., Cagnoni, S., Branke, J., Corne, D.W., Drechsler, R., Jin, Y., Johnson, C.G., Machado, P., Marchiori, E., Rothlauf, F., Smith, G.D., Squillero, G. (eds.) EvoWorkshops 2004. LNCS, vol. 3005, pp. 513–524. Springer, Heidelberg (2004)
12. Kennedy, J., Eberhart, R.C.: Particle Swarm Optimization. In: Proceedings of the IEEE International Conference on Neural Networks, Perth, Australia, pp. 1942–1948. IEEE, Piscataway (1995)
13. Kennedy, J., Eberhart, R.C.: The Particle Swarm: Social Adaptation in Information-Processing Systems. In: Corne, D., Dorigo, M., Glover, F. (eds.) New Ideas in Optimization, ch. 25. McGraw-Hill Publishing Company, England (1999)
14. Parrot, D., Li, X.: A Particle Swarm Model for Tracking Multiple Peaks in a Dynamic Environment using Speciation. In: Proceeding of the IEEE Congress on Evolutionary Computation, pp. 98–103 (2004)
15. Salami, M., Hendtlass, T.: A Fast Evaluation Strategy for Evolutionary Algorithms. Applied Soft Computing 2(3), 156–173 (2003)
16. Salami, M.: Fast Evolutionary Algorithm for Evolvable Hardware. PhD Thesis, Swinburne University of Technology (2003)

A Coevolutionary Particle Swarm Algorithm for Bi-Level Variational Inequalities: Applications to Competition in Highway Transportation Networks

Andrew Koh

Abstract. A climate of increasing deregulation in traditional highway transportation, where the private sector has an expanded role in the provision of traditional transportation services, provides a background for practical policy issues to be investigated. One of the key issues of interest, and the focus of this chapter, would be the equilibrium decision variables offered by participants in this market. By assuming that the private sector participants play a Nash game, the above problem can be described as a Bi-Level Variational Inequality (BLVI). Our problem differs from the classical Cournot-Nash game because each and every player's actions are constrained by another variational inequality describing the equilibrium route choice of users on the network. In this chapter, we discuss this BLVI and suggest a heuristic coevolutionary particle swarm algorithm for its resolution. Our proposed algorithm is subsequently tested on example problems drawn from the literature. The numerical experiments suggest that the proposed algorithm is a viable solution method for this problem.

1 Introduction

Mathematicians have defined a class of problems known as equilibrium problems with equilibrium constraints (EPECs) [33]. A particular subclass of these problems is the Bi-Level Variational Inequalities (BLVI). These are effectively Cournot-Nash games modeled in a hierarchical fashion: The upper level problem is a Cournot-Nash game and the constraints intrinsically define yet another Nash game parameterized by the solutions to the former. In this chapter, a coevolutionary particle swarm algorithm is extended to solve a BLVI that arises from competition in the provision of services within the highway transportation sector.

Andrew Koh
Institute for Transport Studies, University of Leeds, Leeds, LS2 9JT, United Kingdom
e-mail: a.koh@its.leeds.ac.uk

R. Chiong, S. Dhakal (Eds.): Nat. Intel. for Sched., Plan. and Pack. Prob., SCI 250, pp. 195–217.
springerlink.com

Our motivation stems from the observation that in recent years, there has been an increasing amount of private sector participation within areas that are conventionally the privy of the public purse. The driving force behind this change is brought about by the higher efficiency of the private sector coupled with increasing public pressures on governments for accountability and the corresponding need to derive value for money from their various budgetary commitments which are ultimately funded by the tax paying public [30, 53].

In highway transportation, privately operated roads are not novel concepts [56]. However there has been little analysis on this topic in terms of the competition between private sector providers and the equilibrium outcomes, save for theoretical studies by economists (e.g. [39]). In reality, there have already been examples of private sector participation in toll road construction and operation around the world [15]. In return for the private capitalists funding the large initial capital investments for the construction of the road, they are contractually allowed to collect tolls, for some pre-specified duration, on traffic when the road is finally opened to use [12]. In an era when government budgets are becoming increasingly tight and with traffic congestion becoming more of a problem in many major cities, the private sector is recognized as having an increasing role to play in the provision of traditional highway transportation investment. When a private firm is engaged with the provision of such services and in competition with others simultaneously doing the same, the concept of Nash equilibrium [36] can be used to model the equilibrium decision variables offered to the market.

Even though a Cournot-Nash equilibrium problem can be formulated as a variational inequality, the problem that we describe in this chapter differs because the players are constrained by a second variational inequality describing the equilibrium routing behavior of travelers on a transportation network. Hence the problem is formally a "Bi-Level" (denoting the two level hierarchical nature of the problem) Variational Inequality. The objective of our research is to explore the possibilities of employing a multipopulation coevolutionary method based on the particle swarm algorithm [23] to solve the resulting BLVI.

The rest of this chapter is organized as follows. In the next section, the basic traffic assignment concepts are given to provide sufficient background for readers not familiar with the tools of transportation network analysis. The variational inequality that describes the equilibrium of a transportation system is developed. In Section 3 we discuss the concept of Nash equilibrium in further detail with reference to the behavior of players in a Cournot-Nash game and subsequently formulate the BLVI of the problem at hand. Existing algorithms available to solve this problem are briefly described. In Section 4 we describe the coevolutionary particle swarm algorithm, developed based on an analogy with competition between species in natural systems, to solve the BLVI. The objective of the coevolutionary process is to evolve swarms of strategies for each player that are robust against the strategies evolved by other players while maintaining, at each iteration, the system equilibrium of the transportation network. In Section 5 we give some examples to illustrate the performance of our proposed solution algorithm on several problems. Finally in Section 6 the findings are summarized and we provide further directions for research in this developing subject area.

As the subject matter of this chapter transcends both market structures and game theory, we will use the terms "players" and "firms" interchangeably to refer to the private sector participants in this market who are the players in the Nash game governed by the variational inequality constraint. For the same reasons, "profits" and "payoff" are also used synonymously.

2 Transportation Network Analysis

To facilitate understanding of this chapter, we discuss basic concepts associated with traffic assignment and network analysis which are the key focus areas of this chapter. We discuss the traffic assignment problem and the theoretical model explaining the route choice decisions of users on a highway network in a deterministic equilibrium setting and finally show how this can be formulated as a variational inequality. Readers interested in this topic may refer to the texts [37, 49] for further details.

2.1 *Traffic Assignment Problem*

Transportation network analysis seeks to understand factors affecting the route choice decisions of road users so that control policies affecting it may be formulated. The key to understanding this is facilitated by the traffic assignment problem (TAP) which is the methodology of assigning users desiring to travel between origin-destination (OD)[1] pairs onto a traffic network.

In a given transportation network represented by a directed network graph, let:

A: the set of all links in the transportation network
B: the set of all links that are subject to tolls and capacity enhancements $B \subseteq A$,
R: the set of all routes in the network,
H: the set of all OD pairs in the network,
Ω: the set of all feasible flows and demands,
R_h: the set of routes between OD pair $h\,(h \in H)$,
d: the vector of demands between each OD pair $\mathbf{d} = [d_h]\,(h \in H)$,
$\mathbf{D^{-1}}$: the continuous and decreasing inverse demand function giving the travel cost as a function of the number of trips for each OD pair $\mathbf{D^{-1}} = [D_h^{-1}]\ (h \in H)$,
μ: the minimum travel cost between OD pair $\mu = [\mu_h]\,(h \in H)$
E: the route flow on all routes in the network $\mathbf{E} = [E_r]\,(r \in R)$
C: the cost of travel on all routes in the network $\mathbf{C} = [C_r]\,(r \in R)$
v: the vector of link flows, $\mathbf{v} = [v_a]\ (a \in A)$,
δ_{ar} : 1 if the route $r\,(r \in R)$ uses link $a\,(a \in A)$, 0 otherwise,
x: the vector of tolls, $\mathbf{x} = [x_a]\ (a \in B)$,

[1] Origins and destinations are equivalent to sources and sinks in other fields where analogous concepts of network analysis are employed.

K: the vector of capacities, $\mathbf{K} = [k_a]$ $(a \in A)$,
y: the vector of capacity enhancements, $\mathbf{y} = [y_a]$ $(a \in B)$,
$\mathbf{c}(\mathbf{v},\mathbf{x},\mathbf{y})$: the vector of link travel costs as a function of link flows, $\mathbf{c}(\mathbf{v},\mathbf{x},\mathbf{y}) = [c_a(v_a,0,0)]$ $(a \in A, a \notin B)$;$\mathbf{c}(\mathbf{v},\mathbf{x},\mathbf{y}) = [c_a(v_a,x_a,y_a)]$ $(a \in B)$
$t_a(v_a,k_a)$: the monotonically non decreasing travel time on the link *excluding* tolls on the link a. Note that we have $\frac{\partial t_a(v_a,k_a)}{\partial v_a} > 0$ and $\frac{\partial t_a(v_a,k_a)}{\partial k_a} < 0$
$t0_a$: a scalar free flow travel time for link a $(a \in A)$,
ρ_a: a positive scalar for link a $(a \in A)$,
λ_a: a positive scalar for link a $(a \in A)$.

Throughout this section we assume that **x** and **y** have already been exogenously specified. We will discuss how these are chosen in Section 3.

2.1.1 From Link Travel Time to Path Travel Costs

The travel time function on a link gives the travel time of the link as a function of traffic flows on the link and a commonly used functional form is depicted in 1.

$$t_a(v_a,k_a) = \begin{cases} t0_a + \rho_a \left(\frac{v_a}{k_a}\right)^{\lambda_a}, \forall a \in A, a \notin B \\ t0_a + \rho_a \left(\frac{v_a}{k_a + y_a}\right)^{\lambda_a}, \forall a \in B \end{cases} \tag{1}$$

The functional form in 1, known as the Bureau of Public Roads (BPR) function [3], includes link capacity as a determinant of the link travel time. When monetary tolls are converted into a time equivalent amount, then 2 allows for a unique map between the link travel time and the travel cost.

$$c_a(v_a,x_a,y_a) = \begin{cases} t_a(v_a,k_a), \forall a \in A, a \notin B \\ t_a(v_a,k_a) + x_a, \forall a \in B \end{cases} \tag{2}$$

Equation 2 states that if the use of the link is not subject to a toll charge, then the travel cost of using that link is given solely by the travel time taken to traverse it. However, if travel on it requires payment of a toll, then the time equivalent of the toll is added to the time cost of travel.

Note that a route or path between an OD pair comprises all links that constitute that route from O to D. In general, there could possibly be several routes available for use by any single OD pair and that in equilibrium not all these routes would necessarily be used.

2.2 *TAP as a Variational Inequality*

The behavioural principle underlying the highway users' choice of routes in the TAP is founded on Wardrop's Equilibrium Condition [57]. We state this more formally in Lemma 1.

Lemma 1. *Wardrop's Equilibrium Condition of route choice implies that at equilibrium the following conditions are simultaneously satisfied:*

$$\begin{aligned}
&E_{r\in R} \geq 0 \Leftrightarrow C_{r\in R} \leq D_h^{-1}\ \forall h \in H, \forall r \in R;\\
&E_{r\in R} = 0 \Leftrightarrow C_{r\in R} > D_h^{-1}\ \forall h \in H, \forall r \in R;\\
&d_{h\in H} \geq 0 \Leftrightarrow D_h^{-1} \geq \mu_{h\in H}\ \forall h \in H;\\
&d_{h\in H} = 0 \Leftrightarrow D_h^{-1} < \mu_{h\in H}\ \forall h \in H;
\end{aligned}$$

The first two conditions of Lemma 1 state that at equilibrium, all routes used between a given OD pair have equal costs and routes with higher costs will not be used. The next two conditions stipulate that travel occurs between OD pair $h, (h \in H)$ only if the marginal utility derived from travel (given by the inverse demand function) is greater than the minimum travel cost $\mu_h, (h \in H)$. Wardrop's Equilibrium Condition implies therefore that at equilibrium, no user can decrease his OD travel costs by unilaterally changing routes. For a pre-determined vector of tolls $\mathbf{x}$ and capacity enhancement levels $\mathbf{y}$, the following variational inequality (VI) can be used to restate Wardrop's Equilibrium Condition:

Find $(\mathbf{v}^*, \mathbf{d}^*) \in \Omega$ such that:

$$\mathbf{c}(\mathbf{v}^*, \mathbf{x}, \mathbf{y})^T(\mathbf{v} - \mathbf{v}^*) - \mathbf{D}^{-1}(\mathbf{d}^*)^T(\mathbf{d} - \mathbf{d}^*) \geq 0, \quad \forall(\mathbf{v}, \mathbf{d}) \in \Omega \tag{3}$$

Proposition 1. *The solution of the Variational Inequality defined by 3 results in a vector of link flows and demands ($(\mathbf{v}^*, \mathbf{d}^*) \in \Omega$) that satisfies Wardrop's Equilibrium Condition of route choice given by Lemma 1.*

For a proof of Proposition 1, see [9, 50].

2.3 Convex Optimization Reformulation

In the particular instance (and in the cases considered in this chapter) when the travel cost of using a link is dependent only on its own flow[2], there exists an equivalent convex optimization program for the above VI as given by 4-7 [2].

$$\min_{\mathbf{v},\mathbf{d}} \sum_{a\in A} \int_0^{v_a} c_a(z)\,\mathrm{d}z - \sum_{h\in H} \int_0^{d_h} D_h^{-1}(z)\mathrm{d}z. \tag{4}$$

Subject to: $((\mathbf{v}^*, \mathbf{d}^*) \in \Omega)$ where Ω is a closed and convex set defined by 5 to 7.

$$\sum_{r\in R_h} E_r = d_h, \forall h \in H. \tag{5}$$

$$v_a = \sum_{r\in R} E_r \delta_{ar}, \forall a \in A. \tag{6}$$

$$E_r, d_h \geq 0, \forall r \in R, \forall h \in H. \tag{7}$$

[2] This is known as the separability assumption.

The objective function 4 is a mathematical construct to solve for the equilibrium link and demand flow tuple that satisfies Wardrop's Equilibrium Condition [49]. In this programme, the first constraint states that the flow on each route used by each OD pair is equal to the total demand for that OD pair. The second constraint is a definitional constraint which stipulates that the flow on a link comprises flow on all routes that use that link. The last constraint restricts the equilibrium flows and demands to be non negative. As Ω is closed and convex (Ω is in fact a bounded polyhedron), the equilibrium link flows and demands $((\mathbf{v}^*, \mathbf{d}^*) \in \Omega)$ are unique [2].

3 A Model of Competition

With the above definition of the TAP from Section 2 in place, we are now able to consider the problem of a set of pre-defined P (indexed by $i = 1,2,...,P$) private firms, individually and non cooperatively, choosing the toll and the capacity enhancement levels on one and only one of P pre-defined links in a highway network.

To simplify our exposition, but without loss of generality, we have implicitly assumed that the cardinality of set P (i.e. $|P|$) is exactly equal to the cardinality of the set of $B, (B \subseteq A)$ network links which are subject to toll charges and capacity enhancements. In the ensuing discussion, we consider a given toll $(x_{i \in P})$ and capacity enhancement combination $(y_{i \in P})$ of a link to be the only strategic variables available to each of these firms. We formulate the optimization problem facing each firm before appealing to the concept of Nash Equilibrium to determine the strategic combinations offered by these firms in competition. Our problem statement follows that given in [58].

3.1 Optimization Formulation

The profit (or payoff) to firm $i, i \in P$ is the difference between the toll revenue obtained by charging a toll on the link and her investment cost of capacity enhancement, $I(y_i)$. Mathematically, the resulting choice of the strategic variables for each may be represented by the optimization problem in 8:

$$\underset{x_i, y_i}{\text{Max}} \quad \psi_i(\mathbf{x}, \mathbf{y}) = v_i(\mathbf{x}, \mathbf{y})x_i - \theta I(y_i), \forall i \in P \tag{8}$$

where v_i is obtained by solving the variational inequality (3) i.e.

$$\mathbf{c}(\mathbf{v}^*, \mathbf{x}, \mathbf{y})^T(\mathbf{v} - \mathbf{v}^*) - \mathbf{D}^{-1}(\mathbf{d}^*)^T(\mathbf{d} - \mathbf{d}^*) \geq 0, \quad \forall (\mathbf{v}, \mathbf{d}) \in \Omega$$

In 8, the scalar θ, common to all firms, allows for conversion of the enhancement costs from monetary equivalents into time value of money as we work in time units throughout this chapter.

3.1.1 Single Firm Case

If there is only a single firm (i.e. $|P| = 1$), the problem is in fact an instance of a mathematical program with equilibrium constraints (MPEC) [31]. These are hierarchical optimization problems with the key characteristic that the lower level problem describes a variational inequality (such as those in 3) defining an equilibrium in a system. In the parlance of economics, the MPEC is the equivalent of a Stackelberg [51] or "leader follower" game [16] which provides a paradigm for considering the actions of a single leader making decisions in transportation and elsewhere. This hierarchical relationship is illustrated in Figure 1.

This paradigm is developed, in accordance with Stackleberg's proposition, that the sole decision maker (acting as a leader) sets the strategic variables of the system, and the highway users follow by taking into account the firm's decisions in formulating their route choice on the network, manifesting as traffic flows $\mathbf{v}$. As we have shown in 8, the link flows have to satisfy the variational inequality and this serves as a constraint to the leader's optimization problem, fitting therefore into the definition of the MPEC. Evolutionary methods based on Genetic Algorithms [19] and Differential Evolution [52] have been proposed to solve several planning problems formulated as MPECs within transportation network analysis [25, 60].

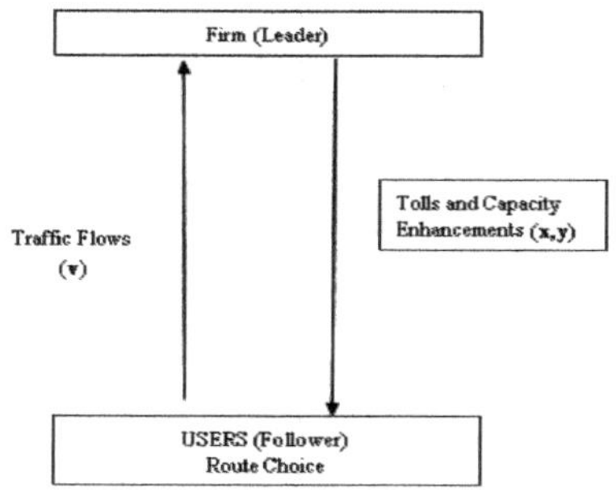

Fig. 1 Single Player Model

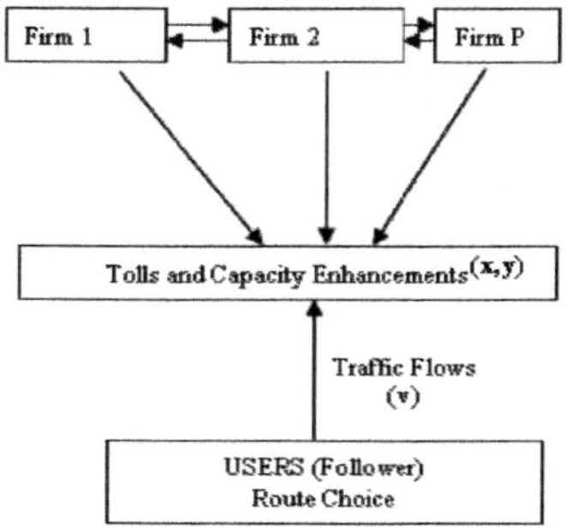

Fig. 2 Multiple Player Model

3.1.2 Extension to Multiple Firms

However, when there are multiple firms, each firm continues to take into consideration the route choices of the users as in the single firm model (i.e. the vertical relationship between leaders and followers in Figure 2) but *additionally* the choices made by other firms as required by the rationality postulate of non-cooperative behavior [18]. The latter is illustrated by the horizontal relationships between the various players at the upper level not present in the single firm model. In other words, whilst these firms play a Cournot-Nash game [8] amongst themselves, the "leader follower"relationship between these firms and the highway users still applies as in the single firm case[3].

3.2 Nash Equilibrium

In our current game context, a firm chooses its strategic combination from a set of feasible toll and capacity enhancement strategies denoted by $X_i \times Y_i$. The common strategy set of tolls and capacity enhancements across all P players may therefore be written as $X(=\prod_{i\in P} X_i)$ and $Y(=\prod_{i\in P} Y_i)$ respectively. When player $i,(i \in P)$, chooses her tolls and capacities, she is faced with the strategic choices of her competitors doing the same simultaneously. Writing her competitors' choices of strategic variables as $\mathbf{x}_{-i} = \{x_1,...,x_{i-1},x_{i+1},...,x_P\}$ and $\mathbf{y}_{-i} = \{y_1,...,y_{i-1},y_{i+1},...,y_P\}$, then a toll and capacity enhancement combination $(x_i{}^*,y_i{}^*)$ is a Nash Equilibrium if the following in 9 is satisfied:

$$\psi_i\left(x_i^*,y_i^*,\mathbf{x}_{-i}^*,\mathbf{y}_{-i}^*\right) \geq \psi_i\left(x_i,y_i,\mathbf{x}_{-i}^*,\mathbf{y}_{-i}^*\right), \forall\left(x_i,y_i\right) \in X_i \times Y_i, \forall i \in P \tag{9}$$

Equation 9 states that a Nash Equilibrium is attained when no player in the game has an incentive to deviate from her current strategy. She is therefore doing the best she can given what her competitors are doing [43].

3.2.1 Variational Inequality

A Nash Equilibrium can be represented as a VI [18, 35]. In our current context, assume the profit function ψ_i is convex in $X_i \times Y_i$ strategies when viewed as a function of (x_i,y_i) alone, then the first order optimality conditions for the optimization problem facing player i given in 8 are the following:

$$-\frac{\partial \psi_i\left(\mathbf{x}^*,\mathbf{y}^*\right)}{\partial x_i}\left(x_i - x_i^*\right) \geq 0 \tag{10}$$

and

$$-\frac{\partial \psi_i\left(\mathbf{x}^*,\mathbf{y}^*\right)}{\partial y_i}\left(y_i - y_i^*\right) \geq 0 \tag{11}$$

Let $f_x(\mathbf{x}^*,\mathbf{y}^*) = -\frac{\partial \psi_i(\mathbf{x}^*,\mathbf{y}^*)}{\partial x_i},(i \in P)$ and $f_y(\mathbf{x}^*,\mathbf{y}^*) = -\frac{\partial \psi_i(\mathbf{x}^*,\mathbf{y}^*)}{\partial y_i},(i \in P)$ then combining 10 and 11 we therefore arrive at the following VI

$$\left(\mathbf{x}-\mathbf{x}^*\right)^{\mathrm{T}} f_x(\mathbf{x}^*,\mathbf{y}^*) + \left(\mathbf{y}-\mathbf{y}^*\right)^{\mathrm{T}} f_y(\mathbf{x}^*,\mathbf{y}^*) \geq 0, \forall \mathbf{x} \in X, \mathbf{y} \in Y \tag{12}$$

Proposition 2. *The solution of the Variational Inequality defined by 12 results in strategies* $(\mathbf{x}^* \in X, \mathbf{y}^* \in Y)$ *satisfying the definition of Nash equilibrium given by 9.*

For a proof of Proposition 2, see [18][4].

[3] Thus the model is sometimes known as a Multi-Leader-Follower Game [29].

[4] Proposition 2.2, p276.

3.3 Bi-Level Variational Inequality Representation

Combining Proposition 2 given by 12 and Proposition 1 relating to the equilibrium condition of the TAP as given in 3, we therefore represent the Bi-Level Variational Inequality (BLVI) for this multi-firm game as follows:

Find $\mathbf{x} = \{x_1^*,...,x_{i-1}^*,x_i^*,x_{i+1}^*,...,x_P^*\}$ and $\mathbf{y} = \{y_1^*,...,y_{i-1}^*,y_i^*,y_{i+1}^*,...,y_P^*\}$ such that

$$(\mathbf{x}-\mathbf{x}^*)^{\mathrm{T}} f_x(\mathbf{x}^*,\mathbf{y}^*,\hat{\mathbf{v}},\hat{\mathbf{d}}) + (\mathbf{y}-\mathbf{y}^*)^{\mathrm{T}} f_y(\mathbf{x}^*,\mathbf{y}^*,\hat{\mathbf{v}},\hat{\mathbf{d}}) \geq 0, \forall \mathbf{x} \in X, \mathbf{y} \in Y \tag{13}$$

where for each $\mathbf{x} \in X$ and $\mathbf{y} \in Y$, $(\hat{\mathbf{v}},\hat{\mathbf{d}})$ is a unique solution to the following VI in 14

$$\mathbf{c}(\hat{\mathbf{v}},\mathbf{x},\mathbf{y})^T(\mathbf{v}-\hat{\mathbf{v}}) - \mathbf{D}^{-1}(\hat{\mathbf{d}})^T(\mathbf{d}-\hat{\mathbf{d}}) \geq 0, \quad \forall (\mathbf{v},\mathbf{d}) \in \Omega \tag{14}$$

3.3.1 Solution Algorithms for BLVI

The BLVI is not only applicable in transportation network analysis[5], but is also a model encountered in the deregulated electricity transmission markets [4, 22, 44] and elsewhere. The solution method proposed in these references, based primarily on [21], amounts to decomposing the problem into a series of inter-related MPECs i.e. one for each player. An outline of the method is given in Algorithm 1.

Algorithm 1. Gauss-Jacobi Fixed Point Iteration

Step 1: Set iteration counter $\delta = 0$. Select a convergence tolerance parameter, $\varepsilon(\varepsilon > 0)$. Choose a strategy for each player. Let the initial strategy set be $\mathbf{x}^\delta = \{x_1^\delta,...,x_{i-1}^\delta,x_i^\delta,x_{i+1}^\delta...x_P^\delta\}$ and $\mathbf{y}^\delta = \{y_1^\delta,...,y_{i-1}^\delta,y_i^\delta,y_{i+1}^\delta...y_P^\delta\}$. Set $\delta = \delta + 1$ and go to Step 2,

Step 2: For the ith player $i \in \{1,2,...,P\}$, solve the following optimization problem:

$$(x_i^{\delta+1},y_i^{\delta+1}) = \max_{x_i \in X, y_i \in Y} \psi_i(\mathbf{x}^\delta,\mathbf{y}^\delta) \quad i \in \{1,2,...,P\},$$

subject to

$$\mathbf{c}(\mathbf{v}^*,\mathbf{x},\mathbf{y})^T(\mathbf{v}-\mathbf{v}^*) - \mathbf{D}^{-1}(\mathbf{d}^*)^T(\mathbf{d}-\mathbf{d}^*) \geq 0, \quad \forall (\mathbf{v},\mathbf{d}) \in \Omega$$

Step 3: If $(\sum_{i=1}^{P} \left\|x_i^{\delta+1} - x_i^\delta\right\|$ and $\sum_{i=1}^{P} \left\|y_i^{\delta+1} - y_i^\delta\right\|) \leq \varepsilon$ terminate, else set $\delta = \delta + 1$ and return to Step 2.

The drawback with Algorithm 1 is that it can fall prey to being trapped in local optima [61] and this depends on the starting point assumed in Step 1.

At the time of writing, the study of BLVIs has only just begun to receive the attention of researchers. Aside from the Gauss-Jacobi Algorithm mentioned above, Mordukhovich [34] has applied tools of non smooth analysis (see [5, 46]) in order

[5] This concept has been used implicitly in modeling competition between transit operators, see[62].

to solve the BLVI. The proposed method is complex, employing advanced tools of variational analysis, and has thus far not been applied within transportation analysis. In his PhD thesis [54], Su proposed alternatively a sequential method for solving EPECs by iteratively relaxing the complementarity conditions in each player's program and solving a sequence of resulting complementarity problems. These two methods are relatively novel and could provide promise for further algorithmic developments for solving general BLVIs.

However the primary drawback of the above algorithms is their requirement for derivative information. For the specific case of transportation network analysis, it was shown [45] that the equilibrium constraint governing the route choice of users is not continuously differentiable. Additional assumptions made to satisfy the continuous differentiability requirement could possibly limit the practical applications of the algorithms to special cases. Thus methods relying on derivatives for a search direction might be problematic to apply. To overcome these constraints, we propose instead a derivative free coevolutionary particle swarm algorithm as a solution heuristic for the BLVI thus described.

4 A Coevolutionary Particle Swarm Approach

Any proposed solution method must take into account three separate but intertwined elements consistent with the modeling framework of the BLVI viz,

1. optimization for each player as given by equation 8,
2. hierarchy as based on the premise that the firms act as leaders and the highway users at the lower level take the leaders' strategic variables as given in optimizing their route choice decisions within a Stackleberg framework, and
3. intra-firm dependency consistent with the assumptions that each firm makes its decisions taking into account what its competitors are doing.

Thus the proposed algorithm is designed with a view to taking into account these three aforementioned elements. In particular,

1. the Particle Swarm Optimization (PSO) Algorithm is used as the global optimization method,
2. the evaluation process used within PSO to determine the fitness of candidate solutions has to be modified to retain the Stackleberg framework, and
3. the use of coevolution to deal with the intra-firm dependency such that strategies evolved by one player are robust to the strategies played by its competitors.

This section discusses how these elements are integrated within our proposed solution approach. A pseudo code description of our algorithm is then given.

4.1 Operation of PSO

PSO is a member of a class of Swarm Intelligence methods. It was first developed by James Kennedy and Richard Eberhart [23] inspired by the simulation of fish schools and bird flocks. Since 1995, it has gained increasing popularity due to its effectiveness in solving difficult optimization tasks, with practical applications in diverse fields which range from chemistry [7, 38], mechanical engineering [1, 17], to civil engineering [41].

Let $\mathbf{x}_{-i} = \{x_1,...,x_{i-1},x_{i+1},...,x_P\}$ and $\mathbf{y}_{-i} = \{y_1,...,y_{i-1},y_{i+1},...y_P\}$ be predetermined strategies for all players excluding player i and $\psi_i : \Re^{2P} \rightarrow \Re$ as given by the objective function in 8. The global optimization problem for player i is to find x_i^* and y_i^* such that $\psi_i(x_i^*,y_i^*,\mathbf{x}_{-i},\mathbf{y}_{-i}) \geq \psi_i(x_i,y_i,\mathbf{x}_{-i},\mathbf{y}_{-i}),\forall x_i,y_i$.

To solve this problem, the iterative PSO algorithm begins with the initialization (random generation) of the positions of a J particle, 2 dimensional swarm. Each vector of the swarm is defined as: $M_j^i = \{m_{jx}^i, m_{jy}^i\}, \forall j \in J$. Here the subscripts x and y indicate the pertinent toll and capacity enhancements for the link each particle defines, thus each particle represents a real number encoding of the potential solution for the optimization problem. Each particle is associated (also generated randomly initially) with a velocity $S_j^i = \{s_{jx}^i, s_{jy}^i\}, \forall j \in J$. The best position ever encountered by the jth particle at each iteration is denoted $Q_j^i = \{q_{jx}^i, q_{jy}^i\}, \forall j \in J$. Define δ as the iteration counter and let g be the index of the particle that attained the lowest function value ever encountered by the entire swarm up to iteration δ. Each particle is flown through the problem space and has its velocity updated by 15 [6]:

$$S_j(\delta+1) = \chi(S_j(\delta) + \alpha\phi_1(Q_j(\delta) - M_j(\delta)) + \beta\phi_2(Q_g(\delta) - M_j(\delta))), \quad \forall j \in J. \tag{15}$$

In 15, χ, the constriction coefficient used to limit velocity explosion and promote convergence, is conventionally set to 0.7298 based on the theoretical analysis of Clerc and Kennedy [6]. The scalars, α and β, are cognitive weights used to represent the attraction of a particle towards the personal and global bests respectively. Finally ϕ_1 and ϕ_2 are pseudo random numbers $\in (0,1)$. Combining the effects of these variables, the velocity update equation given by 15 highlights the fundamental operation of PSO: each of these potential solutions attempt to discover better solutions by "flying" in the search space with a velocity defined as a stochastic combination of its best ever attained position and the entire swarm's best position.

With its new velocity, its particle position can then be updated using 16 :

$$M_j(\delta+1) = M_j(\delta) + S_j(\delta+1), \quad \forall j \in J. \tag{16}$$

Once the particle position is updated, the particle is evaluated again. If the function value is better than the best value encountered by the particle so far, the best ever position for the jth particle is updated and if this value is also better than the highest function value ever encountered by the swarm up to iteration δ, the global best position Q_g is also updated. In other words, when considering player $i, i \in P$ we use 17 to update the personal best particle positions depending on the fitness obtained.

[6] We drop i superscripts from this point to reduce notational clutter and only reintroduce it when necessary to avoid confusion.

$$Q_j^i(\delta+1) = \begin{cases} Q_j^i(\delta) \text{ if } \psi_i(M_j^i(\delta+1),\mathbf{x}_{-i},\mathbf{y}_{-i}) \leq \psi_i(Q_j^i(\delta),\mathbf{x}_{-i},\mathbf{y}_{-i}) \\ M_j^i(\delta+1) \text{ if } \psi_i(M_j^i(\delta+1),\mathbf{x}_{-i},\mathbf{y}_{-i}) > \psi_i(Q_j^i(\delta),\mathbf{x}_{-i},\mathbf{y}_{-i}) \end{cases} \quad (17)$$

Now since $Q_g^i(\delta+1) \in \{Q_1^i(\delta+1),...,Q_J^i(\delta+1)\}$, we choose the global best particle positions from this set such that 18 is satisfied.

$$\psi_i(Q_g^i(\delta),\mathbf{x}_{-i},\mathbf{y}_{-i}) = \max\{\psi_i(Q_1^i(\delta),\mathbf{x}_{-i},\mathbf{y}_{-i}),...,\psi_i(Q_J^i(\delta),\mathbf{x}_{-i},\mathbf{y}_{-i})\} \quad (18)$$

This process is repeated for a number of iterations until some user specified termination criteria is satisfied, usually the maximum number of iterations.

4.1.1 Global Best, Local Best and Unified PSO

The version we have presented above is developed on the assumption that the neighborhood of the particle is defined as the entire swarm [13, 24]. This is often referred to in the literature as the "global best PSO" since in 15, the particle moves towards a combination of its own best position and the *entire* swarm's best position. To elucidate the so called "local best PSO" [13, 24], let l be the index of the particle that attained the best function value ever encountered by the jth particle's *neighborhood* up to the current iteration. The most common neighborhood topology used for this local best variant is known as the ring topology [13] where the neighbors of the jth particle are the $(j-1)^{\text{th}}$ and the $(j+1)^{\text{th}}$ particle. The local best particle position in this case is the best function value attained by either of these 3 particles. In the local best version, the velocity update is performed by 19:

$$S_j(\delta+1) = \chi(S_j(\delta) + \alpha\phi_1(Q_j(\delta) - M_j(\delta)) + \beta\phi_2(Q_l(\delta) - M_j(\delta))), \quad \forall j \in H, \quad (19)$$

with all other terms remaining as previously defined.

In the local best case we therefore have $Q_l^i(\delta+1) \in \{Q_{j-1}^i(\delta+1), Q_j^i(\delta+1), Q_{j+1}^i(\delta+1)\}$ [7] such that neighborhood analogue of 18 is satisfied. It is then trivial to recover the "global best PSO" by defining the neighborhood of the jth particle as the entire swarm [24].

An interesting variation to the basic algorithm was to synergize both the local and global search mechanisms in the Unified Particle Swarm Optimization (UPSO) Algorithm as proposed in [40]. In this model, the velocity update makes use of both the global and local versions simultaneously using 20-22 as follows:

$$G_j(\delta) = \chi(S_j(\delta) + \alpha\phi_1(Q_j(\delta) - M_j(\delta)) + \beta\phi_2(Q_g(\delta) - M_j(\delta))), \quad \forall j \in J \quad (20)$$

$$L_j(\delta) = \chi(S_j(\delta) + \alpha\phi_1(Q_j(\delta) - M_j(\delta)) + \beta\phi_2(Q_l(\delta) - M_j(\delta))), \quad \forall j \in J, \quad (21)$$

$$S_j(\delta+1) = uG_j(\delta) + (1-u)L_j(\delta), \quad \forall j \in J. \quad (22)$$

For purposes of exposition, we have used $G_j(\delta)$ and $L_j(\delta)$ as dummy variables to distinguish the global and local velocity schemes. In 22, $u \in [0,1]$ is a scalar

[7] Obvious modifications are necessary to deal with the first and last particles in the swarm.

unification factor that combines the global and local updating mechanisms of the PSO search algorithm. When u is 0, the UPSO reverts to "local best PSO" and when u is 1, the UPSO reverts to "global best PSO". Particle position updating then follows as previously i.e. using 16.

4.2 Hierarchical Evaluation Framework

The evaluation process to determine the fitness of a particle has to be developed within the Stackleberg framework (recall our discussion in Section 2) since the firms take into account the traveler's behavior for a given toll and capacity perturbation. To do this, we used the method from [25].

Given $\mathbf{x}_{-i}$ and $\mathbf{y}_{-i}$ (strategies of the competitors), a two stage process is used to evaluate the profit function and hence determine the fitness of a particle. In the first stage, the tolls and capacity enhancement vectors are input into the TAP (equations 4 - 7) to solve for the (unique) vector of link flows such that routing satisfies Wardrop's Equilibrium Condition. These link flows serve as an input into the second stage which determines the actual profits using the leaders' objective specified by 8. A summary of the process to determine the fitness of particle $j, j \in J$ for player $i, i \in P$ is given in Algorithm 2.

Algorithm 2. Two Stage Fitness Evaluation Process

Step 1: Combine m^i_{jx} and $\mathbf{x}_{-i}$ to obtain $\mathbf{x}$. Combine m^i_{jy} and $\mathbf{y}_{-i}$ to obtain $\mathbf{y}$.
Step 2: Solve the VI (3) to obtain v_i i.e. Find $(\mathbf{v}^*, \mathbf{d}^*) \in \Omega$ such that:

$$\mathbf{c}(\mathbf{v}^*, \mathbf{x}, \mathbf{y})^T (\mathbf{v} - \mathbf{v}^*) - \mathbf{D}^{-1}(\mathbf{d}^*)^T (\mathbf{d} - \mathbf{d}^*) \geq 0, \quad \forall (\mathbf{v}, \mathbf{d}) \in \Omega$$

Step 3: Evaluate $\psi^i_j(\mathbf{x}, \mathbf{y})$ ($\equiv v_i m^i_{jx} - \theta I(m^i_{jy})$) to determine the profit and hence determine the fitness of the particle j.

4.3 Coevolution

To allow for the players to consider the strategies of their competitors, all strategies are evolved simultaneously. This coevolution can be interpreted as the simultaneous adaptation of fitness from interaction between different species, which in our case corresponds to the different players within the game. The development of coevolutionary algorithms was inspired from an analogy with biological coevolution [11, 55]. In the context of nature inspired optimization algorithms, it is usual to classify coevolutionary algorithms into 2 categories: either cooperative or competitive. Extending the parallel with biological phenomenon in the natural world, the former imitates symbiosis while the latter imitates parasitism. An example of the former class of cooperative type algorithms was exploited in [42] where different species (one species for each problem dimension) were evolved to cooperatively solve an optimization problem using genetic algorithms.

It is however the latter case of competitive coevolution [47, 48] that is the focus of this chapter. In this case, the fitness of a sub-population is "contingent" (i.e. dependent) on the strategies of other species [14, 28]. Hence the key concept of "contingent" fitness evaluation is a necessary ingredient embedded in our proposed search algorithm and it is envisaged that the resulting "evolutionary arms race" [48] would lead to fitter strategies selected by players in response to strategies evolved by their competitors.

4.4 Algorithm Outline

Recall that there are P subpopulations containing J particles each. Combining all the aforementioned elements, the pseudo code describing our PSO based method to tackle the BLVI thus specified is given in Algorithm 3.

Algorithm 3. Coevolutionary Particle Swarm Algorithm

```
Step 1 Initialization: Generate Sub Populations of particles
and velocities randomly for all players decision variables.
Step 2 Randomly select one strategy vector from each player
as its Nash strategy.
Step 3 Evaluate each sub population given the Nash strategy
of others by the Two Stage Evaluation Process to identify
the global best particle; Set this as the Nash strategy for
the player.
REPEAT
 Step 4 Synchronization:
  Announce Nash Strategy to all players
  For each subpopulation i = 1 to P do
  Step 5 Re-evaluate subpopulation i by the Two Stage
  Evaluation Process given the announced Nash Strategies
  in Step 4 and  obtain personal bests and the global best
   For each particle j = 1 to J do
    Step 6 Fly particle j through problem space using the
    velocity update equation.
    Step 7 Update particle j's position using position
    update equation.
    Step 8 Evaluate new particle fitness by the Two Stage
    Evaluation Process, update personal best if fitter,
    update global best if fitter.
   Next j
  Step 9 Identify new global best particle of
  subpopulation i and set this as the Nash strategy
  for the player.
 Next i
UNTIL Termination Criteria
```

In the initialization phase of the algorithm, we randomly generate particle positions and velocities corresponding to the strategies of each player. One strategy from each subpopulation is then randomly selected as the initial Nash strategy for each player. Subsequently each subpopulation is evaluated separately to determine the profit for each player, given the above initial moves of the other players. From this process, the personal best and global best particle for each player can be identified. When all subpopulations have been evaluated, each individual player's global best particles are announced to the whole group. We term this the synchronization phase of the algorithm so that every participant is cognizant of their competitors' actions.

With the new global best strategy of all players announced, each subpopulation is once more evaluated to determine its own global best strategy in the light of the foregoing announcement. For each subpopulation, particles are then flown through the search space and the particle positions are updated using the particle swarm method. New global best positions are produced and these are announced in the next synchronization phase of the algorithm. The algorithm terminates after a number of user specified maximum number of iterations. Any variant of the PSO algorithm e.g. global, local, UPSO or even other variants (see [13]) can be employed in the velocity and position updates in Steps 6 and 7 of the algorithm. In our examples, we utilize UPSO explained earlier as it integrates the global and local search elements of the swarm.

Whilst the proposed algorithm is primarily a heuristic, our numerical examples in the following section suggest that it provides a potential solution algorithm for the BLVI.

5 Numerical Examples

The proposed coevolutionary particle swarm algorithm is applied on two problems drawn from the literature. The first example assumes that the private firms are only allowed to collect tolls for the operation of the road and ignores the costs associated with capacity enhancement. The objective therefore is one of maximizing revenue from the tolls. The second example is a literature problem given in [58] which explicitly considers capacity provision.

Table 1 PSO Parameters Used for Numerical Examples

Parameter	Value
Maximum Number of Iterations	200
Swarm Size (H)	12
Unification Factor (u)	0.5
Constriction Factor (χ)	0.7298
Cognitive Factor (α)	2.05
Cognitive Factor (β)	2.05

The parameters used in our PSO algorithm are shown in Table 1. Our reported results are based on the average of 30 runs of the algorithm with different random seeds. No attempt was made in this work to seek optimal parameters which might improve the performance of our proposed algorithm.

5.1 Example 1: Network 1

The first example is taken from [27]. The link specific parameters and the specific form of the demand functions can be found therein. This network has 18 one way links with 6 OD pairs ((Node) 1 to 5, 1 to 7, 5 to 1, 5 to 7, 7 to 1 and 7 to 5). As mentioned above, the objective of each player is to maximize toll revenue by modifying the toll level $\mathbf{x}$ and we have ignored capacity enhancement. The problem therefore is a modified form of 8 and adopts that shown in 23.

$$\underset{x_i}{\text{Max}} \quad \psi_i(\mathbf{x}) = v_i(\mathbf{x})x_i, \forall i \in P \tag{23}$$

where v_i is obtained by solving the variational inequality (3) i.e.

$$\mathbf{c}(\mathbf{v}^*,\mathbf{x})^T(\mathbf{v}-\mathbf{v}^*) - \mathbf{D}^{-1}(\mathbf{d}^*)^T(\mathbf{d}-\mathbf{d}^*) \geq 0, \quad \forall(\mathbf{v},\mathbf{d}) \in \Omega$$

Two mutually exclusive scenarios are considered for this numerical example. In Scenario A, Links 3 and 4 shown as dashed lines in Figure 3, are the only links in this network that are subject to tolls. In Scenario B, Links 7 and 10, shown in an alternative style of dashed lines, are the only links subject to tolls in the network. In both scenarios, however, there is one private firm operating on each link. The maximum allowable toll was capped at 1000 seconds. Recently, [26] employed a variant of Algorithm 1 for the solution of this problem. We compare the results reported therein with that from application of the Coevolutionary Particle Swarm Algorithm. This is shown in Table 2 and Table 3 for Scenarios A and B respectively.

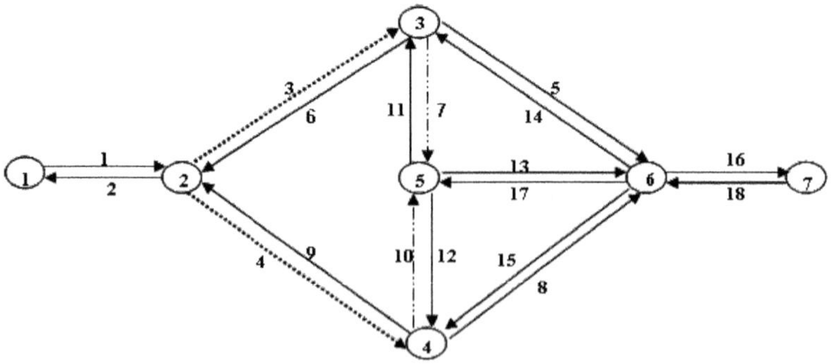

Fig. 3 Highway Network for Example 1, link numbers are indicated on links and travel is in direction indicated by arrows

Table 2 Example 1: Results of PSO algorithm on Scenario A

Firm	Link	Proposed Algorithm Toll (secs)	Revenue (secs/hr)	[26] Toll (secs)	Revenue (secs/hr)
1	3	530.55	461,861	530.63	461,882
2	4	505.62	420,242	505.65	420,293

Table 3 Example 1: Results of PSO algorithm on Scenario B

Firm	Link	Proposed Algorithm Toll (secs)	Revenue (secs/hr)[1]	[26] Toll (secs)	Revenue (secs/hr)
1	7	141.36	105,294	141.37	105,295
2	10	138.29	100,846	138.29	100,848

For Scenario A, the average revenue over 30 runs for Firm 1 and 2 are 461,861 and 420,242 with standard deviations of 0.0023 and 0.0018 respectively. In the case of Scenario B, the average revenue over 30 runs for Firm 1 and 2 are 105,294 and 100,846 with standard deviations of 0.091 and 0.033 respectively.

From Figure 3 we see that when Links 3 and 4 are both subject to tolls, there are no alternative free routes connecting Origin Node 1 to the rest of the network. On the other hand, even when both Links 7 and 10 are both subject to tolls, Link 17 continues to be free to use. This is the intuitive explanation for the much higher toll levels under Scenario A compared to Scenario B. Nevertheless, the proposed Coevolutionary Particle Swarm Algorithm converges to similar solutions to that obtained using Algorithm 1 reported in [26]. The convergence of the algorithm for a typical run under each scenario are shown in Figure 4 and Figure 5.

5.2 *Example 2: Network 2*

The 11 link network for this example is taken from [58] and shown in Figure 6 with 4 OD pairs ((node) 1 to 7, 2 to 7, 3 to 7 and 6 to 7). This model considers capacity enhancement with 3 private operators on this network each optimizing their profits ψ_i as outlined in 8. Specifically, θ, given as 0.114, is common to all players and the enhancement cost functions take the form $I(y_i) = t0_i y_i, \forall i \in P$. In other words, the enhancement cost is proportional to the free flow travel time ($t0_a, a \in B$) for each of these links. The free flow times for the 3 links (9,10 and 11) to be improved (shown as dashed lines in Figure 6) are 11,11 and 15 secs respectively.

Details of the link parameters and the functional forms of the travel demand relationships for the 4 OD pairs can be found in [58] where a heuristic gradient based

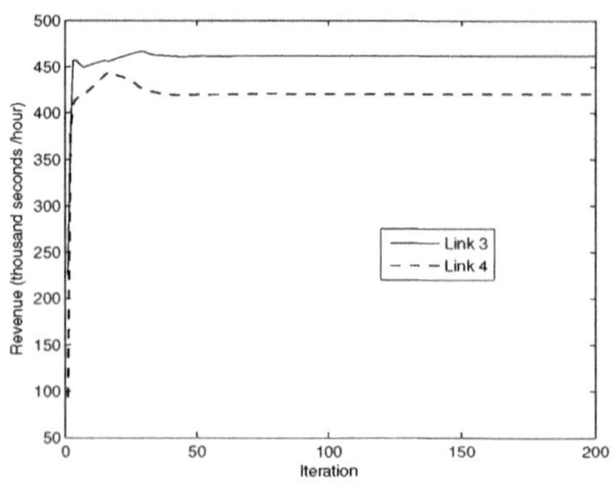

Fig. 4 Convergence of Proposed Algorithm for Scenario A, Example 1

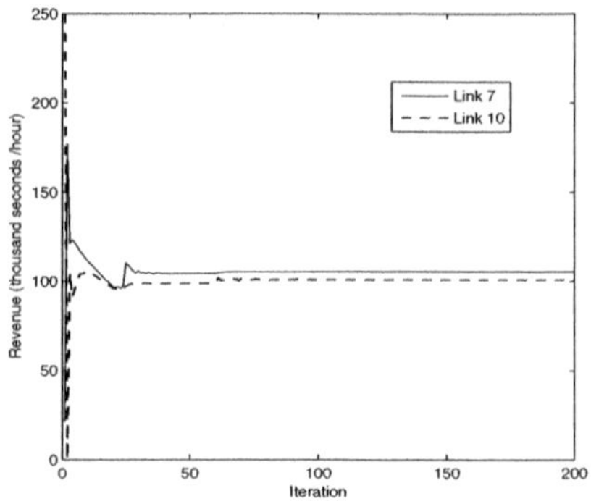

Fig. 5 Convergence of Proposed Algorithm for Scenario B, Example 1

Fig. 6 Highway Network for Example 2, link numbers are indicated on links and travel is in direction indicated by arrows

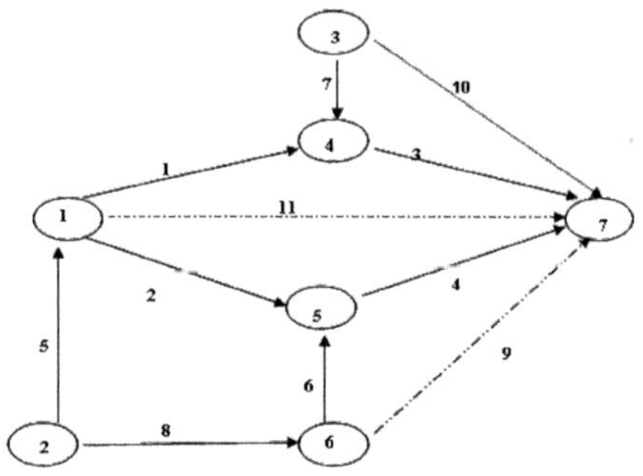

algorithm was applied. We compare the results reported therein with those produced from the Coevolutionary Particle Swarm Algorithm as summarized in Table 4.

Table 4 Example 2: Results of PSO algorithm

		Proposed Algorithm			[58]		
Firm	Link	Toll (secs)	Capacity (vehicles)	Profit (secs/hr)	Toll (secs)	Capacity (vehicles)	Profit (secs/hr)
1	9	4.52	151.64	301.93	4.52	151.60	301.43
2	10	4.76	192.90	417.53	4.76	193.04	417.14
3	11	2.97	61.42	25.98	2.97	61.88	25.92

The average profit for firms 1, 2 and 3 over 30 runs (in seconds per hour) are 301.93, 417.53 and 29.58 with standard deviations 0.0033,0.0041 and 0.0017 respectively. The convergence of the proposed PSO based algorithm is shown in Figure 7.

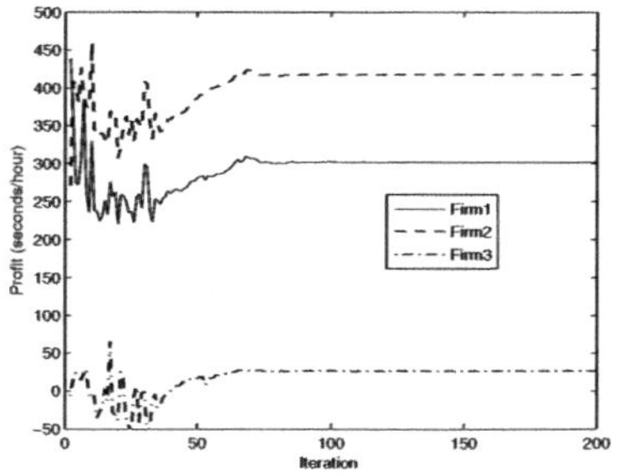

Fig. 7 Convergence of Proposed Algorithm for Example 2

6 Conclusions and Directions for Further Research

In this chapter, we presented a model of competition that arises in a highway network with private sector participants. The private firms on a road network are assumed to act as players in a Cournot-Nash game in choosing their decision variables and their actions are constrained by a variational inequality describing the equilibrium routing pattern of the highway users. The Nash equilibrium paradigm was used to model the behavior of the players in the resulting Bi-Level Variational Inequality (BLVI).

To solve the BLVI and determine the optimal choice of strategic variables of these firms, we employed a coevolutionary particle swarm algorithm to evolve strategies for each player that are robust against strategies evolved by their competitors. The traffic equilibrium constraint is achieved by solving the traffic assignment problem with a view to determining the global best strategy and during the synchronization phase of our proposed algorithm, the Nash strategy is revealed by all players simultaneously. Our algorithm embodies the principle that the fitness of a strategy for one player is contingent on the strategies revealed by all other players which is consistent with the rationality postulate of non-cooperative behavior in games. The proposed method was applied to two problems from the literature. In the examples, we have illustrated that our proposed algorithm easily obtained the solutions provided by others in their earlier research. Thus the proposed algorithm is a potentially useful method for this class of intrinsically non smooth optimization problems.

In our application of the proposed coevolutionary particle swarm algorithm, we utilized the unified particle swarm method which combined both the local and global search elements. There is no compelling reason however to prevent other variants of particle swarm optimization to be employed. Further work will study in greater effect the implications of swarm size and such parameters on convergence of the algorithm. Further research could additionally investigate the impact of these parameters on the convergence of the algorithm to the Nash equilibrium. This includes the possibility of employing fuzzy logic or other algorithms to adaptively tune the parameters of PSO as well as exploring alternative neighborhood topologies in addition to the ring topology that was adopted here.

Implicit in this chapter was the assumption that the private firms did not collude or that regulatory mechanisms prevented them from doing so. If an explicit collusion was possible, then the problem becomes one of monopoly optimization subject to the equilibrium constraint and this reverts to the familiar form of the MPEC. However if collusion was implicit, then the problem is in fact a multiobjective (i.e. vector valued) optimization problem with equilibrium constraints [32],[59]. Further research studying the application of evolutionary multiobjective optimization algorithms [10] to solve this latter category of problems is underway.

Acknowledgements

The author acknowledges the advice of colleagues Professor Mike Maher, Dr Simon Shepherd and Dr Richard Connors without implicating any of them in errors or omissions which remain the author's sole responsibility. The author is grateful to Professor Hai Yang and Dr Feng Xiao (Hong Kong University of Science and Technology) for clarifying their example. The author thanks the editors and 3 anonymous reviewers for comments on a draft version of this chapter.

References

1. Abido, M.: Optimal design of power system stabilizers using particle swarm optimization. IEEE Transactions on Energy Conversion 17(4), 406–413 (2002)
2. Beckmann, M., McGuire, C., Winsten, C.: Studies in the Economics of Transportation. Yale University Press, New Haven (1956)
3. Bureau of Public Roads, Traffic Assignment Manual. Urban Planning Division, US Dept. of Commerce, Washington DC (1964)
4. Cardell, J., Hitt, C., Hogan, W.: Market power and strategic interaction in electricity networks. Resource and Energy Economics 19(1-2), 109–137 (1997)
5. Clarke, F.H.: Optimization and Nonsmooth Analysis. Wiley, New York (1983)
6. Clerc, M., Kennedy, J.: The particle swarm - explosion, stability, and convergence in a multidimensional complex space. IEEE Transactions on Evolutionary Computation 6(1), 58–73 (2002)
7. Cockshott, A., Hartman, B.: Improving the fermentation medium for Echinocandin B production, Part II: Particle swarm optimization. Process Biochemistry 36(7), 661–669 (2001)
8. Cournot, A.A.: Recherches sur les Principes Mathematiques de la Theorie des Richesses. Hachette, Paris (1838); English trans. in: Researches into the Mathematical Principles of the Theory of Wealth. A M Kelley, New York (1971)
9. Dafermos, S.C.: Traffic Equilibrium and Variational Inequalities. Transportation Science 14(1), 42–54 (1980)
10. Deb, K.: Multi-objective Optimization using Evolutionary Algorithms. John Wiley, Chichester (2001)
11. Ehrlich, P.R., Raven, P.H.: Butterflies and plants: a study in coevolution. Evolution-International Journal of Organic Evolution 18(4), 586–608 (1964)
12. Engel, E., Fischer, R., Galetovic, A.: A New Approach to Private Roads. Regulation 25(3), 18–22 (2002)

13. Engelbrecht, A.P.: Computational Intelligence: An Introduction. John Wiley, New York (2002)
14. Ficici, S.G., Pollack, J.B.: Game theory and the Simple Coevolutionary Algorithm. In: Deb, K., Rudolph, G., Lutton, E., Merelo, J.J., Schoenauer, M., Schwefel, H.-P., Yao, X. (eds.) PPSN 2000. LNCS, vol. 1917, pp. 467–476. Springer, Heidelberg (2000)
15. Fisher, G., Babbar, S.: Private Financing of Toll Roads. RMC Discussion Paper Series, vol. 117. World Bank, Washington DC (1996)
16. Fisk, C.: Game theory and transportation systems modeling. Transportation Research Part B 18(4/5), 301–313 (1984)
17. Fourie, P., Groenwold, A.: The particle swarm optimization algorithm in size and shape optimization. Structural and Multidisciplinary Optimization 23(4), 259–267 (2002)
18. Gabay, D., Moulin, H.: On the uniqueness and stability of Nash-equilibria in non cooperative games. In: Bensoussan, A., Dorfer, P.K., Tapiero, C. (eds.) Applied Stochastic Control in Econometrics and Management Science: Contributions to Economic Analysis, vol. 130, pp. 271–293. North-Holland, Amsterdam (1980)
19. Goldberg, D.E.: Genetic Algorithms in Search, Optimization, and Machine Learning. Addison-Wesley, Boston (1989)
20. Henderson, J.M., Quandt, R.E.: Microeconomic Theory: A Mathematical Approach. McGraw Hill, Auckland (1980)
21. Harker, P.T.: A variational inequality approach for the determination of Oligopolistic Market Equilibrium. Mathematical Programmming 30(1), 105–111 (1984)
22. Hobbs, B.F., Metzler, C.B., Pang, J.S.: Strategic gaming analysis for electric power networks: An MPEC approach. IEEE Transactions on Power Systems 15(2), 638–645 (2000)
23. Kennedy, J., Eberhart, R.C.: Particle swarm optimization. In: Proceedings of the IEEE International Conference on Neural Networks, pp. 1942–1948 (1995)
24. Kennedy, J., Eberhart, R.C., Shi, Y.: Swarm Intelligence. Morgan Kaufmann Series in Artificial Intelligence, San Francisco (2001)
25. Koh, A.: Solving Transportation bi-level programs with Differential Evolution. In: Proceedings of the IEEE Congress on Evolutionary Computation, pp. 2243–2250 (2007)
26. Koh, A., Shepherd, S.P.: Tolling, Capacity Selection and Equilibrium Problems with Equilibrium Constraints. In: Proceedings 3rd Kuhmo-Nectar Conference, Free University of Amsterdam, July 3-4 (2008)
27. Koh, A., Shepherd, S.P., Sumalee, A.: Second Best Toll and Capacity Optimisation in Networks: Solution Algorithm and Policy Implications. Transportation 36(2), 147–165 (2009)
28. Koza, J.R.: Genetic evolution and coevolution of computer programs. In: Langton, C.G., Taylor, C., Farmer, J.D., Rasmussen, S. (eds.) Artificial Life II, pp. 603–629. Addison-Wesley, Redwood City (1990)
29. Leyffer, S., Munson, T.: Solving Multi-Leader-Follower Games. Preprint Argonne National Laboratory/MCS-P1243-0405 (2005)
30. Lockwood, S.: Public private partnerships in U.S. highway finance: ISTEA and beyond. Transportation Quarterly 49(l), 5–26 (1995)
31. Luo, Z.Q., Pang, J.S., Ralph, D.: Mathematical Programs with Equilibrium Constraints. Cambridge University Press, Cambridge (1996)
32. Mordukhovich, B.S.: Equilibrium problems with equilibrium constraints via multiobjective optimization. Optimization Methods and Software 19(5), 479–492 (2004)
33. Mordukhovich, B.S.: Optimization and equilibrium problems with equilibrium constraints. Omega-International Journal of Management Science 33(5), 379–384 (2005)

34. Mordukhovich, B.S.: Variational Analysis and Generalized Differentiation. II: Applications. Grundlehren Series in Fundamental Principles of Mathematical Sciences, vol. 331. Springer, Berlin (2006)
35. Nagurney, A.: Network Economics: a Variational Inequality Approach. Kluwer, Boston (1999)
36. Nash, J.: Equilibrium points in N-person games. Proceedings of the National Academy of Science 36(1), 48–49 (1950)
37. Ortuzar, J.D., Willumsen, L.: Modelling Transport. Wiley, Chichester (1998)
38. Ourique, C., Biscala, E., Pinto, J.: The use of particle swarm optimization for dynamical analysis in chemical processes. Computers and Chemical Engineering 26(12), 1783–1793 (2002)
39. de Palma, A., Lindsey, R.: Private roads: competition under various ownership regimes. Annals of Regional Science 34(1), 13–35 (2000)
40. Parsopoulos, K.E., Vrahatis, M.N.: Parameter selection and adaptation in Unified Particle Swarm Optimization. Mathematical and Computer Modelling 46(1-2), 198–213 (2007)
41. Perez, R., Behdinan, K.: Particle swarm approach for structural design optimization. Computers and Structures 85(19-20), 1579–1588 (2007)
42. Potter, M.A., De Jong, K.: A cooperative coevolutionary approach for function optimization. In: Davidor, Y., Männer, R., Schwefel, H.-P. (eds.) PPSN 1994. LNCS, vol. 866, pp. 249–257. Springer, Heidelberg (1994)
43. Pyndyck, R., Rubinfeld, S.: Microeconomics. Macmillan, New York (1992)
44. Ralph, D., Smeers, Y.: EPECs as models for electricity markets. In: Proceedings of IEEE Power Systems Conference and Exposition, pp. 74–80 (2006)
45. Robinson, S.M.: Strong regularity and the sensitivity analysis of traffic equilibria: A comment. Transportation Science 40(4), 540–542 (2006)
46. Rockafellar, R.T., Wets, J.B.: Variational Analysis. Grundlehren Series in Fundamental Principles of Mathematical Sciences, vol. 317. Springer, Berlin (1998)
47. Rosin, C.D., Belew, R.K.: Methods for Competitive Co-Evolution: Finding Opponents Worth Beating. In: Eshelman, L.J. (ed.) Proceedings of the 6th International Conference on Genetic Algorithms, Pittsburgh, USA, July 15-19, 1995, pp. 373–381 (1995)
48. Rosin, C.D., Belew, R.K.: New methods for competitive coevolution. Evolutionary Computation 5(1), 1–29 (1997)
49. Sheffi, Y.: Urban Transportation Networks: Equilibrium Analysis with Mathematical Programming Methods. Prentice Hall, Englewood Cliffs (1985)
50. Smith, M.J.: The existence, uniqueness and stability of traffic equilibria. Transportation Research Part B 13(4), 295–304 (1979)
51. von Stackelberg, H.: The Theory of the Market Economy (1934); English trans. in: Peacock, A. Oxford University Press, New York (1952)
52. Storn, R., Price, K.: Differential Evolution – A Simple and Efficient Heuristic for Global Optimization over Continuous Spaces. Journal of Global Optimization 11(4), 341–359 (1997)
53. Stough, R., Rietveld, P.: Institutional issues in transport systems. Journal of Transport Geography 5(3), 207–214 (1997)
54. Su, C.: Equilibrium Problems with Equilibrium Constraints: Stationarities, Algorithms, and Applications. PhD Thesis, Department of Management Science and Engineering, Stanford University (2005)
55. Thompson, J.N.: The Coevolutionary Process. University of Chicago Press, Chicago (1994)
56. Viton, P.: Private Roads. Journal of Urban Economics 37(3), 260–289 (1995)

57. Wardrop, J.G.: Some theoretical aspects of road traffic research. Proceedings of the Institution of Civil Engineers - Transport 1(2), 325–378 (1952)
58. Yang, H., Xiao, F., Huang, H.: Private Road Competition and Equilibrium with Traffic Equilibrium Constraints. Journal of Advanced Transportation 43(1), 21–45 (2009)
59. Ye, J.J., Zhu, Q.J.: Multiobjective optimization problems with variational inequality constraints. Mathematical Programming Series A 96(1), 139–160 (2003)
60. Yin, Y.: Genetic Algorithms-Based Approach for Bilevel Programming Models. ASCE Journal of Transportation Engineering 126(2), 115–120 (2000)
61. You, S., Baldick, R.: Hybrid coevolutionary programming for Nash equilibrium search in games with local optima. IEEE Transactions on Evolutionary Computation 8(4), 305–315 (2004)
62. Zubeita, L.: A network equilibrium model for oligopolistic competition in city bus services. Transportation Research Part B 32(6), 413–422 (1998)

Ant Colony Optimization and Its Application to the Vehicle Routing Problem with Pickups and Deliveries

Bülent Çatay

Abstract. Ant Colony Optimization (ACO) is a population-based metaheuristic that can be used to find approximate solutions to difficult optimization problems. It was first introduced for solving the Traveling Salesperson Problem. Since then many implementations of ACO have been proposed for a variety of combinatorial optimization problems. In this chapter, ACO is applied to the Vehicle Routing Problem with Pickups and Deliveries (VRPPD). VRPPD determines a set of vehicle routes originating and ending at a single depot and visiting all customers exactly once. The vehicles are not only required to deliver goods but also to pick up some goods from the customers. The objective is to minimize the total distance traversed. The chapter first provides an overview of the ACO approach. Next, VRPPD is described and the related literature is reviewed. Then, an ACO approach for VRPPD is presented. The approach proposes a new visibility function which attempts to capture the "delivery" and "pickup" nature of the problem. The performance of the approach is tested using well-known benchmark problems from the literature.

1 Introduction

Ant Colony Optimization (ACO) is a population-based metaheuristic that can be used to find approximate solutions to difficult optimization problems [16]. It was first introduced for solving the Traveling Salesperson Problem (TSP) [15, 18]. Since then many implementations of ACO have been proposed for a variety of combinatorial optimization problems such as Quadratic Assignment Problem [34], Scheduling Problems [11], Sequential Ordering Problem [21], and various Vehicle Routing Problems [5, 6, 14, 22, 30, 31].

The approach is based on the observation of the behavior of real ant colonies searching for food sources. Real ants deposit an aromatic essence, called pheromone,

Bülent Çatay
Sabanci University, Faculty of Engineering and Natural Sciences,
Tuzla, 34956 Istanbul, Turkey
e-mail: catay@sabanciuniv.edu

R. Chiong, S. Dhakal (Eds.): Nat. Intel. for Sched., Plan. and Pack. Prob., SCI 250, pp. 219–244.
springerlink.com

on the path they walk. Other ants searching for food sense that pheromone and use this information in selecting their path. The quantity of pheromone deposited on a path is based on the length of the path and the quality of the food source. As more ants follow a path the level of pheromone on that path will increase, thus increasing its selection probability by other ants. In ACO, artificial ants are used for searching good solutions to an optimization problem by taking advantage of this cooperative learning process.

In this chapter, we apply the ACO approach to the well-known Vehicle Routing Problem with Pickups and Deliveries (VRPPD). The classical Vehicle Routing Problem (VRP) involves a set of delivery customers to be serviced by a fleet of vehicles housed at a central depot. The objective of the problem is to develop a set of vehicle routes originating and terminating at the depot such that all customers are serviced, the demands of the customers assigned to each route do not violate the capacity of the vehicle that services the route, and the total distance traveled by all vehicles is minimized. VRPPD is a variant of the VRP where the vehicles are not only required to deliver goods to customers but also to pick up some goods from the customers. Customers receiving goods are called linehauls and customers sending goods are called backhauls. VRPPD may be classified into three categories: (i) *Deliveries First, Pickups Second*: the vehicles pick up goods only after they have delivered their goods; (ii) *Mixed Pickups and Deliveries*: the vehicles deliver and pick up goods in any sequence along their routes; and (iii) *Simultaneous Pickups and Deliveries*: the vehicles simultaneously deliver and pick up goods [28].

VRP with delivery first, pickup second is the first VRPPD problem introduced in the literature and is known as the VRP with Backhauls (VRPB). The reason why the vehicles have to finish delivering their load before they start picking up items may be due to the difficulty of rearranging the delivery and pickup items on the vehicles, e.g. rear loaded vehicles. However, it is also possible to perform both tasks in any order or simultaneously when the vehicle is nearly empty or is designed for both rear and side loading and unloading. Hence, several variants of this problem have been proposed over time relaxing the restriction of servicing backhaul customers after the linehauls as well as introducing multiple-depot cases. In this chapter, we consider two of these variants: Mixed VRP with Backhauls (MVRPB) and VRP with Simultaneous Pickups and Deliveries (VRPSPD). In MVRPB and VRPSPD the objective and constraints are the same as in VRPB except the servicing order of the customers, which makes the former two problems more complicated because of the fluctuating loads on the vehicle along the route.

VRPB has been extensively studied in the literature. However, the research on MVRPB is scant and VRPSPD has only recently received some attention. These two problems are more realistic and applicable to real-world situations. This chapter attacks these problems using an ant algorithm and is organized as follows: Section 2 depicts the mechanism of the ACO metaheuristic and summarizes some of the variants proposed in the literature. Section 3 is devoted to the description of VRPPD and the overview of various approaches proposed for solving the problem. Section 4 introduces an ACO approach by proposing a new visibility function and Section 5 presents the computational experiments and numerical results. Finally, concluding remarks are given in the last section.

2 Ant Colony Optimization

ACO is a metaheuristic approach designed for solving hard combinatorial optimization problems. Real ant colonies deposit pheromone on the paths they walk while searching for food sources. If other ants searching for food sense the pheromone on a path, they are likely to follow it rather than traveling at random, thus reinforcing the path. As more and more ants follow a path the level of pheromone on that path will enhance, which in turn will increase its selection probability by other ants. On the other hand, the pheromone evaporates over time, reducing the chance of other ants following the path. The longer the path between the nest and the food source the more the pheromone evaporates. Thus, the pheromone levels remain higher on the shorter paths. As a consequence, the level of pheromone laid is basically based on the path length and the quality of the food source.

The experimental setting given in Figure 1 illustrates the above described behavior of the real ants. Figure 1.(a) shows a path that has been formed by ants walking between the food source A and the nest E. When the path is cut off with an obstacle as shown in Figure 1.(b) the ants located at point B walking from A to E and those located at point D walking from E to A have to choose either the path passing through point C or the path passing through point H. Since there is no previous pheromone trail on any of the two alternative paths, the selection of either path by the first ants reaching these points is equally likely. Since the path BCD is shorter than the path BHD the ant that has selected the path through point C will arrive at point D before the ant that has selected the path through point H. Hence, an ant returning from E to A and located at point D will find a stronger trail on path DCB

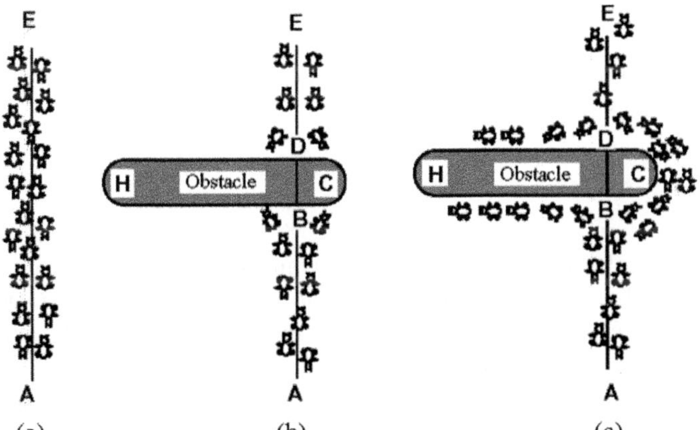

Fig. 1 The natural behavior of real ants [18]

due to the ants that have already selected that path by chance and those walking through BCD. Therefore, the selection probability of path DCB will be larger than that of path DHB. Consequently, the amount of pheromone on path BCD will increase faster than the pheromone on path BHD because of the larger number of ants following path BCD per unit time and the evaporation factor. In time, all ants will select the shorter path [18].

ACO simulates this natural behavior of real ants to solve combinatorial optimization problems by using artificial ants. To apply ACO, the optimization problem is transformed into the problem of finding the best path on a weighted graph. The artificial ants incrementally build solutions by moving on the graph using a stochastic construction process guided by artificial pheromone and heuristic information known as visibility [16]. The amount of pheromone deposited on arcs is proportional to the quality of the solution generated and increases at run-time during the computation.

The Ant System (AS) is the first ACO algorithm which was applied for solving the TSP [15, 18]. Given a number of cities and the costs of traveling from any city to any other city, TSP aims at finding the least-cost round-trip route that visits each city exactly once and then returns to the starting city. In AS, each ant probabilistically chooses the next city to visit based on a heuristic combining the distance to that city and the amount of virtual pheromone deposited on the arc to that city. The ants explore, depositing pheromone on each arc they cross, until they have all completed a tour. At this point the ant which has completed the shortest tour deposits virtual pheromone along its complete tour. The amount of pheromone deposited is inversely proportional to the tour length; i.e., the shorter the tour, the more amount of pheromone the ant deposits on the arcs of the corresponding tour.

Although AS provided competitive results its performance was still inferior in large instances compared to other algorithms specifically designed for the TSP [19]. However, its successful application has led to many extensions for various combinatorial optimization problems utilizing a similar construction mechanism. Some early applications include the elitist strategy for Ant System (EAS) [15, 18], rank-based version of Ant System (ASrank) [6], *MAX-MIN* Ant System (*MM*AS) [35], Ant Colony System (ACS) [17], and Multiple Ant Colony System (MACS) [22].

In the next section we provide a more detailed explanation of the mechanisms of AS approach and its extensions applied to the TSP.

2.1 Ant System

In AS , K artificial ants probabilistically construct tours in parallel exploiting a given pheromone model. Initially, all ants are placed on randomly chosen cities. At each iteration, each ant moves from one city to another, keeping track of the partial solution it has constructed so far. The algorithm has two fundamental components:

- The amount of pheromone on arc (i, j), τ_{ij}
- Desirability of arc (i, j), η_{ij}

where arc (i, j) denotes the connection between city i and city j.

At the start of the algorithm an initial amount of pheromone τ_0 is deposited on each arc: $\tau_{ij} = \tau_0 = K/L_0$, where L_0 is the length of an initial feasible tour and K is the number of ants. In AS, the initial tour is constructed using the nearest-neighbor algorithm; however, another TSP heuristic may be utilized as well. The desirability value (also referred to as visibility or heuristic information) between a pair of cities is the inverse of their distance $\eta_{ij} = 1/d_{ij}$, where d_{ij} is the distance between cities i and j. So, if the distance on the arc (i, j) is long, visiting city j after city i (or vice-versa) will be less desirable.

Each ant constructs its own tour utilizing a transition probability: an ant k positioned at a city i selects the next city j to visit with a probability given by

$$p_{ij}^{k} = \begin{cases} \dfrac{\tau_{ij}^{\alpha}\eta_{ij}^{\beta}}{\sum\limits_{l \in N_i^k} \tau_{il}^{\alpha}\eta_{il}^{\beta}} & \text{, if } j \in N_i^k \\ 0 & \text{, otherwise} \end{cases} \tag{1}$$

Here, N_i^k denotes the set of not yet visited cities; α and β are positive parameters to control the relative weight of pheromone information τ_{ij} and heuristic information η_{ij}. Note that $\tau_{ij}^{\alpha}\eta_{ij}^{\beta}$ is also referred to as the attractiveness and is denoted as φ_{ij}.

After each ant has completed its tour, the pheromone levels are updated. The pheromone update consists of the pheromone evaporation and pheromone reinforcement. The pheromone evaporation refers to uniformly decreasing the pheromone values on all arcs. The aim is to prevent the rapid convergence of the algorithm to a local optimal solution by reducing the probability of repeatedly selecting certain cities. The pheromone reinforcement process, on the other hand, allows each ant to deposit a certain amount of pheromone on the arcs belonging to its tour. The aim is to increase the probability of selecting the arcs frequently used by the ants that construct short tours. The pheromone update rule is the following:

$$\tau_{ij} \leftarrow (1-\rho)\tau_{ij} + \sum_{k=1}^{K} \Delta\tau_{ij}^{k}, \qquad \forall (i, j) \tag{2}$$

In this formulation, ρ $(0 < \rho \leq 1)$ is the pheromone evaporation parameter and $\Delta\tau_{ij}^{k}$ is the amount of pheromone deposited on arc (i, j) by ant k and is computed as follows:

$$\Delta\tau_{ij}^{k} = \begin{cases} \frac{1}{L^k} & \text{, if ant } k \text{ uses arc } (i, j) \text{ on its tour} \\ 0 & \text{, otherwise} \end{cases} \tag{3}$$

where L^k is the length of tour constructed by ant k.

Prior to the pheromone update a local search procedure may be applied on the tours constructed by the ants to reduce the distance traversed. It has been observed that such a procedure enhances the performance of the AS algorithm. In Figure 2 an overview of the steps of the algorithm is provided.

```
compute visibility
initialize pheromone levels
while (max number of iterations is not reached)
   for each ant
     while (not all cities are visited)
       select a not yet visited city
       update tour length and list of not yet visited cities
     end while
     perform local search
   end for
   perform pheromone update
   save the best-so-far solution
end while
```

Fig. 2 Description of AS

2.2 The Extensions of AS

In the EAS [15, 18] an elitist strategy is implemented by further increasing the pheromone levels on the arcs belonging to the best tour achieved since the initiation of the algorithm. That best-so-far tour is referred to as the "global-best" tour. The pheromone update rule is as follows:

$$\tau_{ij} \leftarrow (1-\rho)\tau_{ij} + \sum_{k=1}^{K} \Delta\tau_{ij}^{k} + w\Delta\tau_{ij}^{gb}, \qquad \forall(i,j) \tag{4}$$

Here, w denotes the weight associated with the global-best tour and $\Delta\tau_{ij}^{gb}$ is the amount of pheromone deposited on arc (i, j) by the global-best ant and calculated by the following formula:

$$\Delta\tau_{ij}^{gb} = \begin{cases} \frac{1}{L^{gb}} & \text{, if global best ant uses arc } (i,j) \text{ on its tour} \\ 0 & \text{, otherwise} \end{cases} \tag{5}$$

where L^{gb} is the length of global-best tour.

In the AS_{rank} [6] a rank-based elitist strategy is adopted in an attempt to prevent the algorithm from being trapped in a local minimum. In this strategy, w best-ranked ants are used to update the pheromone levels and the amount of pheromone deposited by each ant decreases with its rank. Furthermore, at each iteration, the global-best ant is allowed to deposit the largest amount of pheromone. The update rule is the following:

$$\tau_{ij} \leftarrow (1-\rho)\tau_{ij} + \sum_{r=1}^{w-1} (w-r)\Delta\tau_{ij}^{r} + w\Delta\tau_{ij}^{gb}, \qquad \forall(i,j) \tag{6}$$

The ACS presented in [17] attempts to improve AS by increasing the importance of exploitation versus exploration of the search space. This is achieved by employing a strong elitist strategy to update pheromone levels and a pseudo-random proportional rule in selecting the next node to visit. The strong elitist strategy is applied by using the global-best ant only to increase the pheromone levels on the arcs that belong to the global-best tour:

$$\tau_{ij} \leftarrow (1-\rho)\tau_{ij} + \rho\Delta\tau_{ij}^{gb}, \qquad \forall(i,j) \tag{7}$$

The mechanism of the pseudo-random proportional rule is as follows: an ant k located at customer i may either visit its most favorable customer or randomly select a customer. The selection rule is the following:

$$j^k = \begin{cases} \arg\max\limits_{j \in N_i^k} \tau_{ij}^{\alpha}\eta_{ij}^{\beta} & \text{, if } z \le z_0 \\ J^k & \text{, otherwise} \end{cases} \tag{8}$$

where z is a random variable drawn from a uniform distribution $U[0,1]$ and z_0 ($0 \le z_0 \le 1$) is a parameter to control exploitation versus exploration. J^k is selected according to the probability distribution (1). ACS also uses local pheromone updating while building solutions: as soon as an ant moves from city i to city j the pheromone level on arc (i, j) is reduced in an attempt to promote the exploration of other arcs by other ants. The local pheromone update is performed as follows:

$$\tau_{ij} \leftarrow (1-\xi)\tau_{ij} + \xi\tau_0 \tag{9}$$

where ξ is a positive parameter less than 1.

Similar to ACS, *MM*AS [35] uses either the global-best ant or the iteration-best ant alone to reinforce the pheromone. It has been observed that using iteration-best ant at the start of the algorithm and then gradually increasing the frequency of using the global-best ant for the pheromone update improves the performance. However, this strategy may cause a rapid convergence to a sub-optimal solution. Thus, maximum and minimum limits on the pheromone levels are imposed to avoid stagnation. The interval in which the pheromone may vary is set to $[\tau_{min}, \tau_{max}]$. The pheromone levels are initialized at τ_{max} to allow the exploration of the search space at the beginning. In addition, the pheromone levels are reinitialized whenever the system approaches stagnation or no improvement has been achieved after a number of consecutive iterations.

Gambardella et al. [22] developed a multiple ACS (MACS) for solving the VRP with Time Windows (VRPTW). VRPTW has two objectives: to minimize the number of vehicles used and the total tour time. The former is considered to be the primary objective, i.e. a solution with less number of vehicles but longer travel time is preferred over a solution with more vehicles but shorter travel time. MACS attempts to minimize both objectives simultaneously by using two parallel ant colonies. The first colony, named as ACS-VEI, reduces the number of vehicles while the second, named as ACS-TIME, minimizes the total tour time by using the number of vehicles provided by ACS-VEI. Although the two ant colonies run in parallel they use independent pheromone trails.

The interested reader is referred to [19] for more details on ACO metaheuristic and its variants.

3 Vehicle Routing Problem with Pickups and Deliveries

In this section, we first describe VRPPD and present a 0-1 mixed integer linear programming model following the formulation of [13]. We next review the existing literature on MVRPB and VRPSPD.

3.1 Problem Description

VRPPD deals with a single depot distribution/collection system servicing a set of customers by means of a homogeneous fleet of vehicles, i.e. all vehicles have the same capacity. The customers may require two types of service: a delivery and/or a pickup. Products to be delivered are loaded at the depot and products picked up are transported back to the depot. The objective is to find the set of vehicle routes servicing all the customers with the minimum total distance. A maximum route length restriction may be imposed on the vehicles.

In VRPB, each customer has either a delivery or a pickup demand to be satisfied and the vehicle services the linehaul customers first. The main reasoning behind visiting linehaul customers before backhaul customers is the fact that linehaul customers have precedence over backhaul customers in many real world cases and vehicles are often rear loaded. The latter causes problems when rearranging the items on the vehicle, thus preventing the mixed routes and simultaneous pickup and delivery. However, the improved design of vehicles allows side loadings, making the mixed routes a more practical option since that would provide shorter routes. Thus, in MVRPB, each customer has *either a delivery or a pickup* demand and backhaul and linehaul customers may be visited in any order. On the other hand, servicing the customers in any order but not allowing simultaneous pickup and delivery is not practical and realistic in many real world situations. As a result, VRPSPD was proposed where each customer has *both a delivery and a pickup* demand and both services are performed simultaneously. Although the customers can be visited in any order along the route in both problems they must be serviced exactly once.

From a practical point of view VRPPD models situations such as distribution of bottled drinks, chemicals, LPG tanks, laundry service of hotels, etc. where the customers are typically visited for a double service. In the case of the distribution of the bottled drinks for instance, full bottles are delivered to customers and empty ones are brought back either for re-use or for recycling. In the distribution of chemicals case, some hazardous materials may need to be returned for safe disposal. Regulations or environmental issues may also force companies to take responsibility for their products throughout their lifetime and to collect them.

In VRPB, the loads of linehaul customers and backhaul customers can be checked separately during the delivery route and pickup route, respectively, to ensure that the vehicle capacity is not exceeded. In MVRPB, however, the decrease or increase on the vehicle load at each customer must be checked depending on whether the customer is a linehaul or backhaul customer, respectively. Similarly, in VRPSPD, the net change (decrease or increase) on the vehicle load at each customer must be monitored. Therefore, in these two problems the vehicle capacity must not be exceeded at any arc along the route.

Since VRPB is out of the scope of this study we omit further discussion on the problem and refer the interested reader to [4, 23, 36] for details.

3.2 Problem Formulation

Mathematically, VRPSPD is described by a set of homogenous vehicles V, a set of customers J, and a complete undirected graph $G(N, A)$. The graph consists of $n+1$ vertices where the customers are denoted by 1, 2, ..., n and the depot is represented by the vertex 0. $A = \{(i, j)\colon i, j \in N, i \neq j\}$ denotes the set of arcs that represents connections between the depot and the customers and among the customers. A cost (time, distance) c_{ij} is associated with each arc (i, j). Each vehicle has capacity Q and each customer (node) i is characterized by its geographical location and its delivery and pickup requests D_i and P_i, respectively. Finally, Q, D_i, P_i, and c_{ij} are assumed to be non-negative integers. The VRPSPD determines a set of paths (routes) such that:

1. each vehicle travels exactly one route;
2. each customer is visited only once by one of the vehicles completely satisfying its demand and supply;
3. the load carried by a vehicle between any pair of adjacent customers on the route must not exceed its capacity; and
4. total distance given by the sum of the arcs belonging to these routes is minimal.

In addition, a maximum route length (maximum time) restriction may be imposed on the vehicles.

Following the model in [13] the 0-1 mixed integer linear programming model of VRPSPD can be formulated as follows:

Decision Variables

L_j load of vehicle after having serviced customer $j \in J$

z_j subtour elimination variable

$$x_{ijv} = \begin{cases} 1 & \text{, if vehicle } v \text{ travels directly from customer } i \text{ to } j \\ 0 & \text{, otherwise} \end{cases}$$

Mathematical Model

$$\text{Minimize} \quad z = \sum_{i\in N}\sum_{j\in N}\sum_{v\in V} c_{ij}x_{ijv} \tag{10}$$

Subject to

$$\sum_{j\in N}\sum_{v\in V} x_{ijv} = 1 \qquad i \in J \tag{11}$$

$$\sum_{i\in N}\sum_{v\in V} x_{ijv} = 1 \qquad j \in J \tag{12}$$

$$\sum_{i\in N} x_{ikv} - \sum_{j\in N} x_{kjv} = 0 \qquad k \in J, v \in V \tag{13}$$

$$L_j \geq \sum_{i\in N}\sum_{j\in J} D_j x_{ijv} - D_j + P_j - M\left(1 - x_{0jv}\right) \quad j \in J, v \in V \tag{14}$$

$$L_j \geq L_i - D_j + P_j - M\left(1 - \sum_{v\in V} x_{ijv}\right) \qquad i,j \in J, i \neq j \tag{15}$$

$$\sum_{i\in N}\sum_{j\in J} D_j x_{ijv} \leq Q \qquad v \in V \tag{16}$$

$$L_j \leq Q \qquad j \in J \tag{17}$$

$$z_j \geq z_i + 1 - n\left(1 - \sum_{v\in V} x_{ijv}\right) \qquad i,j \in J, i \neq j \tag{18}$$

$$z_j \geq 0 \qquad j \in J \tag{19}$$

$$x_{ijv} \in \{0,1\} \qquad i,j \in N, v \in V \tag{20}$$

where M is a sufficiently large number (e.g. $M = \max\left\{\sum_{j\in N_c}(D_j + P_j), \sum_{i\in N}\sum_{\substack{j\in N\\ j\neq i}} C_{ij}\right\}$).

The objective function (10) minimizes the total distance traveled. Constraint sets (11) and (12) assure servicing each customer exactly once. Constraint (13) makes sure that if a vehicle arrives at a customer, then the same vehicle departs from it. The load after servicing the first customer is defined with constraint (14) while the load "en route" is limited with constraint (15). Constraint sets (16) and (17) ensure that the load when leaving the depot and "en route", respectively, does not exceed the vehicle capacity. Constraint (18) is the subtour elimination constraint. Constraint (19) is the non-negativity constraint and constraint (20) defines the binary variables.

MVRPB may be considered as a special case of VRPSPD in which some of the customers require only delivery service while the remaining customers require only pickup service. In other words, we define J_L as the set of linehaul customers, $J_L = \{j: j\in J, D_j > 0, P_j = 0\}$, J_B as the set of backhaul customers, $J_B = \{j: j \in J, D_j = 0, P_j > 0\}$, and $J = J_L \cup J_B$. Therefore, the above VRPSPD model also formulates MVRPB where $P_j = 0$ for $j \in J_L$ and $D_j = 0$ for $j \in J_B$.

We can prove that VRPSPD and MVRPB are *NP*-hard in the following way: Let $J_B = \oslash$. Then MVRPB reduces to VRP, which is known to be *NP*-hard. Hence,

MVRPB is also *NP*-hard. Since MVRPB is a special case of VRPSPD, VRPSPD is *NP*-hard as well.

3.3 Literature Review

Although research on VRPSPD has recently gained momentum, there are only a few papers attacking MVRPB. Golden et al. [24] developed an algorithm which inserts backhaul customers into the routes formed by the linehaul customers. The algorithm utilizes a penalty factor which considers the number of linehaul customers left on the route after the insertion point.

Casco et al. [7] proposed a load-based insertion procedure where the insertion cost for backhaul customers is determined based on the remaining load to be delivered along the route of the vehicle. Salhi and Nagy [33] modified this method by proposing the cluster insertion of backhauls to solve both MVRPB and VRPSPD. In their problem structure, nodes are represented as disjoint delivery or pickup nodes; thus repetitive servicing is allowed. Salhi and Nagy also investigated the case with multiple depots. Nagy and Salhi [28] improved their previous approach using several heuristics in which they first find a solution to the VRP by allowing infeasibilities then modify this solution to make it feasible for the MVRPB and VRPSPD. The proposed approach is capable of solving both single- and multi-depot problems.

Wade and Salhi [38] proposed an ant algorithm which uses the ACS approach of Dorigo and Gambardella [17]. However, the computational results were rather poor compared to those in the literature. Wade and Salhi [39] further enhanced their ant algorithm by using different mechanisms, and hence, improved their earlier results. Recently, Ropke and Pisinger [32] developed a unified heuristic for a large class of VRPPD based on a large neighborhood search (LNS). The proposed heuristic provides competitive results for both MVRPB and VRPSPD.

VRPSPD was first introduced by Min [26] as a book distribution and collection problem between a central library and 22 remote libraries in Ohio using two vehicles. Min utilizes a cluster-first route-second approach and solves the TSP to optimality as sub-problems. Halse [25] proposed a cluster-first route-second approach for VRPSPD as well as for several other variants of VRPPD. In this approach the nodes are first distributed to vehicles and then the problem is solved using the 3-opt algorithm. Halse also utilizes Lagrangean relaxation and column generation techniques and discusses the results for single depot instances with 22 to 150 customers.

Angelelli and Mansini [2] addressed the VRPSPD with time windows constraints. They developed a branch-and-price approach based on a set covering formulation for the master problem. A relaxation of the elementary shortest path problem with time windows and capacity constraints is used as the pricing problem. Branch-and-bound is applied to obtain integer solutions.

Dethloff [13] presented insertion-based heuristics using four different criteria. He developed 40 instances to test his algorithm. He also compared his results with those of Salhi and Nagy [33] and reported an improvement on Min's problem. Vural et al. [37] reported improvements on the results of Dethloff problems employing a dual genetic algorithm approach. In this approach first tours are created and partitioned into sub-tours, then a local search is performed, and finally crossover and mutation operations are executed.

Crispim and Brandão [12] proposed a hybrid tabu search (TS)-variable neighborhood search approach. The approach uses the sweep algorithm to construct the initial solution and then performs arc exchanges in the TS procedure. If it is faced with any overloads it exchanges the order of the customers on the route until it achieves feasibility. For the improvement phase it uses "insert" and "swap" moves by penalizing the overloads. Other TS approaches for VRPSPD include Chen and Wu [8] which presented an insertion-based procedure followed by a hybrid heuristic based on the record-to-record travel, tabu lists, and improvement procedures; Montané and Galvão [27] which developed a TS algorithm using "insert", "exchange", "split and splice" (on two routes), and 2-opt; Bianchessi and Righini [3] which proposed constructive algorithms, local search algorithms, and TS algorithms to obtain approximate solutions fast; and Wassan et al. [40] which designed a reactive tabu search metaheuristic. The last approach provides competitive results.

In what follows we propose an ACO algorithm for efficiently solving MVRPB and VRPSPD and test its performance against the approaches presented in the above mentioned papers.

4 An Ant Algorithm for VRPPD

An initial solution is first obtained using the nearest-neighbor heuristic: start at the depot and then select the not yet visited closest feasible customer as the next customer to be visited. A customer is feasible if visiting her next does not violate the capacity constraint (and the maximum route length restriction, if any). If no feasible customer is available then the route is terminated at the depot and a new route is initiated. The procedure is repeated until all customers are serviced. This solution is used to initialize the pheromone trails on the arcs as follows: $\tau_0 = n/L_0$, where L_0 is the length of the nearest-neighbor heuristic solution. Note that this heuristic does not guarantee feasibility if a limit on the number of vehicles is imposed.

4.1 Heuristic Information

In the classical ant approaches developed for solving TSP and VRP the visibility value between a pair of customers is the inverse of their distance. On the other hand, [6, 14] employed the savings function as the visibility function for solving the VRP. While the latter utilized the classical Clarke and Wright savings function ([10]) the former used the Paessens' parametrical savings function ([29]). The classical savings function calculates the savings in distance achieved by serving two customers i and j on the same route instead of serving them on different routes using the following formula:

$$S_{ij} = (c_{0i} + c_{i0} + c_{0j} + c_{j0}) - (c_{0i} + c_{ij} + c_{j0}) = c_{i0} + c_{0j} - c_{ij} \tag{21}$$

where c_{i0} (c_{0j}) is the distance of customer i (j) to the depot and c_{ij} represents the distance between the customer i and j. Since a high value of savings indicates that visiting customer j after customer i is a desired choice the tour length is expected to be shorter if the probability of moving from customer i to customer j increases with the savings value.

Paessens' formulation aims at collecting more information about the distribution of the customers in an attempt to avoid the circumferenced formation of routes. The proposed parameterized formulation is as follows:

$$S_{ij} = c_{i0} + c_{0j} - \lambda c_{ij} + \mu \left| c_{i0} - c_{0j} \right| \tag{22}$$

where λ and μ are non-negative constants.

In our approach, the visibility function consists of two components. The first component is an enhanced savings function developed by [20]:

$$S_{ij} = c_{i0} + c_{0j} - \lambda c_{ij} + \mu cos(\theta_{ij}) \left| c_{0i} - c_{j0} \right| \tag{23}$$

where θ_{ij} is the angle formed by the two rays originating from the depot and crossing the customers i and j, and λ and μ are non-negative parameters. The proposed savings function (23) can be regarded as a more general enhancement to Paessens' savings formulation and was shown to perform better on various problem sets.

The second component takes into consideration the load of the vehicle on its route. This component is equal to the largest of the ratio of delivery to customer j to the average value of all deliveries and the ratio of pickup from customer j to the average value of all pickups if total deliveries or total pickups so far have exceeded half of the vehicle capacity; and is equal to 1 otherwise. The idea is to basically give more chance of selection to customers requiring larger delivery or pickup quantities. Our motivation in doing so stems from the "put first larger items" approach used in [1]. The reason why we start employing this approach after half of the vehicle capacity is used up is to let the first component determine the selection of the customers at the early stages of the route construction in an attempt to not adversely affect the influence of this heuristic information on building a shorter route. In other words, the first component acts as a primary heuristic information whereas the second component starts playing a role after we have already constructed a partial tour using only the distance criterion. The computation of the second visibility value is as follows:

$$R_j = \begin{cases} \max\left(\frac{P_j}{\overline{P}}, \frac{D_j}{\overline{D}}\right) & \text{, if } \min\left(\sum\limits_{k \in V_q} P_k, \sum\limits_{k \in V_q} D_k\right) > \frac{Q}{2} \\ 1 & \text{, otherwise} \end{cases} \tag{24}$$

Here, $\overline{D}$ ($\overline{P}$) is the average delivery (pickup) and V_q is the set of customers already visited by the associated vehicle q. Note that the first component is static whereas the second depends on the current load of the vehicle. The visibility function is then the following:

$$\eta_{ij} = S_{ij} \times R_j \tag{25}$$

4.2 Route Construction

The route construction process uses the pseudo-random proportional rule of ACS depicted in Section 2.2. In addition, a candidate list is used in selecting a customer to visit, i.e. N_{ik} consists of a number of customers that have not been visited yet and have the largest attractiveness values.

After an ant has constructed its tour, a local search is performed in an attempt to further improve the solution. In our algorithm we use the "swap" and "move" procedures sequentially. In swap two customers are exchanged whereas in move a customer is removed and inserted into another arc. These procedures are applied both within routes and between different routes.

4.3 Pheromone Update

Our pheromone update consists of a rank-based *MMAS* strategy. In this strategy, w best-ranked ants of each iteration along with the best-so-far ant are used to update the pheromone trails. The pheromone reinforcement of each ant is proportional to its rank. Our pheromone update rule is as follows:

$$\tau_{ij} \leftarrow (1-\rho)\tau_{ij} + \sum_{r=1}^{w}(w-r)\Delta\tau_{ij}^{r} + w\Delta\tau_{ij}^{gb} \tag{26}$$

Here, $\Delta\tau_{ij}^{r}=1/L^{r}$ for all arcs (i, j) belonging to the tour built by the rth best ant where L^{r} is the length of the corresponding tour. gb denotes the global-best ant. If the pheromone level on any arc drops below an explicit lower limit or exceeds an explicit upper limit it is set equal to that limit. In other words, if any $\tau_{ij} < \tau_{min}$ $(\tau_{ij} > \tau_{max})$ then $\tau_{ij} = \tau_{min}$ $(\tau_{ij} = \tau_{max})$. The aim in using this *MMAS* approach is to reduce the risk of a premature convergence.

5 Experimental Study

The proposed algorithm is coded using C++ and executed on an Intel Pentium T2130 1.86 GHz processor with 1 Gb RAM. The parameters in the savings function are $\lambda = \mu = 1$. The parameters of the ACO algorithm are set according to initial experimental runs as: $z_0 = 0.7, \alpha = 1, \beta = 4, \rho = 0.1, \tau_{max} = n/\rho L^{gb}$, and $\tau_{min} = \rho\tau_{max}/50$. For consistency, the same parameter values are used for solving both VRPSPD and MVRPB. The number of best ants used for the pheromone reinforcement and the size of the candidate list used in the selection of the next customer to be visited are proportional to the number of ants and the number of customers, respectively, and their values are set to $w = n/10$ and $s = n/5$, respectively. Since setting the number of ants equal to the number of customers has been observed to perform well in the literature (see e.g. [19]) we also adopted the same strategy. For each problem instance we performed 10 runs, each carried on for 100 iterations.

Table 1 Results for the VRPSPD data set of Dethloff [13]

Problem	**MG**	**ACO Avg**	**ACO Best**	**%Gap**
SCA3-0	640.55	640.47	636.06	**-0.71**
SCA3-1	697.84	708.59	700.50	0.38
SCA3-2	659.34	662.72	659.86	0.08
SCA3-3	680.04	685.13	680.04	0.00
SCA3-4	690.50	691.26	690.50	0.00
SCA3-5	659.90	673.27	662.75	0.43
SCA3-6	653.81	661.08	653.69	**-0.02**
SCA3-7	659.17	668.83	659.17	0.00
SCA3-8	719.70	724.62	720.06	0.05
SCA3-9	681.00	687.91	682.33	0.20
SCA8-0	981.47	992.60	978.10	**-0.34**
SCA8-1	1077.44	1085.45	1079.92	0.23
SCA8-2	1050.98	1058.43	1046.20	**-0.46**
SCA8-3	983.34	1027.30	1006.59	2.31
SCA8-4	1073.46	1098.11	1077.01	0.33
SCA8-5	1047.24	1081.44	1067.29	1.88
SCA8-6	995.59	1004.30	990.44	**-0.52**
SCA8-7	1068.56	1088.35	1079.20	0.99
SCA8-8	1080.58	1100.73	1086.72	0.57
SCA8-9	1084.80	1091.19	1074.40	**-0.97**
CON3-0	631.39	619.30	617.98	**-2.17**
CON3-1	554.47	562.89	558.69	0.76
CON3-2	522.86	522.55	519.11	**-0.72**
CON3-3	591.19	591.63	591.19	0.00
CON3-4	591.12	597.08	590.49	**-0.11**
CON3-5	563.70	572.44	564.88	0.21
CON3-6	506.19	504.51	501.34	**-0.97**
CON3-7	577.68	591.48	585.51	1.34
CON3-8	523.05	524.51	523.14	0.02
CON3-9	580.05	591.57	588.84	1.49
CON8-0	860.48	873.67	861.40	0.11
CON8-1	740.85	772.33	753.81	1.72
CON8-2	723.32	729.46	725.54	0.31
CON8-3	811.23	852.86	835.77	2.94
CON8-4	772.25	795.59	775.67	0.44
CON8-5	756.91	781.35	769.20	1.60
CON8-6	678.92	701.69	695.43	2.37
CON8-7	814.50	819.04	811.96	**-0.31**
CON8-8	775.59	795.40	775.56	0.00
CON8-9	809.00	834.63	826.95	2.17
Average	764.25	776.64	767.58	0.39

5.1 Benchmark Problems and Results for the VRPSPD

The performance of the algorithm for VRPSPD is tested using two well-known benchmark problem sets from the literature. The first problem set was proposed by Salhi and Nagy [33] based on the 14 VRP problems proposed in [9]. The number of customers in this data varies from 50 to 199 and problems CMT6-10 and CMT13-14 impose a maximum route length restriction for the vehicles. For each VRP instance in [9], Salhi and Nagy generated a VRPSPD problem by splitting the original demand between demand and pickup loads. Another instance was obtained by exchanging these demand and pickup loads of every other customer. Thus, two classes of problems were generated and referred to as problem classes X and Y.

The second problem set was presented by Dethloff [13] where random instances with 50 customers were generated considering two different geographical scenarios: In scenario SCA, the coordinates of the customers are uniformly distributed over the interval [0,100]. In scenario CON, half of the customers are distributed in the same way as in SCA while the coordinates of the other half are uniformly distributed over the interval [100/3,200/3]. The delivery demand D_j of the customers is uniformly distributed over the interval [0,100]. The pickup demand P_j are computed by using a random number r_j that is uniformly distributed over the interval [0,1] such that P_j $=(0.5+ r_j)D_j$. Instances with different vehicle capacities were generated by choosing the minimal number of vehicles μ. Then, the corresponding capacity was set to $C = \sum_{s\in J} D_s/\mu$ where μ was chosen to be 3 or 8.

Table 1[1] compares the best-so-far distances for Dethloff's problems to the average and best distance values found using ACO. Avg and Best columns denote the average distance and best distance, respectively. %Gap column shows the percentage difference between the best-so-far distances and ACO distances and calculated as (Best-so-far/ACO Best)-1. We note that only Ropke and Pisinger [32] and Montané and Galvão [27] utilize this data set in their experiments; however, the former only reports the average results. Thus, we compare our results to those of [27] which are better than those of [13] in all instances. We observe that the ACO improves the solutions of 10 problems and matches 5. Although the overall performance of [27] is better the average gap is only 0.39%. On the other hand, ACO requires more computational effort: 18 seconds versus 3.7 seconds (using Athlon 2.0 GHz processor).

Table 2 provides a comparison of the best solutions published by Dethloff [13], Chen and Wu [8], Montané and Galvão [27], Wassan et al. [40] and those obtained by ACO for Salhi and Nagy problems. In this table column n denotes the number of customers and Veh is the number of vehicles. Note that Nagy and Salhi [28] and Ropke and Pisinger [32] reported only their average results (the former for all problems and the latter for class X only). Hence, we cannot make a detailed comparison.

However, we include their results in Table 3 where we make a comparison of the average solutions. Note also that none of the results in [12] is any better than those published in the literature. Thus, we do not use them as benchmarks. Furthermore,

[1] In all the tables presented in this section, the first letter of the names of the corresponding authors are used for referencing.

Table 2 Comparison of results for the VRPSPD data set of Salhi and Nagy [33]

		D		CV		MG		WWN		ACO	
Problem	***n***	**Veh**	**Best**	**Veh**	**Best**	**Veh**	**Best**	**Veh**	**Best**	**Veh**	**Best**
CMT1X	50	3	501	3	478.59	3	472	3	**468.30**	3	479.94
CMT2X	75	7	782	6	688.51	7	695	6	**668.77**	6	707.87
CMT3X	100	5	847	5	744.77	5	**721**	4	729.63	5	729.27
CMT4X	150	7	1050	7	887.00	7	880	7	**876.50**	7	914.65
CMT5X	199	11	1348	10	1089.22	11	1098	9	**1044.51**	11	1133.92
CMT6X	50	6	584	-	-	-	-	6	**556.06**	6	556.68
CMT7X	75	11	961	-	-	-	-	11	903.05	11	**901.22**
CMT8X	100	9	928	-	-	-	-	9	879.60	9	**865.51**
CMT9X	150	15	1299	-	-	-	-	15	1220.00	**14**	**1184.34**
CMT10X	199	19	1571	-	-	-	-	19	1464.58	**18**	**1444.90**
CMT11X	120	4	959	4	**858.57**	4	900	4	861.97	4	898.35
CMT12X	100	6	804	6	678.46	6	675	5	**644.70**	6	678.08
CMT13X	120	11	**1576**	-	-	-	-	12	1647.51	11	1596.01
CMT14X	100	10	871	-	-	-	-	10	823.95	10	**821.75**
CMT1Y	50	3	501	3	480.78	3	470	3	**458.96**	3	475.37
CMT2Y	75	7	782	6	679.44	7	700	6	**663.25**	6	699.89
CMT3Y	100	5	847	5	723.88	5	**719**	4	745.46	5	733.82
CMT4Y	150	7	1050	7	**852.35**	7	878	7	870.44	7	894.11
CMT5Y	199	11	1348	10	1084.27	10	1083	9	**1054.46**	10	1112.66
CMT6Y	50	6	584	-	-	-	-	6	558.17	6	**555.43**
CMT7Y	75	11	961	-	-	-	-	11	903.36	11	**901.22**
CMT8Y	100	9	936	-	-	-	-	10	917.42	9	**865.50**
CMT9Y	150	15	1299	-	-	-	-	15	1213.11	**14**	**1184.05**
CMT10Y	199	19	1571	-	-	-	-	18	**1419.79**	18	1437.07
CMT11Y	120	4	1070	5	859.77	5	910	4	**830.39**	4	933.59
CMT12Y	100	5	825	6	676.23	6	689	6	**659.52**	5	**676.21**
CMT13Y	120	11	**1576**	-	-	-	-	11	1647.04	11	1597.03
CMT14Y	100	10	871	-	-	-	-	10	823.34	10	**821.75**
Average			1010.79						912.64		921.43

Table 3 Comparison of the average results for the VRPSPD data set of Salhi and Nagy [33]

Problem	**Type**	**D**	**NS**	**CV**	**RP**	**MG**	**WWN**	**ACO**
CMT-X	**no distance restriction**	899	-	775	-	777	792	756
	distance restriction	1113	-	-	-	-	1053	1071
	All	1006	991	-	919	-	922	914
CMT-Y	**no distance restriction**	918	-	765	-	778	789	755
	distance restriction	1114	-	-	-	-	1052	1069
	All	1016	989	-	-	-	921	912

since the maximum route length constraints were not considered in [8, 27] we only compare the unrestricted problems. The results show that ACO improves 10 best-so-far distances. In addition, ACO finds best-so-far distance and number of vehicles in

Table 4 Improvements on the best-so-far solutions for the VRPSPD data set of Salhi and Nagy [33]

	Best-so-far			ACO	
Problem	**Reference**	**Veh**	**Dist**	**Veh**	**Dist**
CMT7X	WWN	11	903.05	11	**901.22**
CMT8X	WWN	9	879.60	9	**865.51**
CMT9X	WWN	15	1220.00	**14**	**1184.34**
CMT10X	WWN	19	1464.58	**18**	**1444.90**
CMT14X	WWN	10	823.95	10	**821.75**
CMT6Y	WWN	6	558.17	6	**555.43**
CMT7Y	WWN	11	903.36	11	**901.22**
CMT8Y	WWN	10	917.42	10	**865.50**
CMT9Y	WWN	15	1213.11	**14**	**1184.05**
CMT12Y	WWN	6	659.52	**5**	676.21
CMT14Y	WWN	10	823.34	10	**821.75**

problems CMT9X, CMT10X, and CMT9Y. Although the distance is not improved in CMT12Y, the best-so-far distance is obtained using 5 vehicles. The improvements on the best-so-far solutions are summarized in Table 4. We also observe in Table 2 that the average gap in the performance of ACO is inferior to that of Wassan et al. [40] but the difference is less than 1%. On the other hand, we note that ACO requires more computational effort: an average of 615 seconds versus 221 seconds in [40] using UltraSPARC-IIIi 1062 MHz Solaris 9.

5.2 Benchmark Problems and Results for the MVRPB

To test the performance of our algorithm for MVRPB we used the benchmark problems generated by Salhi and Nagy [33] based on 14 VRP problems proposed in [9] and the problems proposed by Goetschalckx and Jacobs-Blecha [23]. For each VRP instance Salhi and Nagy generated three MVRPB problems replacing every second, forth, and tenth delivery customer with a backhaul customer and assigning a pickup quantity equal to its original delivery quantity. Thus, three classes of problems were generated for 50%, 25%, and 10% of backhauls, respectively, and were referred to as problem classes H, Q, and T, respectively. The data set of [23] consists of 63 instances with the number of customers varying from 25 to 150.

Since only Salhi and Nagy [33] reported the individual results we base our comparisons on those results in Table 5. The results show that the performance of ACO is significantly better. However, the average computation time in [33] is less than 4 seconds on a VAX 4000-500 computer whereas our average is 672 seconds. If we investigate the average results of each type of data given in Table 6, we see that ACO outperforms the more recent results of Nagy and Salhi [28]; however, the average results of Ropke and Pisinger [32] are better.

Table 5 Results for the MVRPB data set of Salhi and Nagy [33]

Problem	SN	ACO			%Gap
		Avg	Best	Veh	
CMT1T	541	522.66	520.93	5	-3.71
CMT2T	839	817.50	802.75	9	-4.32
CMT3T	903	831.36	826.20	7	-8.51
CMT4T	1111	1045.28	1023.20	11	-7.90
CMT5T	1423	1306.66	1295.34	15	-8.97
CMT6T	571	556.66	555.43	6	-2.73
CMT7T	-	906.92	903.05	11	-
CMT8T	911	869.81	865.54	9	-4.99
CMT9T	1164	1201.22	1181.34	14	1.49
CMT10T	1418	1469.37	1447.59	18	2.09
CMT11T	1075	1060.00	1042.46	7	-3.03
CMT12T	827	818.65	804.89	9	-2.67
CMT13T	1600	1608.66	1587.17	11	-0.80
CMT14T	866	840.18	829.76	10	-4.18
CMT1Q	557	497.64	492.79	4	-11.53
CMT2Q	860	756.82	745.64	8	-13.30
CMT3Q	918	778.05	764.88	6	-16.68
CMT4Q	1164	969.08	947.19	9	-18.63
CMT5Q	1477	1228.70	1213.41	13	-17.85
CMT6Q	594	558.86	556.68	6	-6.28
CMT7Q	-	903.37	900.69	11	-
CMT8Q	918	871.30	865.50	9	-5.72
CMT9Q	1178	1202.60	1187.16	14	0.78
CMT10Q	1477	1461.64	1452.52	18	-1.66
CMT11Q	1075	990.50	976.04	6	-9.21
CMT12Q	843	755.57	747.94	7	-11.28
CMT13Q	1613	1610.48	1588.68	11	-1.51
CMT14Q	873	827.75	823.11	10	-5.71
CMT1H	594	469.17	465.02	3	-21.71
CMT2H	873	672.76	664.82	6	-23.85
CMT3H	915	748.75	735.43	4	-19.63
CMT4H	1164	896.41	875.15	7	-24.82
CMT5H	1509	1095.15	1080.69	9	-28.38
CMT6H	594	558.40	556.68	6	-6.28
CMT7H	-	901.28	901.22	11	-
CMT8H	915	867.82	865.51	9	-5.41
CMT9H	1164	1200.56	1191.53	14	2.37
CMT10H	1509	1458.58	1446.11	18	-4.17
CMT11H	1120	868.44	852.62	4	-23.87
CMT12H	850	686.47	671.51	5	-21.00
CMT13H	1546	1616.59	1598.75	11	3.41
CMT14H	866	827.69	821.75	10	-5.11
Average	1036.28	955.60	944.63		-8.85

Table 6 Comparison of the average results for the MVRPB data set of Salhi and Nagy [33]

Problem Set	SN	NS	RP	ACO
10% (T)	2008	1011	955	978
25% (Q)	2050	1034	922	947
50% (H)	2088	1045	881	909

Table 7 compares our average distances to those of Wade and Salhi [39]. In this table, n_1 and n_2 columns denote the number of linehaul and backhaul customers, respectively, and column Q denotes the vehicle capacity. The results show that our ACO approach outperforms the ant system algorithm of [39] in 28 instances out of 46. The average improvement is 0.22.

In Table 8, we report our best distances and compare them to the best distances reported by Halse [25] and Wade and Salhi [39]. The results reveal that ACO improves 17 best-so-far solutions. In addition, ACO finds best-so-far distance and number of vehicles in problems f1, l4, and h1. We also observe that ACO provides very competitive results: the average improvement of ACO on Wade and Salhi's results is 0.03% and on Halse's results is 0.65%. On the other hand, ACO's performance is surprisingly poor on problem instance j4.

5.3 Sensitivity of the Results to the Parameter Setting

In the initial experimental study that we conducted for determining suitable parameter values we observed that the algorithm was robust in the sense that fairly good solutions could be obtained for varying parameter values. This is mostly due to the contribution of the local search strategy to the overall solution quality. As mentioned in Section 4.2, we apply the local search at each iteration after each ant has constructed its tour. This exhaustive local search procedure has a significant contribution in achieving relatively short distances fast. To illustrate the effect of the local search mechanism we provide in Figure 3 the results for 5 sample runs of a medium size problem instance with 100 customers, namely CMT3X. This figure also shows the convergence pattern with regard to the 100 iterations. Note that all parameter values are kept same as in the earlier experimental study. We see that the local search procedure provides a significant improvement in the total distance traversed.

To investigate the robustness of the proposed method with respect to the parameter set we show the variations in the solutions of problem CMT3X for various parameter values in Figure 4. The charts (a)-(d) in this figure report the average distance of 10 runs for different values of the (a) heuristic information parameter β, (b) evaporation rate ρ, (c) pseudo-random proportional rule parameter z_0, and (d) number of best ants used for the pheromone update w, respectively — ceteris paribus. We observe that z_0 and w have more effect on the solution quality compared to β and ρ: the maximum deviations from the best distance are 2.6% and 3.4% in the case of

Table 7 Comparison of the average distances of MVRPB data set of Goetschalckx and Jacobs-Blecha [23]

Problem	n_1	n_2	Q	WS Avg	ACO Avg	%Gap
a1	20	5	1550	223374	225002	0.73
a2	20	5	2550	169797	169500	**-0.17**
a3	20	5	4050	142126	142034	**-0.06**
b1	20	10	1600	234514	231667	**-1.21**
b2	20	10	2600	179760	181391	0.91
b3	20	10	4000	145702	145702	0.00
c1	20	20	1800	240126	237979	**-0.89**
c2	20	20	2600	197287	199002	0.87
c3	20	20	4150	165710	166858	0.69
d1	30	8	1700	307383	309454	0.67
d3	30	8	2750	224598	223616	**-0.44**
d4	30	8	4075	-	184171	-
e1	30	15	2650	225927	223172	**-1.22**
e2	30	15	4300	191342	191240	**-0.05**
e3	30	15	5225	184621	184136	**-0.26**
f1	30	30	3000	251364	244671	**-2.66**
f3	30	30	4400	218818	212939	**-2.69**
f4	30	30	5500	202280	199993	**-1.13**
g1	45	12	2700	311057	304590	**-2.08**
g2	45	12	4300	236623	235416	**-0.51**
g3	45	12	5300	215133	213154	**-0.92**
g5	45	12	6400	203777	204068	0.14
g6	45	12	8000	190572	190128	**-0.23**
h1	45	23	4000	241850	239730	**-0.88**
h2	45	23	5100	220532	215634	**-2.22**
h4	45	23	6100	208693	206046	**-1.27**
h5	45	23	7100	203647	199589	**-1.99**
i1	45	45	3000	336501	332321	**-1.24**
i2	45	45	4000	281864	283803	0.69
i3	45	45	5700	241102	245197	1.70
j1	75	19	4400	341272	337791	**-1.02**
j2	75	19	5600	301779	300712	**-0.35**
j3	75	19	8200	261264	265131	1.48
j4	75	19	6600	278866	284264	1.94
k1	75	38	4100	370134	369437	**-0.19**
k2	75	38	5200	327324	326044	**-0.39**
k4	75	38	6200	308065	302954	**-1.66**
l1	75	75	4000	418644	412875	**-1.38**
l2	75	75	5000	377962	374397	**-0.94**
l4	75	75	6000	339020	350558	3.40
m1	100	25	5200	381006	382016	0.26
m3	100	25	6200	354797	345619	**-2.59**
m4	100	25	8000	311410	312310	0.29
n1	100	50	5700	387926	392479	1.17
n3	100	50	6600	349555	366509	4.85
n5	100	50	8500	332754	336372	1.09
Average				263064	260906	**-0.22**

Table 8 Comparison of results for the MVRPB for the data set Goetschalckx and Jacobs-Blecha [23]

	ACO		WS			H		
Problem	**Best**	**Veh**	**Best**	**Veh**	**%Gap**	**Best**	**Veh**	**%Gap**
a1	223088	8	223088	8	0.00	227725	8	-2.04
a2	169500	5	169500	5	0.00	169497	5	0
a3	142034	3	142034	3	0.00	142032	3	0
b1	229403	7	233001	7	-1.54	233950	7	-1.94
b2	179194	4	179258	4	-0.04	182326	4	-1.72
b3	145702	3	145702	3	0.00	145699	3	0
c1	233087	7	239192	7	-2.55	242931	7	-4.05
c2	197278	5	196883	5	0.20	197276	5	0
c3	164891	3	164891	3	0.00	167663	4	-1.65
d1	307110	11	307110	11	0.00	307875	11	-0.25
d3	220751	7	224598	7	-1.71	222195	7	-0.65
d4	182496	5	-	-	-	-	-	-
e1	220742	7	223774	7	-1.35	222518	7	-0.80
e2	191135	4	190559	4	0.30	190048	4	0.57
e3	183197	4	182804	4	0.21	187793	4	-2.45
f1	243496	6	248333	7	-1.95	254977	6	-4.50
f3	210629	4	217317	5	-3.08	215575	5	-2.29
f4	198709	4	200964	4	-1.12	203448	5	-2.33
g1	298882	10	301235	10	-0.78	304106	10	-1.72
g2	234242	6	235920	6	-0.71	235220	6	-0.42
g3	212841	5	214534	5	-0.79	213757	5	-0.43
g5	202282	4	203233	4	-0.47	202610	4	-0.16
g6	188696	3	189922	3	-0.65	201875	4	-6.53
h1	237631	6	241619	6	-1.65	235269	6	1.00
h2	213756	5	220305	5	-2.97	215649	5	-0.88
h4	203441	4	208412	4	-2.39	202971	4	0.23
h5	197875	3	203193	4	-2.62	201896	4	-1.99
i1	326706	10	327168	10	-0.14	329237	10	-0.77
i2	279396	7	278727	7	0.24	289501	7	-3.49
i3	240974	5	238626	5	0.98	244782	5	-1.56
j1	331175	10	332471	10	-0.39	337800	10	-1.96
j2	296918	8	292698	8	1.44	298432	8	-0.51
j3	260619	6	259243	6	0.53	280070	7	-6.95
j4	280140	7	261066	7	7.31	257895	6	8.63
k1	365664	10	360954	10	1.30	361287	10	1.21
k2	322150	8	323979	8	-0.56	320012	8	0.67
k4	297570	7	298518	7	-0.32	296766	7	0.27
l1	405412	10	416167	11	-2.58	412278	11	-1.67
l2	371012	8	360018	8	3.05	362399	8	2.38
l4	343574	7	337620	7	1.76	341304	7	0.67
m1	377846	10	370920	10	1.87	372840	11	1.34
m3	343133	9	335486	9	2.28	336011	9	2.12
m4	307990	7	310567	7	-0.83	305118	7	0.94
n1	386780	10	370690	10	4.34	385978	10	0.21
n3	357552	9	349516	9	2.30	352992	9	1.29
n5	329462	7	323698	7	1.78	319811	7	3.02
Average	257743		259011		**-0.03**	260698		**-0.65**

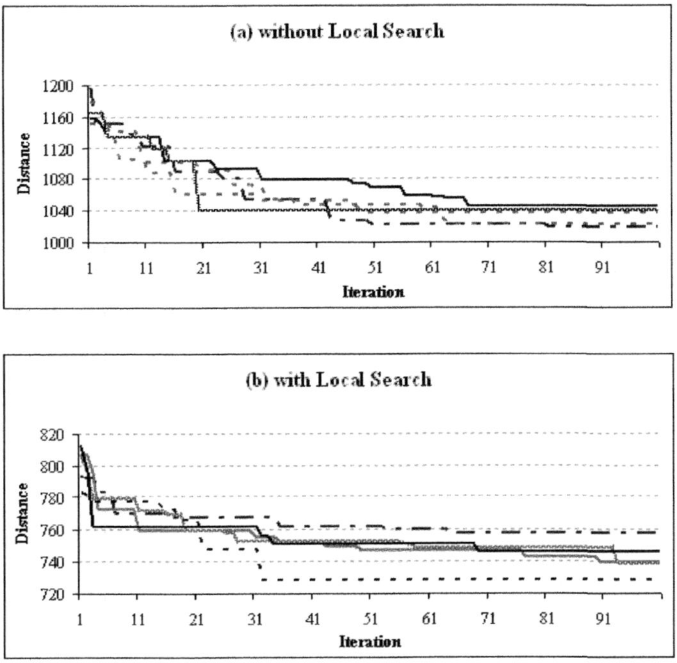

Fig. 3 Sample results of 5 runs for problem CMT3X (a) without local search, (b) with local search

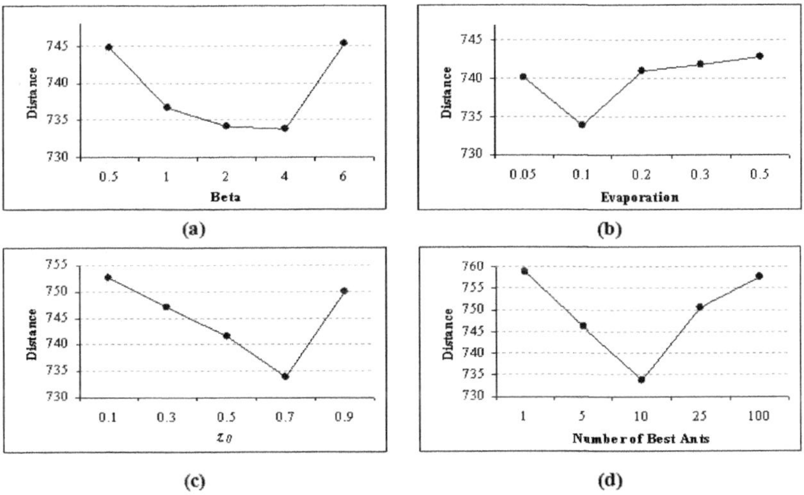

Fig. 4 Average distance of 10 runs of CMT3X for different parameter values of the (a) heuristic information parameter β, (b) evaporation rate ρ, (c) pseudo-random proportional rule parameter z_0, and (d) number of best ants used for the pheromone update w

z_0 and w, respectively, whereas the maximum deviations are 1.2% and 1.6% in the case of β and ρ. Note that these solutions are obtained by varying the value of one parameter only and keeping the remaining parameters at the already determined values. As expected, changing the values of multiple parameters simultaneously would lead to inferior solutions. These results reveal that a different selection of parameter values would not deteriorate the solution quality much due to the contribution of the local search. Nevertheless, the best distances are achieved by using the values we employed in our experimental study: $\beta = 4, \rho = 0.1, z_0 = 0.7$, and w = 100/10=10.

6 Conclusions

In this chapter, we addressed two types of VRPPD, namely MVRPB and VRPSPD which have a growing practical relevance in the reverse logistics literature. The computational complexity of these problems necessitates good heuristic solution procedures. We developed an ant colony algorithm equipped with a new visibility function in an attempt to capture the delivery and pickup nature of these two problems. The experimental analysis reveals promising results compared to the results published in the literature. Furthermore, improvements on some of the best-so-far solutions are obtained as well.

Although a fair comparison of the computational efforts cannot be made because of the use of different processors we observed that our computation times are fairly long compared to other heuristics presented in the literature. We observed that the local search procedure takes a significant amount of time. Thus, further research may focus on reducing the computation times, for instance by decreasing the ant colony size, using a candidate list, and/or by carrying out an elitist local search using only some best-performing ants. More efficient pheromone trail update procedures and visibility functions specific to the problem type may also be investigated to improve the solution quality.

References

1. Altınel, İ.K., Öncan, T.: A new enhancemant of the Clarke and Wright savings heuristic for the capacitated vehicle routing problem. Journal of the Operational Research Society 56, 954–961 (2005)
2. Angelelli, E., Mansini, R.: The vehicle routing problem with time windows and simultaneous pick-up and delivery. In: Klose, A., Speranza, M.G., Van Wassenhove, L.N. (eds.) Quantitative approaches to distribution logistics and supply chain management series. Lecture Notes in Economics and Mathematical Systems, vol. 519, pp. 249–267. Springer, Berlin (2002)
3. Bianchessi, N., Righini, G.: Heuristic algorithms for the vehicle routing problem with simultaneous pick-up and delivery. Computers and Operations Research 34, 578–594 (2007)
4. Brandão, J.: A new tabu search algorithm for the vehicle routing problem with backhauls. European Journal of Operational Research 173, 540–555 (2006)

5. Bullnheimer, B., Hartl, R.F., Strauss, C.: Applying the ant system to the vehicle routing problem. In: Voss, S., et al. (eds.) Meta-heuristics: Advances and trends in local search paradigms for optimization, pp. 285–296. Kluwer Academic Publishers, Boston (1998)
6. Bullnheimer, B., Hartl, R.F., Strauss, C.: An improved ant system algorithm for the vehicle routing problem. Annals of Operations Research 89, 319–328 (1999)
7. Casco, D.O., Golden, B.L., Wasil, E.A.: Vehicle routing with backhauls: models, algorithms, and case studies. In: Golden, B.L., Assad, A.A. (eds.) Vehicle routing: Methods and studies, pp. 127–147. Elsevier, Amsterdam (1988)
8. Chen, J.F., Wu, T.H.: Vehicle routing problem with simultaneous deliveries and pickups. Journal of the Operational Research Society 57, 579–587 (2006)
9. Christofides, N., Mingozzi, A., Toth, P.: The vehicle routing problem. In: Christofides, N., et al. (eds.) Combinatorial optimization, pp. 315–338. Wiley, Chichester (1979)
10. Clarke, G., Wright, W.: Scheduling of vehicles from a central depot to a number of delivery points. Operations Research 12, 568–581 (1964)
11. Colorni, A., Dorigo, M., Maniezzo, V., Trubian, M.: Ant system for job-shop scheduling. Belgian Journal of Operations Research, Statistics and Computer Science 34, 39–45 (1994)
12. Crispim, J., Brandão, J.: Metaheuristics applied to mixed and simultaneous extensions of vehicle routing problems with backhauls. Journal of the Operational Research Society 56, 1296–1302 (2005)
13. Dethloff, J.: Vehicle routing and reverse logistics: the vehicle routing problem with simultaneous delivery and pick-up. OR Spektrum 23, 79–96 (2001)
14. Doerner, K.F., Gronalt, M., Hartl, R.F., Reimann, M., Strauss, C., Stummer, M.: SavingsAnts for the vehicle routing problem. In: Cagnoni, S., Gottlieb, J., Hart, E., Middendorf, M., Raidl, G.R. (eds.) EvoWorkshops 2002. LNCS, vol. 2279, pp. 11–20. Springer, Heidelberg (2002)
15. Dorigo, M.: Optimization, learning and natural algorithms (in Italian), PhD Thesis, Dipartimento di Elettronica, Politecnico di Milano, Italy (1992)
16. Dorigo, M.: Ant colony optimization. Scholarpedia 2(3), 1461 (2008)
17. Dorigo, M., Gambardella, L.M.: Ant colony system: A cooperative learning approach to the traveling salesman problem. IEEE Transactions on Evolutionary Computation 1, 53–66 (1997)
18. Dorigo, M., Maniezzo, V., Colorni, A.: The ant system: Optimization by a colony of cooperating agents. IEEE Transactions on Systems, Man, and Cybernetics-Part B 26, 29–41 (1996)
19. Dorigo, M., Stützle, T.: Ant colony optimization. MIT Press, Cambridge (2004)
20. Doyuran, T., Çatay, B.: Two enhanced savings functions for the Clark-Wright algorithm. In: Blecker, T., Kersten, W., Gertz, C. (eds.) Management in logistics networks and nodes: Concepts, technology and applications. Series on Operations and Technology Management, vol. 8, pp. 245–258. Erich Schmidt Verlag, Berlin (2008)
21. Gambardella, L.M., Dorigo, M.: Ant colony system hybridized with a new local search for the sequential ordering problem. INFORMS Journal on Computing 12, 237–255 (2000)
22. Gambardella, L.M., Taillard, E., Agazzi, G.: MACS-VRPTW: a multiple ant colony system for vehicle routing problems with time windows. In: Corne, D., Dorigo, M., Glover, F. (eds.) New ideas in optimization, pp. 63–76. McGraw-Hill, London (1999)
23. Goetschalckx, M., Jacobs-Blecha, C.: The vehicle routing problem with backhauls. European Journal of Operational Research 42, 39–51 (1989)
24. Golden, B., Baker, A.J., Schaffer, J.: The vehicle routing problem with backhauling: two approaches. In: Hammesfahr, R.D. (ed.) Proceedings of the 21st Annual Meeting of S.E., TIMS, Myrtle Beach, USA, pp. 90–92 (1985)

25. Halse, K.: Modeling and solving complex vehicle routing problems. PhD Thesis, Department of Mathematical Modeling, Technical University of Denmark (1992)
26. Min, H.: The multiple vehicle routing problem with simultaneous delivery and pick-up points. Transportation Research 23A, 377–386 (1989)
27. Montané, F.A.T., Galvão, R.D.: A tabu search algorithm for the vehicle routing problem with simultaneous pick-up and delivery service. Computers and Operations Research 33, 595–619 (2006)
28. Nagy, G., Salhi, S.: Heuristic algorithms for single and multiple depot vehicle routing problems with pickups and deliveries. European Journal of Operational Research 162, 126–141 (2005)
29. Paessens, H.: The savings algorithm for the vehicle routing problem. European Journal of Operational Research 34(3), 336–344 (1988)
30. Reimann, M., Doerner, K.F., Hartl, R.F.: Insertion based ants for vehicle routing problems with backhauls and time windows. In: Dorigo, M., Di Caro, G.A., Sampels, M. (eds.) Ant Algorithms 2002. LNCS, vol. 2463, pp. 135–148. Springer, Heidelberg (2002)
31. Reimann, M., Doerner, K., Hartl, R.F.: Analyzing a unified ant system for the VRP and some of its variants. In: Raidl, G.R., Cagnoni, S., Cardalda, J.J.R., Corne, D.W., Gottlieb, J., Guillot, A., Hart, E., Johnson, C.G., Marchiori, E., Meyer, J.-A., Middendorf, M. (eds.) EvoWorkshops 2003. LNCS, vol. 2611, pp. 300–310. Springer, Heidelberg (2003)
32. Ropke, S., Pisinger, D.: A unified heuristic for vehicle routing problems with backhauls. European Journal of Operational Research 171, 750–775 (2006)
33. Salhi, S., Nagy, G.: A cluster insertion heuristic for single and multiple depot vehicle routing problems with backhauling. Journal of the Operational Research Society 50, 1034–1042 (1999)
34. Stützle, T., Dorigo, M.: ACO algorithms for the quadratic assignment problem. In: Corne, D., Dorigo, M., Glover, F. (eds.) New ideas in optimization, pp. 33–50. McGraw-Hill, London (1999)
35. Stützle, T., Hoos, H.H.: The MAX-MIN ant system and local search for the traveling salesman problem. In: Bäck, T., Michalewicz, Z., Yao, X. (eds.) Proceedings of the 1997 IEEE International Conference on Evolutionary Computation (ICEC 1997), pp. 309–314. IEEE Press, Piscataway (1997)
36. Toth, P., Vigo, D.: An exact algorithm for the vehicle routing problem with backhauls. Transportation Science 31, 372–385 (1997)
37. Vural, A.V., Çatay, B., Eksioglu, B.: A dual GA approach to capacitated vehicle routing problem with simultaneous pick-up and deliveries. In: Proceedings of the IIE Research Conference (CD-ROM), Atlanta, GA (2005)
38. Wade, A., Salhi, S.: An ant system algorithm for the vehicle routing problems with backhauls. In: De Sousa, J.P. (ed.) Proceedings of the 4th Metaheuristics International Conference (MIC 2001), Porto, Portugal, pp. 99–203 (2001)
39. Wade, A., Salhi, S.: An ant system algorithm for the mixed vehicle routing problem with backhauls. In: Resende, M.G., de Sousa, J.P. (eds.) Metaheuristics: Computer decision-making, pp. 699–719. Kluwer, New York (2003)
40. Wassan, N.A., Wassan, A.H., Nagy, G.: A reactive tabu search algorithm for the vehicle routing problem with simultaneous pickups and deliveries. Journal of Combinatorial Optimization 15, 368–386 (2008)

Evolutionary and Ant Colony Optimization Based Approaches for a Two-Dimensional Strip Packing Problem

Carolina Salto, Guillermo Leguizamón, Enrique Alba, and Juan M. Molina

Abstract. In the last few years, metaheuristic approaches have shown an important development in many application areas. This situation has turned them into one of the more appropriate candidates when dealing with difficult real-world problems for which timely, good-quality solutions are necessary. Furthermore, the class of metaheuristic approaches includes a large number of variants and designs which mainly depend on the concepts from which they are inspired. This chapter aims at giving an overview of Evolutionary Algorithms and Ant Colony Optimization when applied to the two-dimensional strip packing problem. The respective performance of these two metaheuristics are analyzed and compared from different perspectives by implementing a Genetic Algorithm and an Ant Colony System.

1 Introduction

The term metaheuristic involves *solution methods that orchestrate an interaction between local improvement procedures and higher level strategies to create a process capable of escaping from local optima and performing a robust search of a solution space* (see Glover and Kochenberg [17]). This definition is probably one

Carolina Salto
Facultad de Ingeniería - Universidad Nacional de La Pampa, Calle 110 y 9,
General Pico, La Pampa (6360), Argentina
e-mail: saltoc@ing.unlpam.edu.ar

Guillermo Leguizamón
LIDIC - Universidad Nacional de San Luis, Ejército de Los Andes 950,
San Luis (5700), Argentina
e-mail: legui@unsl.edu.ar

Enrique Alba and Juan M. Molina
E.T.S.I. Informática, Universidad de Málaga, 29071 Málaga, España
e-mail: {eat, jmmb}@lcc.uma.es

R. Chiong, S. Dhakal (Eds.): Nat. Intel. for Sched., Plan. and Pack. Prob., SCI 250, pp. 245–266.
springerlink.com

of the more widely accepted and used by the research community in the field of optimization; and certainly, there are a variety of methods and techniques that perfectly fit under the above definition. However, metaheuristics can also be classified in different ways, *e.g.*, according to their source of inspiration for the main ideas incorporated as part of their design.

One important class of metaheuristics are those whose main source of inspiration is the nature, *e.g.*, annealing processes, insects' behavior, evolutionary processes, swarm behavior, just to name a few. From this class of metaheuristics, the development achieved by the Evolutionary Algorithms (EAs) [5, 30], which have been successfully applied to different problems and domains, is remarkable. On the other hand, a more recent but similarly successful metaheuristic is that represented by the Ant Colony Optimization (ACO) [15] which is inspired by the behavior of some species of real ant colonies. EAs and ACO algorithms are clear examples of population-based metaheuristics, however, they differ with respect to the manner in which each of them explore the search space. EAs can be considered an instance-based approach whereas ACO algorithms follow a model-based approach. For the above metaheuristics we have respectively chosen the following specific algorithms for further description: Genetic Algorithms (GAs) and Ant Colony System (ACS).

Certainly, there exist a large number of real-world problems for which near-optimal solutions can be achieved by means of metaheuristics. Our main interest here is in the class of problems that are particularly crucial for some industrial application, more precisely the cutting and packing (CP) problems which usually represent a real challenge when looking to achieve an adequate administration of highly constrained resources as seen in paper and textile industries. In this work, we particularly consider the so-called two-dimensional strip packing problem (2SPP).

According to the above considerations, this chapter aims at presenting a general overview of EAs and ACO metaheuristics when applied to a class of CP problems. In the literature it is hard to find such comparison for the problem under consideration, besides the fact that there are few approaches of ACO metaheuristic used to solve CP problems in two dimensions. This overview is also intended to highlight similarities and differences between the selected metaheuristics. In addition, we will provide a source of alternatives for hybridization to develop more advanced algorithms for the 2SPP. The reason for choosing EAs and ACO for our study is that they are one of the more representative and widely used population-based metaheuristics in the respective class, *i.e.*, instance-based and model-based. As a result, this chapter aims to provide the amateur and experienced researchers a global and unifying vision of two different population-based approaches: EAs and ACO algorithms. The particular characteristics of both metaheuristics are analyzed considering their application to some instances of 2SPP.

We show in Section 2 the formulation and background of the 2SPP. Some considerations about the respective search process of EAs and ACO metaheuristics are given in Section 3. Next, we describe the two algorithms studied here and applied to 2SPP: a GA (Section 4) and an ACS (Section 5). In order to assess the performance of GA and ACS for 2SPP, we describe, in Section 7, some implementation details and present in Section 8 an extensive experimental study on a set of 2SPP instances.

Finally, in Section 9, we summarize the more relevant results achieved in our study and we also analyze some future research areas.

2 Strip Packing Problem and Background

The two-dimensional Strip Packing Problem (2SPP), an NP-hard problem [21], is present in many real-world applications in the glass, paper, textile, and other industries, with some variations on its basic formulation. Typically, the 2SPP consists of a set of M rectangular pieces, each defined by a width w_i and a height h_i, $(i = 1...M)$, which have to be packed in a larger rectangle, the *strip*, with a fixed width W and unlimited length. The objective is to find a layout of all the pieces in the strip that minimizes the required strip length, with the constraint that the pieces have to be packed with their sides parallel to the sides of the strip and without overlapping.

In this study some additional constraints are imposed: pieces must not be rotated and they have to be packed into 3-stage level packing patterns. In these patterns, pieces are packed by horizontal levels (parallel to the bottom of the strip). Inside each level, pieces are packed bottom left justified and, when there is enough room in the level, pieces with the same width are stacked one above the other. 3-stage level patterns are used in many real-world applications in the glass, wood, and metal industries and thus the incorporation of this restriction in the problem formulation. Puchinger and Raidl [33] developed a formal description of the problem at hand by linear programming model. A graphical example of two typical layouts is shown in Figure 1.

Heuristic methods are commonly applied to solve the 2SPP. In the case of level packing, three strategies [26] have been derived from popular algorithms for the one-dimensional case. In each case, the pieces are initially sorted by decreasing height and packed by levels following the obtained sequence. Let i and l denote respectively the current piece and the last created level:

- Next-Fit Decreasing Height (NFDH) strategy: piece i is packed left justified on level l, if it fits. Otherwise, a new level $(l = l + 1)$ is created, and i is placed left justified into it.
- First-Fit Decreasing Height (FFDH) strategy: piece i is packed left justified on the first level where it fits, if any. If no level can accommodate i, a new level is initialized as in NFDH.
- Best-Fit Decreasing Height (BFDH) strategy: piece i is packed left justified on that level, among those where it fits, for which the unused horizontal space is minimal. If no level can accommodate i, a new level is initialized.

The main difference between these heuristics lies in the level-selection mechanism for the piece to be placed in. NFDH only uses the current level and the levels are considered in a greedy way, while both FFDH and BFDH analyze all levels built to accommodate the new piece.

Exact approaches are also used [16, 28, 29]. Regarding the existing surveys of metaheuristics in the literature, Hopper and Turton [20, 21] review the approaches developed to solve 2D packing problems using GAs, Simulated Annealing, Tabu

Search, and Artificial Neural Networks. They conclude that EAs are the most widely investigated metaheuristics in the area of cutting and packing. Lodi et al. [27] consider several methods for the 2SPP in their survey and also discuss mathematical models. Specifically, the case where the items have to be packed into rows forming levels is discussed in detail. Few authors restrict themselves to problems involving guillotine packing [6, 31] and n-stage level packing [32, 33, 35, 36]. However, Puchinger and Raidl [32, 33], in particular, deal with the bin packing problem where the objective is to minimize the number of bins needed (raw material with fixed dimensions) to pack all the pieces – an objective which is different, but related, to the one considered in the present work.

Regarding the ACO metaheuristic, it is worth remarking that several ant algorithms have been successfully applied to a wide range of optimization problems. However, only a few have handled packing problems (see [14]), in particular in two dimensions. Boschetti and Maniezzo [8] present a basic ACS for the two-dimensional Finite Bin Packing Problem, which is able to devise solutions of comparable quality to those reported by Boschetti et al. [7]. Levine and Ducatelle [25] use a pure ACO algorithm, as well as an ACO algorithm augmented with local search, for a one-dimensional bin packing problem and cutting stock problem. The results presented are comparable to the ones in the literature. Moreover, their hybrid approach (ACO + local search) could outperform the existing results for the problems under consideration.

It is necessary to note that what have been described above is only a snapshot of what is available in the field. Consequently, many important publications are not mentioned. Interested readers should refer to the works of Wäscher et al. [38], Hopper et al. [19, 20, 22], Burke et al. [9, 10, 11] and Alvarez-Valdes et al. [2, 3]. Apart from these works, some of the latest contributions to the field can be found in [4, 23, 37, 40].

3 A Conceptual Comparison of EAs and ACO

This section is mainly dedicated to researchers new to the area of optimization by metaheuristic approaches, particularly to those interested in EAs and ACO. In the following sections we present a conceptual comparison of these metaheuristics applied to 2SPP which could help better understand the details about the respective algorithms used in this work and described in the following sections.

The more relevant characteristic that EAs and ACO metaheuristics share is the following: they are iterative population-based approaches. However, the "population" that each of them processes and the way it is processed are very different.

As an instance-based approach, EAs process a population of *competitive* individuals that represent a set of possible and complete solutions to the problem at hand. Accordingly, different operators should be defined to explore the problem search space to achieve good quality solutions. However, ACO algorithms are representative of a model-based approach. Their population, instead of solutions representations, consists of a set of *cooperative* ants (agents could be an alternative term) which iteratively build a solution in a step-by-step manner, *i.e.*, they start from an empty solution and proceed by adding problem components, one at a time, to create a complete solution to the problem.

For example, the packing problem studied here (2SPP) accepts, as described further in this chapter, a permutation of natural numbers as a possible solution. Therefore, with a 2SPP instance of size M, the size of the search space for this problem is $M!$ (assuming a permutation representation for the solutions).

First, we analyze the exploration search space from the EA's perspective. The following expression shows a general overview of the iterative evolutionary process followed by an EA when exploring the search space:

$$P(0) \xrightarrow{\mathscr{T}} P(1) \xrightarrow{\mathscr{T}} \cdots \xrightarrow{\mathscr{T}} P(t_{max}) \tag{1}$$

where $P(t) \subset S, t = 0, 1, \ldots, t_{max}$, represents the population at generation t, t_{max} is the maximum number of iterations, and S the whole search space of all possible permutations of size M. The successive populations are obtained by applying a transformation $\mathscr{T}$. This transformation can be conceived as a composition of several specific genetic operators aimed at exploring the search space, with $\mathscr{T} : 2^S \rightarrow 2^S$ where $\mathscr{T}(P(t)) = P(t+1)$. Combination of good genetic material from different solutions determines the building-block of the evolutionary approach to achieve solutions of a higher quality by exploring the respective search space.

The situation described above is different when considering the general ACO algorithms, since they consist of an iterative process that involves a population of ants $\mathscr{A}$ that step-by-step build a set of solutions (let us say $P(t)$, for $t = 0, 1, \ldots, t_{max}$) according to the pheromone ($\tau(t)$) and heuristic (η) values:

$$\begin{array}{ccccccccc} (\mathscr{A}, \tau(0), \eta) & \rightarrow & (\mathscr{A}, \tau(1), \eta) & \rightarrow & \cdots & \rightarrow & (\mathscr{A}, \tau(t_{max}), \eta) \\ & \overset{\mathscr{U}}{\nearrow} & & \overset{\mathscr{U}}{\nearrow} & & \overset{\mathscr{U}}{\nearrow} & \\ \downarrow & & \downarrow & & & & \downarrow \\ P(0) & & P(1) & & \cdots & & P(t_{max}) \end{array} \tag{2}$$

In the above sequence, $\mathscr{U}$ represents the updating process on τ structure which represents the building block globally shared by the whole colony. It is important to note that η values depend, for some problems, on the iteration number as well as the pheromone values. In this case, expression 2 should be modified by changing η to $\eta(t)$ for $t = 0, 1, \ldots, t_{max}$. In addition, it can be observed that set of ants $\mathscr{A}$ remains the same throughout the search process. The pheromone trail updating $\mathscr{U}$ can be defined as $\mathscr{U} : 2^\tau \times 2^S \rightarrow 2^\tau$ such that $\mathscr{U}(\tau(t), P(t)) = \tau(t+1)$.

However, the process of solution generation, either by an evolutionary or ant algorithm, is different when they are hybridized with, for example, a local search procedure as applied in our work. In general, the added local search procedure is intended to exploit good solutions found by the basic algorithm when exploring the search space, and simultaneously providing it, when possible, with a positive feedback about the exploration of promising regions of the search space. Usually the local search (denoted here by $\mathscr{L}$) is applied between two consecutive iterations. Thus, the sequences in Eq. 1 and 2 are respectively changed by:

$$P(0) \xrightarrow{\mathscr{T}} \hat{P}(1) \xrightarrow{\mathscr{L}} P(1) \xrightarrow{\mathscr{T}} \hat{P}(2) \xrightarrow{\mathscr{L}} \cdots \xrightarrow{\mathscr{T}} \hat{P}(t_{max}) \xrightarrow{\mathscr{L}} P(t_{max}) \tag{3}$$

and,

$$\begin{array}{cccccccc}
(\mathscr{A},\tau(0),\eta) & \rightarrow & (\mathscr{A},\tau(1),\eta) & \rightarrow & (\mathscr{A},\tau(1),\eta) & \cdots \rightarrow & (\mathscr{A},\tau(t_{max}),\eta) \\
\downarrow & \overset{\mathscr{U}}{\nearrow} & \downarrow & \overset{\mathscr{U}}{\nearrow} & \downarrow & & \downarrow \\
\hat{P}(0) & \xrightarrow{\mathscr{L}} P(0) & \hat{P}(1) & \xrightarrow{\mathscr{L}} P(1) & \hat{P}(2) & \cdots & P(t_{max})
\end{array} \tag{4}$$

where $\hat{P}(i)$ with $i = 1,\ldots,t_{max}$ represents, for both the EA and ACO approaches, an intermediate set of solutions on which the local search procedure is applied before the next algorithm iteration takes place. Thus, process $\mathscr{L}$ intensifies the search around some regions of the search space and provides a feedback to the algorithm with extra help when diversifying the search. The next two sections provide a detailed description of the algorithms implemented. Following that, the respective hybrid versions are given.

4 A Genetic Algorithm for the 2SPP

In this section we present a steady-state GA for solving the 2SPP (see the structure in Algorithm 1). This algorithm creates an initial population of μ solutions in a random (uniform) way, and then evaluates these solutions. The evaluation uses a layout algorithm to arrange the pieces in the strip to construct a feasible packing pattern. After that, the population goes into a cycle where it undergoes evolution. This cycle involves the selection of two parents by binary tournament and the application of some genetic operators to create a new solution. The newly generated individual replaces the worst one in the population only if it is fitter. The stopping criterion for the cycle is to reach a maximum number of evaluations (*max_evaluations*). The best solution is identified as the best individual ever found which minimizes the required strip length.

Algorithm 1. Genetic algorithm

```
GA
t = 0; {current generation}
initialize(P(t));
evaluate(P(t));
while (t < max_evaluations) do
    P'(t) = evolve(P(t)); {recombination and mutation}
    evaluate (P'(t));
    P(t+1) = select(P'(t),P(t));
    t = t + 1;
end while
```

A chromosome is a permutation $\pi = (\pi_1, \pi_2, ..., \pi_M)$ of M natural numbers (piece identifiers) that defines the input for the layout algorithm (represented by the fitness function) as explained in the following section. For example, let us consider

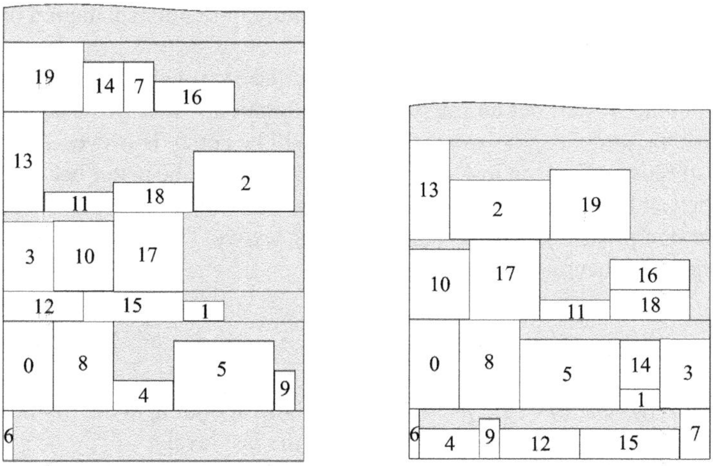

Fig. 1 Example of a layout obtained after applying the MFF strategy (left) and an improved layout (right)

the following problem with eight pieces ($M = 20$), and a strip of width W and an unlimited length. A possible cutting pattern is π = (6 0 8 4 5 9 12 15 1 3 10 17 13 11 18 2 19 14 7 16), where 6 stands for piece 6, 0 for piece 0, 8 for piece 8, and so on. The chromosome gives an order for considering the pieces, which is used afterwards by a layout algorithm. This algorithm will arrange the pieces to be cut on the strip to build a cutting pattern (see Figure 1 left). The GA will devise the best possible permutation so that the layout algorithm finds an optimal cutting pattern, which minimizes the required length of the strip.

The initial population is seeded with a set of greedy rules including information related to the problem (such as the pieces' width, the pieces' area, etc.), resulting in a more specialized initial population. See [34] for more detail.

4.1 The Fitness Function

To assess the quality of the solutions, we first need to achieve the corresponding pieces' layout. In order to generate a 3-stage level pattern for the pieces' layout, we adopt a modified NFDH heuristic —hereafter referred to as *modified next-fit*, or *MNF*— which was proven to be very efficient in [32, 36].

This heuristic gets a sequence of pieces as its input and constructs the packing pattern by placing pieces into stacks, and then stacks into horizontal levels (parallel to the bottom of the strip) bottom left justified in a greedy way, *i.e.*, once a new stack or a new level is started, previous ones are never reconsidered. When there is enough room in the level, neighbor pieces with the same width are stacked one above the other. See Figure 1 (left) for an illustrative example. A more in-depth explanation of the MNF procedure can be found in [36].

Therefore, the fitness of a solution π is defined as the strip length needed to build the corresponding layout, but an important consideration is that two packing patterns could have the same length —so their fitness will be equal. However, from the point of view of reusing the trim loss, one of them can actually be better because the trim loss in the last level (which is still connected to the remainder of the strip) is greater than the one present in last level in the other layout. Therefore we have used the following more accurate fitness function:

$$F(\pi) = strip.length - \frac{l.waste}{strip.length \times W} \tag{5}$$

where $strip.length$ is the length of the packing pattern corresponding to the permutation π and $l.waste$ is the area of reusable trim loss in the last level l.

4.2 *Genetic Operators for Solving the 2SPP*

In this section we describe the recombination and mutation operators, which were incorporated in our GA.

4.2.1 The Recombination Operator

The operator considered for our GA, called Best Inherited Level Recombination (BILX) , is a recently tailored operator for this problem, which was introduced in [36] as BIL. This operator transmits the best levels (groups of pieces which define a level in the layout) of one parent to the offspring. In this way, the inherited levels could be kept or they could even capture some pieces from their neighboring levels depending on how compact the levels are.

The BILX operator works as follows. Let nl be the number of levels in one parent ($parent_1$). In the first step the waste values of all nl levels of $parent_1$ are calculated. After that, a probability of selection, inversely proportional to its waste value, is assigned to each level and a number $(nl/2)$ of levels are selected from $parent_1$. The pieces π_i belonging to the selected levels are placed in the first positions of the offspring, and then, the remaining positions are filled with the pieces which do not belong to those levels, in the order they appear in another parent ($parent_2$).

Actually, the BILX operator differs from the one proposed by Puchinger and Raidl in [32] in that our operator transmits the best levels of one parent and the rest of the pieces are taken from the other parent. However Puchinger's recombination packs all the levels from the two parents sorted according to decreasing values, and repairing is usually necessary in order to guarantee feasibility (*i.e.*, to eliminate piece duplicates and to consider branching constraints), resulting in a more complex and expensive mechanism.

Algorithm 2. ACS algorithm

```
ACS Algorithm
Initialize pheromone values
repeat
   for (each ant k {k = 1,...,a}) do
      repeat
         Choose a random piece i from the non selected pieces
         Open an empty level l (l.height = h_i)
         repeat
            J = {1,...,j} {set of pieces that still fit in l}
            for (w =1 to |J|) do
               T = sum of the pheromone between J[w] and all pieces i already
                  in the level l
               Divide T by the number of pieces in l (|l|)
            end for
            Choose a piece i from J according to the calculated probabilities
            J = J − {i}
            Update the pheromone trail {local update}
         until no remaining piece fits in the level
      until all pieces are placed
      Calculate the objective value for that solution
   end for
   Find the iteration best ant
   Replace the globally best ant if the iteration best was fitter
   Actualize the pheromone trail {global update}
until the maximum number of iterations is reached
```

4.2.2 The Mutation Operator

Stripe (level) Exchange (SE) is the mutation operator selected for our GA implementation. SE operator selects two levels of the packing pattern, at random, and exchanges (swaps) these levels in the chromosome.

5 An Ant Colony System to the 2SPP

In this section we present how the ACS was adapted to solve the 2SPP (see Algorithm 2 for the general structure). The section describes the heuristic information considered, pheromone trail definition, state transition rule, and the objective function which is calculated considering the layout obtained from a permutation of pieces.

5.1 Heuristic and Pheromone Definition

It is well known that the search process of the ACO algorithms is biased and guided by two main components, the heuristic and pheromone values. For many optimization problems, problem-dependent heuristic information can be used to give the ants additional hints for their decisions. For the 2SPP, the heuristic value for a piece j is simply defined as $\eta_j = h_j$, *i.e.*, the height of piece j. On the other hand, the meaning of the pheromone trail is important to the quality of an ACS. The pheromone information should reflect the most relevant information for the process of solution construction. Following the ideas of Boschetti and Maniezzo [8], we maintain solutions in the form of permutations of the set of pieces, which will be directly translated into the corresponding packing pattern by a layout algorithm. The idea here is to favor the piece/position pair because there is a correspondence between the position in the permutation and the level where the corresponding piece will be allocated to. Trail τ_{ij} thus encodes the desirability of having a piece i and j in the same level [25]. Building blocks consisting of a partial packing in a solution (subset of several pieces to be allocated together) can emerge by reinforcing contiguous positions of the piece subset. The pheromone matrix has M rows and M columns.

Once all ants have completed their packing patterns, the pheromone is updated, *i.e.*, a global pheromone updating rule is applied. In this case, only the best ant (which is represented by π^{best}) is allowed to place pheromone after each iteration. This is done according to the following expression:

$$\tau_{ij} = \rho \times \tau_{ij} + (1/F(\pi^{best})) \tag{6}$$

where $0 < \rho < 1$ is the pheromone decay parameter, and $F(\pi^{best})$ is the objective value (described later) of π^{best} in the current iteration. Using only the best ant for updating makes the search much more aggressive. Global updating is intended to provide a greater amount of pheromone to good packing patterns.

Moreover, while ants construct a solution, a local pheromone updating rule is applied . The effect of local updating is to make the desirability of edges change dynamically. The local updating is conducted according to the following expression:

$$\tau_{ij} = (1 - \xi) \times \tau_{ij} + \xi \times \Delta\tau_{ij} \tag{7}$$

where $0 < \xi < 1$ is the parameter and $\Delta\tau_{ij}$ is set to τ_0. Dorigo and Gambardella [13] used this expression to run their experiments with good results.

Another way to promote exploration is by defining a lower limit (τ_{min}) for the pheromone values. The following formula evaluates τ_{min} [25] as:

$$\tau_{min} = \frac{(1/(1-\rho))(1 - \sqrt[M]{pbest})}{(avg - 1)\sqrt[M]{pbest}} \tag{8}$$

where *pbest* is the approximation of the real probability to construct the best solution, *avg* is the average number of pieces to choose from at every decision point when building a solution, defined as $M/2$. Also an evaporation phase occurs at each iteration by updating the pheromone trail by:

$$\tau_{ij} = \gamma \times \tau_{ij} \tag{9}$$

where $0 < \gamma < 1$ is a parameter.

5.2 State Transition Rule Definition

The state transition rule gives the probability with which ant k will choose a piece j as the next piece for its current level l in the partial solution $\pi' = (\pi'_1, \pi'_2, \ldots, \pi'_{M'})$ with $0 \leq M' < M$, which is given by [25]:

$$j = \begin{cases} \max_{j \in J_k(\pi',l)} [\tau_l(j)] \times [\eta_j]^\beta & \text{if } q \leq q_0 \\ S & \text{otherwise} \end{cases} \tag{10}$$

where $\tau_l(j)$ is the pheromone value for piece j in level l, η_j is the heuristic information guiding the ant, and β is a parameter which determines the relative importance of pheromone information versus heuristic information. q is a random number uniformly distributed in [0..1], q_0 is a constant parameter ($0 < q_0 < 1$) which determines the relative importance of exploitation versus exploration. S is a random variable selected according to the probability distribution given in Eq. 11.

$$p_k(\pi',l,j) = \begin{cases} \frac{[\tau_l(j)] \times [\eta_j]^\beta}{\sum_{g \in J_k(\pi',l)} [\tau_l(g)] \times [\eta_g]^\beta} & \text{if } j \in J_k(\pi',l) \\ 0 & \text{otherwise} \end{cases} \tag{11}$$

In Eqs. 10 and 11, $J_k(\pi',l)$ is the set of pieces that qualify for inclusion in the current level by ant k. The set includes those pieces that are still left after partial solution π' is formed, and are small enough to fit in the level l. The pheromone value $\tau_l(j)$, for a piece j in a level l, is given by:

$$\tau_l(j) = \begin{cases} \frac{\sum_{i \in A_l} \tau_{ij}}{|A_l|} & \text{if } A_l \neq \emptyset \\ 1 & \text{otherwise} \end{cases} \tag{12}$$

where A_l is the set of current pieces allocated in level l. In other words, $\tau_l(j)$ is the sum of all the pheromone values of pieces already in level l, divided by the number of pieces in that level. This approach is similar to the one followed by Levine and Ducatelle [25].

5.3 *The Objective Function*

The solution construction process followed by each ant of the ACS gives as result a solution that represents a permutation $\pi = (\pi_1, \pi_2, \ldots, \pi_M)$ of M natural numbers, *i.e.*, the same structure as the chromosomes used in our GA. For that reason, the objective function F defined here is the same as the GA's fitness function defined in Eq. 5. Although the more appropriate term is "objective function" for the ACS, we decided to use "fitness function" in order to unify the terminology with GA in the rest of this chapter, regarding the evaluation of the solutions' quality.

6 The Hybrids GA and ACS

Recently, there have been very different attempts to combine ideas and methods from both local search approaches and metaheuristics, resulting in hybrid algorithms. One of the most common forms of hybrid metaheuristic is the incorporation of local optimization as an extra add-on to the metaheuristic cycle. With the hybrid approach, local optimization is applied to each newly generated solution to move it to a local optimum. Thus the GA and ACS are respectively used to perform global exploration, while additional local search algorithms are used to perform local exploitation around a solution. Because of the properties of the metaheuristics considered here and the local search algorithms, the hybrid approach often outperforms either approaches operating alone.

The local search procedure used in this work consists of the application of a modified version of FFDH described in Section 2 (hereafter referred to as *modified first-fit*, or MFF), which is similar to the adjustment operator presented in a preliminary work [36].

When considering the Hybrid ACS, MFF is applied after a solutions are generated by the ants (after the outer **for** in Algorithm 2). In this way, the local search starts from a solution created by the ACS towards the nearest local optimum for that solution. After this improvement phase, the pheromone trail is updated, so it is in fact a form of Lamarckian search. More specifically, MFF is applied to the best $a/2$ ants' solutions with the aim of improving the trim loss of all levels. MFF considers the pieces in the order given by the solution π (permutation of pieces). The piece π_i is packed into the first level it fits, in an existing stack or in a new one. If no space is found, a new stack containing π_i is created and packed into a new level in the remaining length of the strip. The above process is repeated until no pieces remain in π. The possible new layouts obtained in this way have to be transmitted to the ant colony by updating the pheromone trail in such a way that ants build further solutions biased towards the improved layouts. An example of an improved layout is shown in Figure 1 (right) which will enforce a different pheromone trail updating from the layout in the same figure (left).

In the case of the Hybrid GA, MFF is applied to improve the solution obtained after recombination and mutation, with a predefined probability. In this case, the improvement phase starts from a solution created by the genetic operators.

In this work, another local search technique, Simulated Annealing (SA), was considered to hybridize both the GA and ACS, with a predefined probability after a solution is created. SA [12, 24] is an analogy of an algorithm used in statistical physics, which simulates the cooling of a solid previously heated past its melting point back to its solid state. SA goes through a number of iterations. At each iteration there is a current packing pattern as well as the best packing pattern obtained so far. While searching for an optimal packing pattern, the process moves from one pattern to another. The major difference with regular neighborhood searches is that moves to worse solutions are allowed. The reason for allowing these moves is to give the procedure the opportunity to move away from a local minimum and find a better solution later on.

7 Implementation

In order to evaluate the effectiveness of the different described algorithms, we will analyze the following GAs: *i*) a GA hybridized with SA (GA+SA), *ii*) a GA augmented with the FF heuristic (GA+FF) and finally, *iii*) a GA using both SA and

FF (GA+SA+FF). Regarding the ACS algorithms we carry out the same hybridization experiments, yielding ACS+SA, ACS+FF, and ACS+SA+FF algorithms, respectively. All these algorithms have been compared in terms of the quality of their results. In addition, the analysis of their performance was also addressed in order to understand the relationship between fitness values and diversity.

First, the different parameter values are defined, and then the pure ACS and GA approaches are compared to the hybrid ones in a numerical way, to show the numerical advantages of the hybrid algorithms. We then present the comparative results for the hybrid ACSs and GAs.

Since the results of these algorithms vary depending on the different parameter settings, we previously performed complete analysis to study how the parameters affect the performance of the algorithms. From these previous analyzes, we conclude that the best settings for our problem instances of the restricted strip packing problem are the following.

GA parameters. The population size is set to 512 individuals. The initial population is generated using problem aware rules. The BILX operator is applied with a probability of 0.8, while the mutation probability is set to 0.1. These parameters (population size, stop criterion, probabilities, etc.) were chosen after an examination of some values previously used with success (see [35]).

ACS parameters. The number of ants is set to 50. Regarding their initial positioning, ants are placed randomly. The parameter β, which defines the relative importance of the heuristic information as opposed to the pheromone information, was analyzed in order to determine its influence in a hybrid ACS. The parameter values are the following: $q_0 = 0.9$, $\rho = 0.8$, γ=0.96, ξ=0.1, and $\beta = 1,2,5,10$. The initial pheromone value, $\tau(0)$, was set to τ_{min}.

SA parameters. The number of iterations between two consecutive changes of the temperature is given by the parameter *Markov_Chain_length*, whose name alludes to the fact that the sequence of accepted solutions is a Markov chain (a sequence of states in which each state only depends on the previous one). The *Markov_Chain_length* is set to 30. The maximum number of evaluations performed by the algorithm is set to 100. We decrease the temperature T after every generation by a proportion α (cooling rate), set to 0.9, so that after N iterations, the temperature becomes $T1 = \alpha^N T0$, with $T0 = 1000$.

All the proposed algorithms were run for the maximum number of evaluations 2^{16}.

The algorithms were implemented with MALLBA [1], a C++ software library fostering rapid prototyping of hybrid and parallel algorithms. The platform was an Intel Pentium 4 at 2.4 GHz and 1 GB RAM, linked by Fast Ethernet, under SuSE Linux with 2.4.19-4GB kernel version.

We have considered five randomly generated problem instances with M equal to 100, 150, 200, 250, and 300 pieces and a known global optimum equal to 200 (the minimum length of the strip). These instances belong to the subtype of level packing patterns but the optimum value does not correspond to a 3-stage guillotine pattern. They were obtained by an implementation of a data set generator, following the ideas proposed in [39] with the length-to-width ratio of all M

rectangles in the range $1/3 \leq h/w \leq 3$. These instances are publicly available at `http://mdk.ing.unlpam.edu.ar/~lisi/ documentos/2spp.htm`.

8 Computational Analysis

In this section we will summarize the results of applying the proposed algorithms on all the problem instances. For each algorithm variant we performed 30 independent runs per instance using the parameter values described in the previous section. In order to obtain meaningful conclusions, we have performed an analysis of variance of the results. When the results followed a normal distribution, we used the t-test for the two-group case, and the ANOVA test to compare differences among three or more groups (multiple comparison test). We have considered a level of significance of $\alpha = 0.05$, in order to indicate a 95% confidence level in the results. When the results did not follow a normal distribution, we used the non-parametric Kruskal Wallis test (multiple comparison test), to distinguish meaningful differences among the means of the results for each algorithm.

Also, the evaluation considers two important measures for any search process: the capacity for generating new promising solutions and the rate of the progress in the surroundings of the best found solution (fine tuning or intensification). For a meaningful analysis we have considered the average best solution values and the entropy measure (as proposed in [18]), which is computed as follows:

$$entropy = \frac{\sum_{i=1}^{M}\sum_{j=1}^{M}(\frac{n_{ij}}{\mu})\ln(\frac{n_{ij}}{\mu})}{M\ln M} \qquad (13)$$

where n_{ij} represents the number of times the piece i is at position j in a set of solutions of size s (where s=μ in the GA and s=a in the ACS). This function takes values in [0..1] and a value of 0 indicates that all the individuals in the population are identical.

8.1 Hybrid ACS Approaches

Table 1 shows the results for the ACS approaches. For every instance, the approach that finds the fittest individuals is always an approach using MFF. The worst results are obtained when ACS is augmented with SA. ACS+FF obtains the best packing patterns for all instances, except for instance 250, where ACS+SA+FF was the best approach. The good packing patterns reported by ACS+SA+FF is owing to the local improvement offered by MFF. Using the test of multiple comparisons, we verified that ACS and ACS+SA both have means significantly different from ACS using MFF, while the differences between the two approaches are not significant for instances with M=100, 150, 250, and 300. For the instance with M=200, ACS+FF and ACS+SA+FF present similar best packing patterns. The same holds for ACS and ACS+SA, but the differences between the two groups are significant.

Table 1 Experimental results for the ACS approaches

Inst	ACS		ACS+FF		ACS+SA		ACS+SA+FF	
	best	*avg*$_{\pm\sigma}$	*best*	*avg*$_{\pm\sigma}$	*best*	*avg*$_{\pm\sigma}$	*best*	*avg*$_{\pm\sigma}$
100	252.47	257.54 $\pm$1.36	215.65	218.76 $\pm$0.92	256.24	259.41 $\pm$1.87	217.72	219.59 $\pm$1.02
150	258.72	265.68 $\pm$3.31	216.75	218.44 $\pm$0.82	265.13	271.17 $\pm$3.34	217.71	219.13 $\pm$0.74
200	251.35	260.35 $\pm$3.13	210.15	214.83 $\pm$1.23	254.65	263.24 $\pm$2.95	211.60	215.92 $\pm$1.35
250	238.53	248.72 $\pm$2.95	208.63	209.80 $\pm$0.54	246.52	252.80 $\pm$2.32	208.52	210.91 $\pm$0.94
300	253.31	260.04 $\pm$2.34	213.46	215.28 $\pm$0.62	259.64	262.67 $\pm$1.68	214.65	215.83 $\pm$0.56

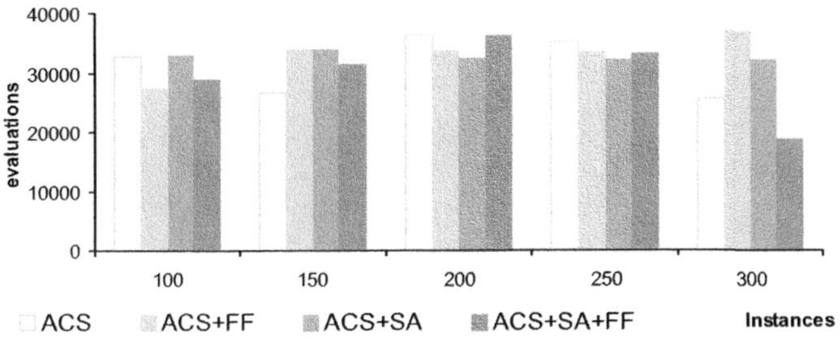

Fig. 2 Mean number of evaluations to reach the best value for each instance

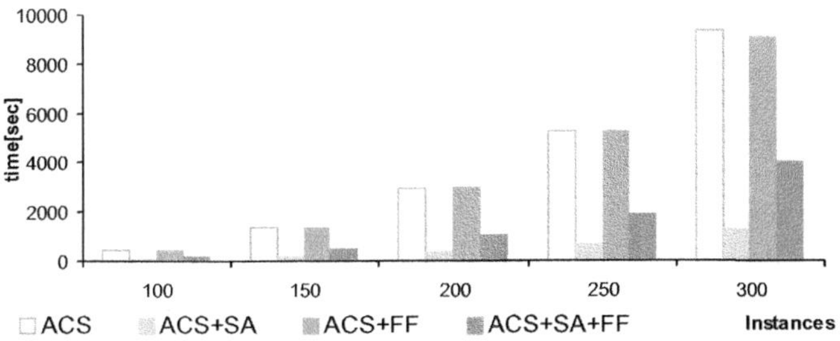

Fig. 3 Mean execution time by the different versions of the ACS

Looking at the average objective values of the best found solutions, ACS+FF produces best values and also small deviations. In terms of the average number of evaluations, Figure 2 shows that, in general, all the approaches present a similar effort to locate their best solutions (statistically corroborated). In particular for the instance of size M=250, ACS and ACS+FF use a similar average number of evaluations to reach the best value while the differences between the other two approaches are significant.

A remarkable point arises while analyzing the mean time spent in the whole search by the different versions of the ACS (see Figure 3). The two hybrid ACSs that include SA perform faster than both the pure ACS and ACS+FF. The slowest one is the pure ACS. The main reason for this result is the following: in the pure

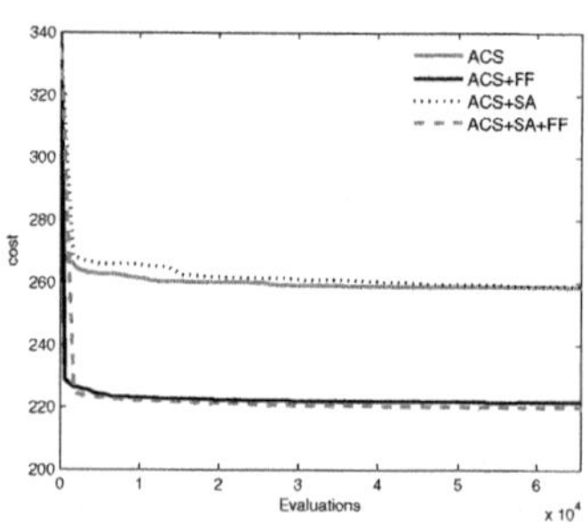

Fig. 4 Best fitness evolution for M=100

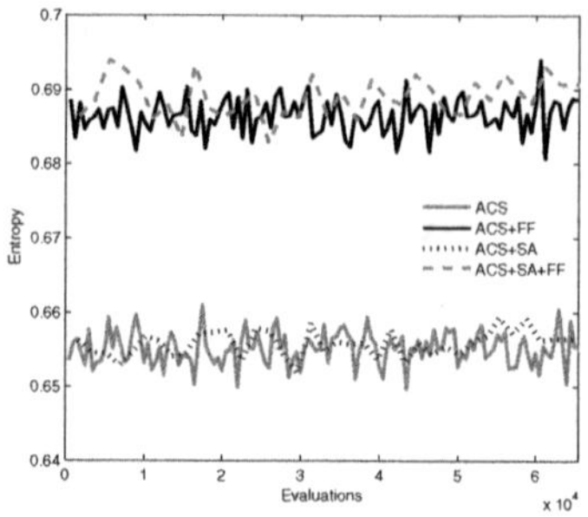

Fig. 5 Entropy colony for M=100

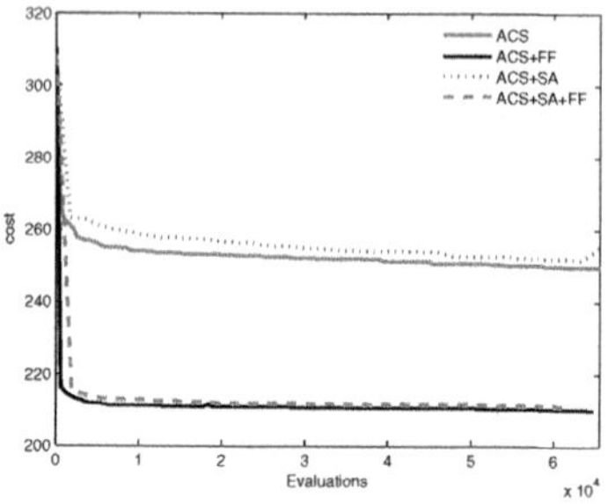

Fig. 6 Best fitness evolution for M=250

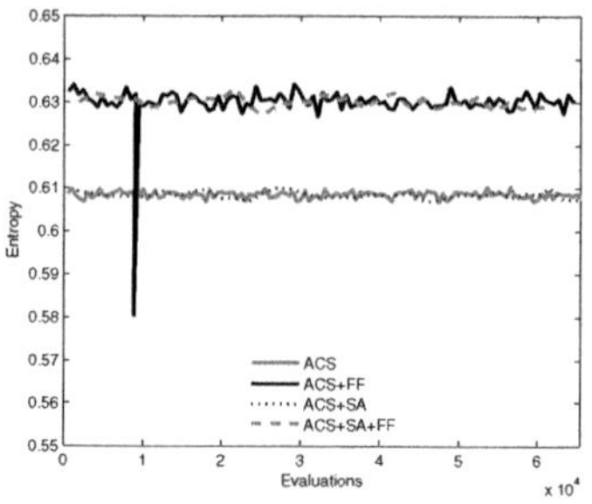

Fig. 7 Entropy colony for M=250

ACS, all the solutions are generated by the ants while in ACS+SA with or without MFF, some are generated by ants and the rest by the local search heuristic. The time that SA spends on improving an existing solution is minimal compared to the one consumed by an ant to generate a new one.

In order to analyze the trade-off between exploration and exploitation, the behavior of the different ACS algorithms is illustrated (for M=100 and M=250, but similar results are obtained for the rest of the instances) in Figure 4 through 7. They present the evolution of the colony entropy [1] and the best solutions, respectively (in y axis), with respect to the number of evaluations (x axis). From these figures we can see that entropy is higher for the ACSs hybridized with MFF than the others, while average population fitness is always lower for the ACSs hybridized with MFF. Therefore the best trade-off between exploration and exploitation is obtained when using MFF.

8.2 *Hybrid GA Algorithms*

In this section we focus on GAs results. Figures in Table 2 clearly show that the hybrid GAs outperform the pure GA, in terms of solution quality, for all instances. In

[1] By "the colony entropy" we mean the entropy of the set of solutions generated by the ant colony.

Table 2 Experimental results for the GA approaches

Inst	GA		GA+FF		GA+SA		GA+SA+FF	
	best	*avg*$_{\pm\sigma}$	*best*	*avg*$_{\pm\sigma}$	*best*	*avg*$_{\pm\sigma}$	*best*	*avg*$_{\pm\sigma}$
100	234.66	240.21 $\pm$4.81	214.48	217.00 $\pm$1.16	215.56	218.88 $\pm$2.18	211.34	215.78 $\pm$1.54
150	238.43	247.94 $\pm$5.30	213.54	214.75 $\pm$1.16	219.89	224.08 $\pm$2.71	212.74	214.32 $\pm$0.62
200	243.52	250.94 $\pm$4.82	207.79	211.02 $\pm$1.16	215.66	217.04 $\pm$0.83	208.87	212.02 $\pm$1.01
250	241.68	249.32 $\pm$4.19	211.34	212.47 $\pm$1.16	218.72	221.04 $\pm$0.94	211.71	212.68 $\pm$0.56
300	241.04	258.10 $\pm$6.10	209.93	211.99 $\pm$1.16	227.44	230.40 $\pm$1.47	209.83	213.05 $\pm$1.11

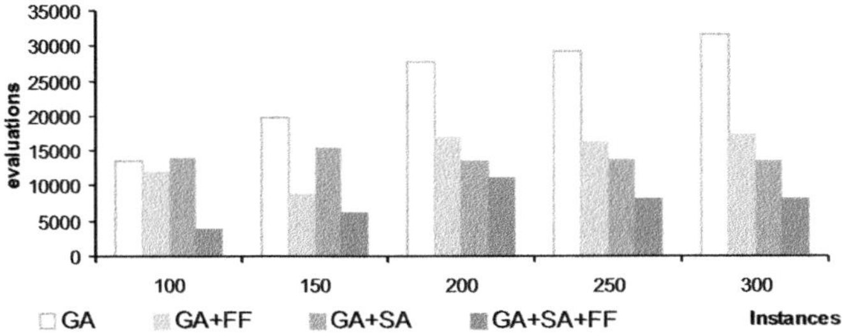

Fig. 8 Mean number of evaluations to reach the best value for each instance

all runs, the best packing patterns reached using hybridization have a low cost and are more robust than the best ones obtained with the GA (see the avg$_{\pm\sigma}$ column), and the differences are more evident as the instance dimension increases. Using the test of multiple comparison, we have verified that the differences among the results of GA+FF and GA+SA+FF are not significant. The reason for this performance is the improvement, in the pieces' layout, obtained after the application of MFF, which rearranges the pieces in order to reduce the trim loss inside each level. Nevertheless, GA+SA+FF presents best packing patterns in means, but there are significant differences with GA and GA-SA.

The hybrid algorithms significantly reduce the number of evaluations required to find good solutions (see Figure 8). As the instance dimension increases, the difference between GA and GA+SA is more notable. We carried out statistical test, which indicates that the difference among the algorithms is significant under this metric.

In particular, GA+SA+FF exploits the capacity of GA for quickly finding and exploiting promising regions of the search space and the effective combination of SA and FF to find local optimum of a small region of the search space.

Figure 10 shows that the GAs algorithms, except the hybrid GA+FF, have a similar decrease in entropy. An interesting point is how GA+FF maintains the genotypic diversity to quite high levels throughout the search process. From Figure 9 we observe that the curves of GA and GA+SA are very close during the whole search process, showing the worst performance. GA+FF presents a fast decrease of the average population fitness. In fact, we can infer that GA+FF seems to provide the best

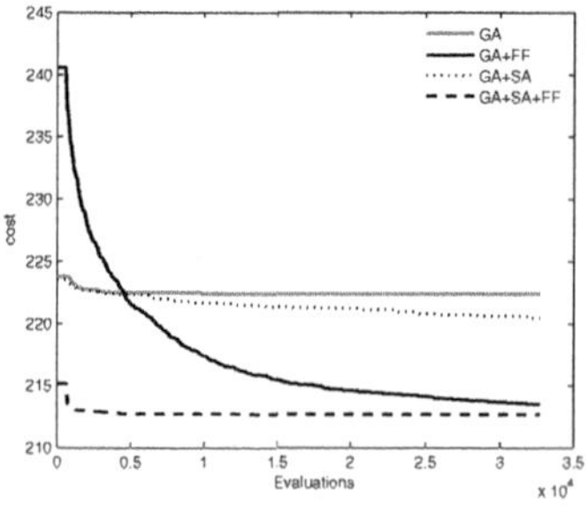

Fig. 9 Best fitness evolution for M=250

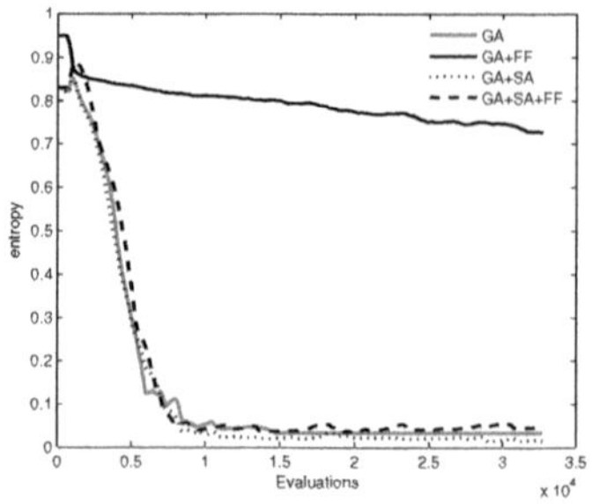

Fig. 10 Entropy colony for M=250

trade-off between exploration and exploitation; in this case MFF provides a feedback to the algorithm as extra help when maintaining a diverse pool of solutions.

8.3 *Hybrid ACS versus Hybrid GA Algorithms*

In this section, we perform a comparison between the best hybrid ACS (ACS+FF) and the best hybrid GA (GA+FF+SA). The results of the comparison are summarized in Table 3. We notice that the performance of ACS+FF is inferior to that of GA+SA+FF, except for instance M=250 (the differences are statistically significant). MFF works properly for both ACS and GA, intensifying the search around some regions of the search space.

These hybrid algorithms were compared with respect to the entropy (Figures 12 and 14) as well as the evolution of the average population fitness (Figures 11 and 13). These comparisons are illustrated over the instances with M=250 and M=150 (as an illustrative example of the rest). From Figure 11, we observe that the hybrid GA has a fast decrease in the average population fitness (resulting in faster convergence) meanwhile the opposite situation is observed in Figure 13. On the other hand, the hybrid ACS approach presents acceptable entropy values throughout the search process, a remarkable quality of the ACS.

Table 3 Experimental results for the GA approaches

Inst	ACS+FF			GA+SA+FF		
	best	***avg***$\pm\sigma$	***eval***	***best***	***avg***$\pm\sigma$	***eval***
100	215.65	218.76 ± 0.92	27287.63	211.34	215.78 ± 1.54	20624.00
150	216.75	218.44 ± 0.82	33853.73	212.74	214.32 ± 0.62	20314.00
200	210.13	214.83 ± 1.23	33725.43	208.87	212.02 ± 1.01	20719.00
250	208.63	209.80 ± 0.52	33437.50	211.71	212.68 ± 0.56	20521.00
300	214.66	215.46 ± 0.57	28671.60	209.83	213.05 ± 1.11	20725.00

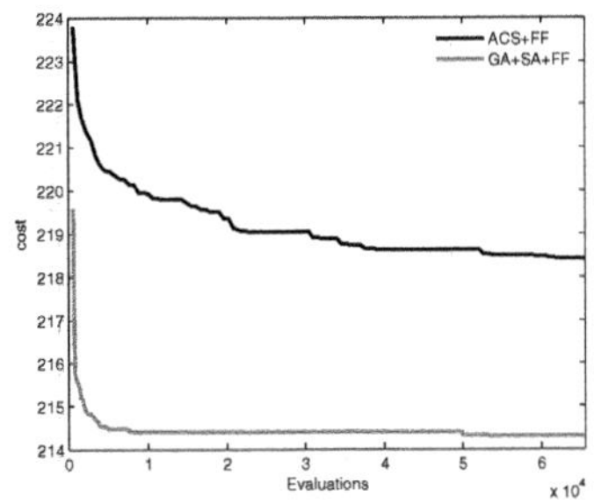

Fig. 11 Best fitness evolution for M=150

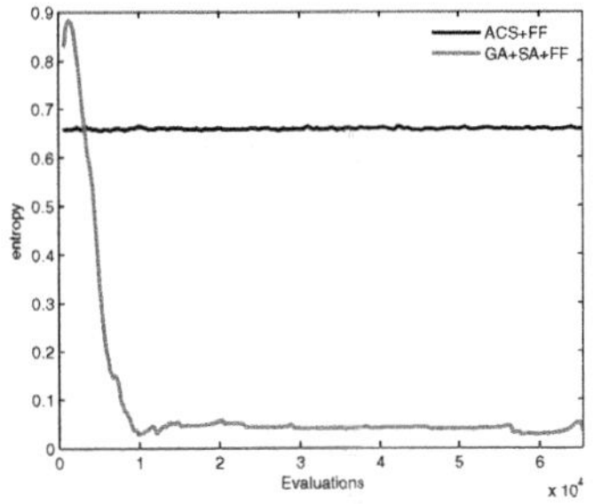

Fig. 12 Entropy colony for M=150

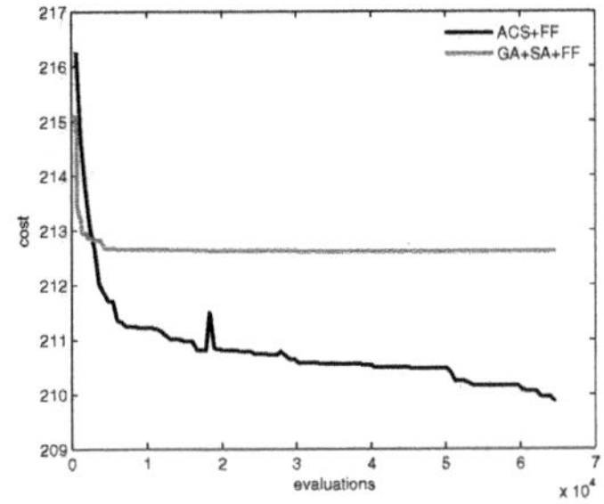

Fig. 13 Best fitness evolution for M=250

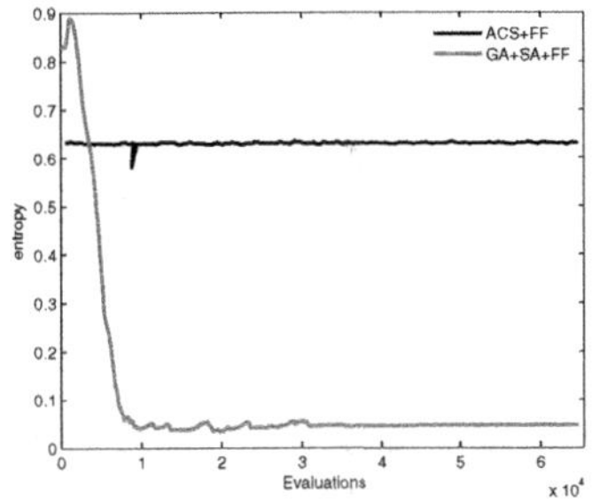

Fig. 14 Entropy colony for M=250

9 Conclusions

In this chapter, we have analyzed the behavior of two metaheuristics for solving a constrained 2SPP. In particular, we considered GAs and ACSs as representative algorithms of two population-based metaheuristics. Moreover, we presented different hybridization alternatives to achieve more advanced algorithms for the 2SPP. The study, validated from a statistical point of view, analyzed the capacity of the two approaches to generate new, potentially promising individuals and the ability to maintain a diversified population.

The first part of our study was related to the ACS. During the process of solution construction, the ants make the best possible choice, as indicated by the pheromone trail and heuristic information. The entries in the pheromone trail matrix express the desirability to have two pieces in the same level.

The hybrid ACS combines an ACS with local search. Each ant's solution is improved by moving some of the pieces around, and the improved solutions are used to update the pheromone trail. ACS performs a rather coarse-grained search, providing good starting points for local search to refine the results. In this case two local search procedures were considered: MFF (a modified version of the well known FFDH heuristic) and SA. Computational results of the ACS with MFF local search are much better than those obtained without local search and also using SA. Additionally, the computation times with local search are much smaller.

On the other hand, the incorporation of local search procedure into the evolutionary process makes a significant improvement to the 2SPP algorithms' results. It provides a faster sampling of the search space in all studied instances and exhibits a satisfactory trade-off between exploration and exploitation. In particular, the combination of SA and MFF in the search procedure maintains the population diversity for longer than in the case of a plain GA.

Finally, the GA using the two local search procedures achieves better results than the hybrid ACS with MFF, regarding both efficiency (effort) and quality of the solutions found.

There are several issues which seem to be worth further investigation. Firstly, it is necessary to consider additional analysis of larger numbers and types of 2SPP instances. To do that, we devise a more sophisticated instance generator capable of controlling the complexity of the instances generated. In addition, it is worth remarking that the application of the ACO approaches for CP problems is a relatively new area of research. We plan to go into the analysis of correlation between mean population fitness and genetic diversity in any depth, regarding ACO and GA approaches. Also we propose to construct parallel versions of the algorithms studied in this work. Consequently, there are still several aspects of these algorithms that can be greatly improved to cope with 2SPP as well as others CP problems.

Acknowledgements

This work has been partially funded by the Spanish Ministry of Science and Technology and the European FEDER under contract TIN2005-08818-C04-01 (the OPLINK project). We acknowledge the Universidad Nacional de La Pampa, Universidad Nacional de San Luis, and the ANPCYT in Argentina from which the first and second authors receive continuous support.

References

1. Alba, E., Luna, J., Moreno, L.M., Pablos, C., Petit, J., Rojas, A., Xhafa, F., Almeida, F., Blesa, M.J., Cabeza, J., Cotta, C., Díaz, M., Dorta, I., Gabarró, J., León, C.: MALLBA: A Library of Skeletons for Combinatorial Optimisation. In: Monien, B., Feldmann, R.L. (eds.) Euro-Par 2002. LNCS, vol. 2400, pp. 927–932. Springer, Heidelberg (2002)
2. Alvarez-Valdés, R., Parreño, F., Tamarit, J.M.: A tabu search algorithm for a two-dimensional non-guillotine cutting problem. European Journal of Operational Research 183(3), 1167–1182 (2007)
3. Alvarez-Valdés, R., Parreño, F., Tamarit, J.M.: Reactive GRASP for the strip-packing problem. Computers and Operations Research 35(4), 1065–1083 (2008)
4. Araya, I., Neveu, B., Riff, M.: An efficient hyperheuristic for strip-packing problems. In: Cotta, C., Sevaux, M., Sörensen, K. (eds.) Adaptive and Multilevel Metaheuristics. SCI, vol. 136, pp. 61–76. Springer, Heidelberg (2008)

5. Bäck, T., Fogel, D., Michalewicz, Z.: Handbook of evolutionary computation. Oxford University Press, New York (1997)
6. Bortfeldt, A.: A genetic algorithm for the two-dimensional strip packing problem with rectangular pieces. European Journal of Operational Research 172(3), 814–837 (2006)
7. Boschetti, M., Maniezzo, V.: The two-dimensional finite bin packing problem. Part II. 4OR, 1(2):135–148 (2003)
8. Boschetti, M., Maniezzo, V.: An ant system heuristic for the two-dimensional finite bin packing problem: preliminary results. In: Multidisciplinary Methods for Analysis Optimization and Control of Complex Systems, ch. 7, pp. 233–247. Springer, Heidelberg (2005)
9. Burke, E.K., Kendall, G., Whitwell, G.: A new placement heuristic for the orthogonal stock-cutting problem. Operations Research 52(4), 697–707 (2004)
10. Burke, E.K., Hellier, R., Kendall, G., Whitwell, G.: A new bottom-left-fill heuristic algorithm for the two-dimensional irregular packing problem. Operations Research 54(3), 587–601 (2006)
11. Burke, E.K., Kendall, G., Whitwell, G.: A simulated annealing enhancement of the best-fit heuristic for the orthogonal stock cutting problem. INFORMS Journal on Computing (2009) doi:10.1287/ijoc.1080.0306
12. Cerney, V.: A thermodynamical approach to the travelling salesman problem: an efficient simulation algorithm. Optimization Theory an Applications 45, 41–51 (1985)
13. Dorigo, M., Gambardella, L.M.: Ant colony system: a cooperative learning approach to the traveling salesman problem. IEEE Transactions on Evolutionary Computation 1(1), 53–66 (1997)
14. Dorigo, M., Stützle, T.: The ant colony optimization metaheuristic: Algorithms, applications, and advances. In: Glover, F., Kochenberger, G. (eds.) Handbook of Metaheuristics (2002)
15. Dorigo, M., Stützle, T.: Ant Colony Optimization. MIT Press, Cambridge (2004)
16. Fekete, S.P., Schepers, J.: On more-dimensional packing III: Exact algorithm. Technical Report ZPR97-290, Mathematisches Institut, Universität zu Köln (1997)
17. Glover, F., Kochenberg, G.A. (eds.): Handbook of Metaheuristics. Kuwler Academic Publishers, London (2003)
18. Grefenstette, J.J.: Incorporating problem specific knowledge into genetic algorithms. In: Davis, L. (ed.) Genetic Algorithms and Simulated Annealing, pp. 42–60. Morgan Kaufmann Publishers, San Francisco (1987)
19. Hopper, E., Turton, B.: A genetic algorithm for a 2D industrial packing problem. Computers and Industrial Engineering 37(1-2), 375–378 (1999)
20. Hopper, E.: Two-dimensional packing utilising evolutionary algorithms and other metaheuristic methods. PhD thesis, University of Wales, Cardiff, U.K (2000)
21. Hopper, E., Turton, B.: A review of the application of meta-heuristic algorithms to 2D strip packing problems. Artificial Intelligence Review 16, 257–300 (2001)
22. Hopper, E., Turton, B.: An empirical investigation of meta-heuristic and heuristic algorithms for a 2D packing problem. European Journal of Operational Research 128(1), 34–57 (2001)
23. Imahori, S., Yagiura, M.: The best-fit heuristic for the rectangular strip packing problem: An efficient implementation and the worst-case approximation ratio. In: Computers and Operations Research (2009) doi:10.1016/j.cor.2009.05.008
24. Kirkpatrick, S., Gelatt, C.D., Vecchi, M.P.: Optimization by simulated annealing. Science 4598, 671–680 (1983)
25. Levine, J., Ducatelle, F.: Ant colony optimization and local search for bin packing and cutting stock problems. Journal of the Operational Research Society (55), 705–716 (2004)

26. Lodi, A., Martello, S., Monaci, M.: Recent advances on two-dimensional bin packing problems. Discrete Applied Mathematics 123, 379–396 (2002)
27. Lodi, A., Martello, S., Monaci, M.: Two-dimensional packing problems: a survey. European Journal of Operational Research 141, 241–252 (2002)
28. Martello, S., Monaci, S., Vigo, D.: An exact approach to the strip-packing problem. Informs Journal on Computing 15, 310–319 (2003)
29. Martello, S., Vigo, D.: Exact solution of the two-dimensional finite bin packing problem. Management Science 44, 388–399 (1998)
30. Michalewicz, M.: Genetic Algorithms + Data Structures = Evolution Programs, 3rd revised edn. Springer, Heidelberg (1996)
31. Mumford-Valenzuela, C.L., Vick, J., Wang, P.Y.: Heuristics for large strip packing problems with guillotine patterns: An empirical study. In: Metaheuristics: Computer Decision-Making, pp. 501–522. Kluwer Academic Publishers, BV (2003)
32. Puchinger, J., Raidl, G.: An evolutionary algorithm for column generation in integer programming: An effective approach for 2d bin packing. In: Yao, X., Burke, E.K., Lozano, J.A., Smith, J., Merelo-Guervós, J.J., Bullinaria, J.A., Rowe, J.E., Tiňo, P., Kabán, A., Schwefel, H.-P. (eds.) PPSN 2004. LNCS, vol. 3242, pp. 642–651. Springer, Heidelberg (2004)
33. Puchinger, J., Raidl, G.R.: Models and algorithms for three-stage two-dimensional bin packing. European Journal of Operational Research 183(3), 1304–1327 (2007)
34. Salto, C., Alba, E., Molina, J.M.: Greedy Seeding and Problem-Specific Operators for GAs Solving Strip Packing Problems. In: Optimization Techniques for Solving Complex Problems, pp. 361–378. John Wiley & Sons, Inc., Chichester (2008)
35. Salto, C., Molina, J.M., Alba, E.: Analysis of distributed genetic algorithms for solving cutting problems. International Transactions in Operational Research 13(5), 403–423 (2006)
36. Salto, C., Molina, J.M., Alba, E.: Evolutionary algorithms for the level strip packing problem. In: Proceedings of NICSO, pp. 137–148 (2006)
37. Soke, A., Bingul, Z.: Hybrid genetic algorithm and simulated annealing for two-dimensional non-guillotine rectangular packing problems. Engineering Applications of Artificial Intelligence 19, 557–567 (2006)
38. Wäscher, G., Haußner, H., Schumann, H.: An improved typology of cutting and packing problems. European Journal of Operational Research 183(3), 1109–1130 (2007)
39. Wang, P.Y., Valenzuela, C.L.: Data set generation for rectangular placement problems. EJOR 134, 378–391 (2001)
40. Zhang, D., Chen, S., Liu, Y.: An improved heuristic recursive strategy based on genetic algorithm for the strip rectangular packing problem. Acta Automatica Sinica 33(9), 911–916 (2007)

Diagnosis, Configuration, Planning, and Pathfinding: Experiments in Nature-Inspired Optimization

W.D. Potter, E. Drucker, P. Bettinger, F. Maier, M. Martin, D. Luper,
M. Watkinson, G. Handy, and C. Hayes

Abstract. We present experimental results of applying various nature-inspired optimization techniques to real-world problems from the areas of diagnosis, configuration, planning, and pathfinding. The optimization techniques we investigate include the traditional Genetic Algorithm (GA), discrete (binary and integer-based) Particle Swarm Optimization (DPSO), relatively new Extremal Optimization (EO), and recently developed Raindrop Optimization (RO); all inspired by different aspects of the natural world. We present algorithm setup, issues with adapting the various optimization methods to the selected problems, and the emerging results produced by the methods. We consider the GA to be the baseline technique because of its robustness and widespread application. The major contribution of this chapter deals with the fact that DPSO, EO, and RO have never been applied to the majority of these selected problems, making this the first time most of these results have appeared in the literature.

1 Introduction

A given search algorithm might perform exceptionally well for certain optimization problems while relatively poorly for others, and it is often quite difficult to explain why this is the case. It is therefore difficult to know beforehand which algorithms should be applied to which problems. In this chapter we investigate a number of nature-inspired optimization techniques and present preliminary experimental results of applying these techniques to real-world problems from the areas of diagnosis, configuration, planning, and pathfinding. The optimization methods we investigate include two population-based search schemes: the Genetic Algorithm (GA)

W.D. Potter, E. Drucker, P. Bettinger, F. Maier, M. Martin, D. Luper,
M. Watkinson, G. Handy, and C. Hayes
Institute for Artificial Intelligence, The University of Georgia, Boyd GSRC,
Room 111 Athens, GA 30602-7415, USA
e-mail: potter@uga.edu

R. Chiong, S. Dhakal (Eds.): Nat. Intel. for Sched., Plan. and Pack. Prob., SCI 250, pp. 267–294.
springerlink.com

and Discrete Particle Swarm Optimization (DPSO); and two single-element search schemes: Raindrop Optimization (RO) and Extremal Optimization (EO). Results show a comparison among the methods based on quality of outcome.

For the diagnosis problem, we use an instance of multiple fault diagnosis where we have 25 diseases each with 10 symptoms. The fitness of any single diagnosis is based on how well the diseases in the diagnosis explain the symptoms presented by the patient, whether or not the diseases explain symptoms not exhibited by the patient, and favoring the more common diseases over more rare diseases. The goal is to find the diagnosis containing the fewest diseases that explains all and only the patient's symptoms. Known optimal results are available for comparison with search technique results.

For the configuration problem, we apply our optimization techniques to the now outdated U.S. Army Mobile Subscriber Equipment network configuration problem. Many years ago, the U.S. Army used Mobile Subscriber Equipment to provide primary and backup communications for large-scale military operations. Once the requirements for an operation were determined, mission planners would prepare a "shopping list" of communications components including type of component and quantity. Components were integrated according to component constraints, requirements support, and Army doctrine (connectivity constraints) to form the communications network to support the mission. Reliable configurations are available for comparison with optimization results.

For the planning problem, we address spatial forest planning, which attempts to optimize timber harvest yields by selecting and scheduling specific forest stands for harvesting. Difficulty arises from the adjacency restrictions and timing constraints associated with harvesting the stands. We use a management unit containing 73 stands representing a typical forest from the western United States. Target results used for comparing the optimization techniques are available.

A pathfinding (also known as route finding) problem we have been investigating for some time is called the Snake In The Box problem. The idea behind this problem is to find the longest path, called the Snake, which has certain characteristics in a special type of graph called a hypercube. An *n-dimensional* hypercube has 2^n vertices where each vertex is connected to n other vertices. Using binary numbers to represent the vertex numbers shows that adjacent vertices differ by only one bit (e.g., vertex 101 is connected to vertices 001, 111, and 100 in a 3-dimensional hypercube). We focus on an 8-dimensional hypercube (containing 256 vertices and 1024 edges). The special path constraints require that each vertex on the snake path be connected to a successor and predecessor snake vertex (except the tail and head vertices which each have only one adjacent snake vertex), and that no snake vertex be adjacent to any other snake vertices (i.e., no chords). The optimal solution or longest snake in dimension 8 has yet to be proven, but the current longest known snake has length 97.

Of the optimization techniques examined, DPSO and the GA are the population based search heuristics. The convergence of DPSO is influenced by the overall movement of the swarm as well as relationships among swarm particles with respect to local and best experiences. The GA uses evolution-inspired operators to move through the search space. Genetic material from mating parents is passed on to the resulting offspring that are typically more effective at surviving the environmental pressure applied by the associated fitness function. Individuals more adept at handling the pressure tend to contribute more genetic material to future generations.

EO and RO are relatively new optimization techniques that are not population based but rather start from single points in the search space and use unique mechanisms for moving about within the search space. EO, which has roots in statistical physics, focuses on removing weak components of the solution rather than favoring good components. RO mimics the radiating waves generated by falling raindrops in a pond. In search heuristic terms, a feasible solution is perturbed which may result in an infeasible solution. As the infeasible solution is fixed, movement away from the initial search space location occurs. Perturbation and fixing continues to radiate from the initial location.

In the next section, we present overviews of the problem areas in more detail: multiple fault diagnosis, network configuration, forest planning, and pathfinding. Sections 3 and 4 describe the population-based and single element search schemes, respectively. Included in each search scheme discussion are characteristics of the scheme and how they map to the specific problem areas. Section 5 presents the results from each search scheme when applied to each problem area. These results are compared and discussed in order to provide insights into the appropriateness of applying a particular search technique to a specified problem. We conclude with some general observations regarding the search techniques.

2 Problem Areas

2.1 *Multiple Fault Diagnosis*

In electronics as well as medicine and other domains, multiple fault diagnosis is the identification of a set of problems (diseases) that best explains some observed abnormal behavior (symptoms). This type of problem solving is commonly referred to as abductive inference, and automating this approach has been the focus of numerous research efforts [19, 25]. A closer look at multiple fault diagnosis reveals three major obstacles between a diagnostic problem and a "reasonable" automated solution: the number of possible diagnoses, the "goodness" of a diagnosis, and the search scheme used.

Our goal is to find the diagnosis that best explains the observed symptoms. This best explanation is determined and wholly dependent on the calculation of the goodness of a diagnosis.

Typically, diagnosis quality is based on the notion of parsimoniously covering a set of observable symptoms [27], that is, finding a minimal set covering of diseases that explains a given set of symptoms. The model we adopt is the probabilistic causal model (PCM), introduced by Peng and Reggia [22, 23]. The PCM integrates "symbolic cause-effect inference with numeric probabilistic inference" to solve multiple fault diagnosis problems.

In the PCM, a multiple fault diagnosis problem is characterized as a 4-tuple $\langle D, M, C, M^+ \rangle$, where:

- D is a finite nonempty set of disorders (i.e., diseases).
- M is a finite nonempty set of manifestations (i.e., symptoms).
- C is a relation, a subset of $D \times M$. This relation pairs diseases with associated symptoms such that (d,m) in C means that disease d may cause symptom m.
- M^+ is a subset of M that identifies the observed symptoms (the symptoms that the patient presents). Note that symptoms not identified in M^+ are assumed to be absent.

A diagnosis DI (a subset of D) identifies the diseases that are possibly responsible for the symptoms in M^+. This leads directly to a bit-string representation with one bit per disease which is either part of the diagnosis ("on") or not ("off"). Diagnosis DI covers M^+ if each of the individual symptoms in M^+ is associated with at least one of the diseases in DI as determined using C. As with M^+, diseases not identified in DI are assumed to be absent.

Associated with each disease in D is a prior probability p indicating how rare or common the disease is. Associated with each causal association in C is a causal strength c such that $0 \leq c \leq 1$ and represents how frequently a disease causes a symptom. Note that c is not equivalent to the conditional probability $P(m|d)$ used in earlier Bayesian approaches. The causal strength represents the conditional probability $P(d\ causes\ m|d)$, and does not change, that is, we may expect the frequency with which d causes m to remain stable. An additional assumption stipulates that no symptom may exist in M^+ unless it is actually caused by some disease in D.

Now, we have $|D|$ prior probabilities and $|D| \times |M|$ causal strengths. Using these values, Peng and Reggia derive a formula for calculating the "relative likelihood", denoted by $L(DI, M^+)$, of a diagnosis DI given observable symptoms M^+. The likelihood is the product of three factors:

$$L(DI, M^+) = L1 \times L2 \times L3 \qquad (1)$$

where:

- $L1$ is the likelihood that diseases in DI cause the symptoms in M^+. For diagnoses that do not cover M^+, $L1$ evaluates to 0 thus forcing L to 0. Unfortunately, this denies any analysis of non-cover diagnoses.
- $L2$ is the likelihood that diseases in DI do not cause symptoms outside of M^+ (e.g., in $M - M^+$). That is, $L2$ gives us a measure based on the expected symptoms that are actually absent. Ideally, we prefer $L2$ values that are close to 1.
- $L3$ is the likelihood that a highly probable (very common) disease d contributes significantly in the overall likelihood of a diagnosis DI containing d.

To summarize, $L1$ forces L to focus on only diagnoses that cover or explain all symptoms in M^+, $L2$ forces L to focus on "irredundant" and "relevant" covers, and $L3$ forces L to focus on more likely or common diseases than on rare or less likely diseases. Irredundant covers do not contain any excess diseases that could be removed and leave us with a smaller cover. Relevant covers (which contain the set of irredundant covers) ensure that only diseases associated with symptoms in M^+ are seriously considered.

One aspect of the MFD problem that makes it unique relative to the other problems discussed here is that we base the results on diagnosing all possible symptom sets, not just one patient's symptoms. Since we have 10 symptoms, that amounts to 1023 diagnosis trials per experiment, excluding the healthy case. The quality of the search technique when applied to the MFD problem is based on the number of diagnosis trials (of the 1023 total trials) in which the search technique found the optimal solution. For example, if 763 trials resulted in the optimal diagnosis out of the 1023 trials, then the reliability of the search technique was determined to be 763/1023 or 75% reliable.

2.2 Network Configuration

In the early 1990s, U.S. Army communications networks were configured using Mobile Subscriber Equipment (MSE), which provided communications support for a typical five-division corps in an area of up to 15,000 square miles [21]. The communications system was characterized as an area-switched system that provided both voice and data communications. An MSE network supported radio subscribers using VHF radio links, and supported wire subscribers using hardwired links and connectivity to other NATO forces and to commercial networks.

The backbone of an MSE network was composed of up to 42 Node Centers. Attached to the backbone were various types of Extension Nodes, Control Centers, NATO Interfaces, and Remote Access Units. The differences between the various types of Extension Nodes were directly related to the number of wire subscribers that could be supported and the actual fashion in which they were supported. A Large Extension Node could support nearly 200 wire subscribers, for example. Radio subscribers were supported by Remote Access Units. Rounding out the completion of a network configuration were the other various components including NATO interface units and System Control Centers.

Configuring an MSE network amounted to determining which components to use for a specified mission, and how many of each component to use [7, 24]. The mission requirements identified only the number of wire subscribers and radio subscribers to support using the network. Military mission planners would spend many hours developing various scenarios that incorporated all aspects of the mission. The communications aspect of a mission — the development of a battlefield communications network that supports both wire and radio telephone communications requirements — is a particularly difficult task due to the numerous dimensions involved, including the types of components and their quantities. We focus our attention on the "shopping list" of components needed for the mission. The shopping list corresponds to a network configuration and is a sequence of integers representing the number of components needed for the mission, with each position in the list corresponding to a different component. Consequently, we have a discrete integer-sequence problem representation that can be used directly by each of our heuristic search schemes. Of course, the network must function properly, support the current mission, and satisfy other configuration constraints (such as constraints imposed by U.S. Army doctrine, the terrain, and certain limitations of the components, such as proper connectivity).

Configuring an optimal network from various differing components and component quantities is the type of problem that is affected by combinatorial explosion; as the number of network components increases, the number of possible network configurations increases drastically. Using a familiar example, N network components may be configured into $N!$ networks (having N components in each network). This situation worsens if we allow varying network sizes.

Because of the combinatorial explosion in the number of possible networks, configuration algorithms that are guaranteed to find the optimal network structure via an exhaustive or near exhaustive search are of little value when dealing with larger networks. Since a typical MSE five-division corps contains over 40 Node Centers (not to mention the other types of components), it is clear that we need good heuristic approaches to solve the MSE configuration problem.

2.3 Forest Planning

In the forest planning problem, the goal is to obtain an optimal timber harvest schedule for a multiple-stand management unit forest subject to certain constraints. Specifically, in a viable harvest schedule, each stand is harvested in at most one of n harvest periods, and no two geographically adjacent stands are harvested in the same period. Among other things, the constraint prevents possible erosion and/or esthetic problems (this is called the green-up issue). For our specific tests, a problem from [4] was used. The objective was to find an optimal harvest schedule for a management unit of 73 stands. Optimality is measured according to the function:

$$\mathit{Minimize} \sum_{i=0}^{n} (H_i - T)^2$$

where i is the harvest period, H_i the total timber harvested in period i across all stands, and T denotes a per period target harvest volume for the entire forest. The total harvest cycle consisted of 30 years, with periods spaced 10 years apart (for a total of $n = 3$ periods). The goal is to achieve an even-flow harvest as close as possible to the target volume T, which for this management unit was given as 34,467 MBF (thousand board feet). The most reasonable representation for a harvest schedule is a 73-element integer array where each position corresponds to a stand and contains the stand's harvest period.

Forest planning problems that involve the location and timing of harvest units usually require integer decision variables and can be difficult combinatorial problems to solve. Interest in using these types of forest planning techniques has recently increased, motivated by voluntary and regulatory programs and facilitated by advancements in computer technology [2, 3].

2.4 The Snake in the Box Problem

The snake-in-the-box problem is a difficult problem in mathematics and computer science that was first described by Kautz in the late 1950's [14]. Snake-in-the-box codes have many applications in electrical engineering, coding theory, and computer network topologies. Generally, the longer the snake for a given dimension, the more useful it is in these applications [15].

Hunting for snakes (achordal-induced paths in an n-dimensional hypercube) deals with finding the longest path, following the edges of the hypercube, that obeys certain constraints. First, every two nodes, or vertices, that are consecutive in the path must be adjacent to each other in the hypercube, and second, every two nodes that are not consecutive in the path must not be adjacent to each other in the hypercube. A third constraint, whether the first node in the path is adjacent or non-adjacent to the last node in the path, determines if the path is a 'coil' or a 'snake.' While coils have received the most attention in the literature [12], both snakes and coils have useful applications. For this chapter we deal primarily with searching for open paths, or snakes, where the start node is not adjacent to the end node.

An n-dimensional hypercube contains 2^n nodes that can be represented by the 2^n n-tuples of binary digits of the hypercube's Gray code. By labeling each node of the hypercube with its Gray code representation, adjacencies between nodes can be detected by the fact that their binary representations differ by exactly one coordinate. In Figure 1, a path through the three-dimensional hypercube is highlighted. This path, $\{0,1,3,7,6\}$ in integer representation and $\{000,001,011,111,110\}$ in binary representation, is one specific example of a longest-possible snake for a three-dimensional hypercube. The length of this snake is four.

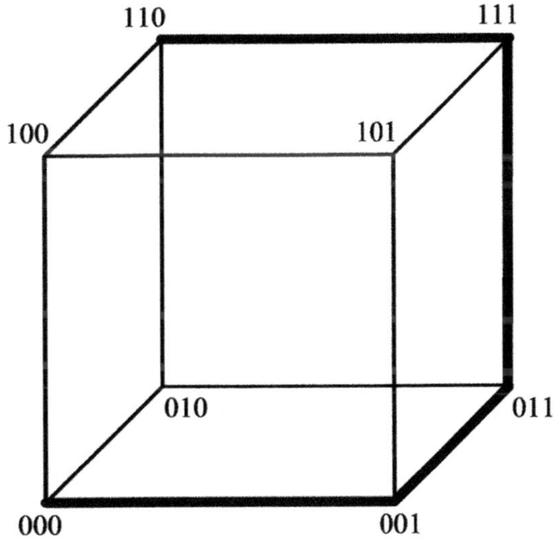

Fig. 1 A Three-Dimension Hypercube with Embedded Snake

Common representations for the snake path include the integer sequence of nodes on the path (node sequence), an integer sequence of bit positions changed on the path (transition sequence), and a bit string representation where each bit represents a hypercube node and the snake path is identified by "on" bits.

3 Population-Based Search Schemes

3.1 The Genetic Algorithm

The Genetic Algorithm (GA) models the process of evolution by natural selection [11, 13]. The driving force of natural evolution is selection pressure, which is applied by competition between organisms for passing on their genetic material. In the GA, potential solutions are treated as individuals in a population, and pressure is exerted through the use of selection schemes that choose individuals for mating. A common mate selection scheme is k-ary tournament selection, where k individuals are selected at random and the competitor with the highest fitness goes on to mate. After two individuals are selected, they exchange genetic material through a process known as crossover. A very common example is two-point crossover, where two random points within the representation are selected and material between the points is swapped to create offspring with genetic material from the parents. Crossover combines elements of separate individuals to exploit good areas of the fitness landscape. The newly created solutions then undergo mutation, which randomly modifies a (randomly selected) component of an individual. Numerous variations exist for the GA. However, we try to stay with the Simple GA, which uses just the basic evolutionary operators described above.

Regardless of the flavor of GA used, the most important features by far affecting its performance are: (1) the way the problem is represented within the GA individuals, and (2) the fitness function, which is the mechanism used to compare the strength of individuals. GA operators go hand in hand with the problem representation, and can lead to different levels of implementation difficulty. The problems addressed in this chapter lend themselves to different representation schemes. All but the Snake In The Box problem allow for straightforward representations and the use of standard operators. The SIB problem, however, has at least three quite different GA representations with corresponding repercussions on performance (e.g., the node sequence representation has adjacency issues, the transition sequence representation has length determination issues, and the bit string representation has both adjacency and length determination issues).

3.1.1 GA-MFD Setup

In the early days of genetic algorithms, the primary representation scheme was a bit string corresponding to a solution in the problem search space. Multiple Fault Diagnosis lends itself directly to the bit string representation with each bit corresponding to a disease. The intuitive or natural interpretation of an individual bit string is a diagnosis where an "on" bit means the particular disease is included in the patient's diagnosis and an "off" bit means the particular disease is not present in the diagnosis.

For our MFD scenario, we incorporate a minor modification to the Relative Likelihood discussed earlier in order to allow it to work smoothly with the GA. Specifically, in the original relative likelihood scheme, two non-covering diagnoses would both evaluate to a zero fitness, whereas in the modified scheme it is possible for each diagnosis to have a very low yet distinct fitness value. This allows the GA some small measure of variance between non-covering diagnoses.

In addition to using the modified relative likelihood to compute the fitness, we have the causal association matrix which, based on practical knowledge of the domain, provides the measure of the relationship between the set of diseases and the set of symptoms. For our scenario, we have 25 diseases in the disease set, and 10 symptoms in the symptom set. Also, related to each disease is the notion of its occurrence within the population, that is, whether it is a common or rare disease. Consequently, each disease has associated with it a measure of a patient's likelihood of having the disease. During flu season, the probability that a patient has the flu is higher than the probability of the patient having the bubonic plague, for example. Finally, we need to input the patient's symptoms before we can begin the search for the proper diagnosis.

As might be expected, reliability (i.e., the percent of optimal diagnostic trials) was generally better with larger population sizes (we typically used populations of 100, 200, and 300). Also, within a given population size, higher crossover rates tended to produce better reliabilities (we typically used 0.6, 0.7, and 0.8 for crossover probabilities). Finally, increased reliabilities were also observed with increased mutation rates within given population and crossover parameter settings (we typically used 0.005, 0.01, 0.015, and 0.02). Continuing to increase population size would allow us to continue to increase the diagnosis reliability but at the expense of runtime. Continuing to increase either crossover or mutation probabilities led to decreases in reliability. The GA stopped each diagnostic search when either a total of 60 generations elapsed or there was no improvement for 20 consecutive generations (indicating a measure of convergence). In general, the GA performed fairly well giving reliabilities in the upper 80s (e.g., out of 1023 symptom sets, the GA found the optimal diagnosis 87% of the time).

3.1.2 GA-MSE Setup

Configuring an MSE network involves determining the proper number of each type of equipment component to use for the given mission. We refer to the list of components and quantities as the mission "shopping list". The shopping list is represented by a sequence of integers corresponding to the number of components needed for the mission, with each position in the list corresponding to a different component. The set of available MSE components includes Node Centers (NCs), Large Extension Nodes (LENs), two types of Small Extension Nodes (SEN1s and SEN2s), System Control Centers (SCCs), NATO interface units (NAIs), and Radio Access

Units (RAUs). The network must function properly, support the current mission, and satisfy other configuration constraints (such as constraints imposed by U.S. Army doctrine, the terrain, and certain limitations of the components, such as proper connectivity).

Each position in the sequence corresponds to a particular component type, and the values of the integers range between the minimum and maximum Corps complement for the component type. Because of the MSE representation, the typical selection and crossover operators within the genetic algorithm can be applied directly. However, the mutation operator was modified for integer sequences rather than binary sequences. In addition, in the scheme the mutation rate decreases over the evolutionary cycle allowing more exploration in the early generations and more exploitation in the later generations.

The MSE fitness function takes into consideration several important issues in determining the fitness of an individual (solution)[24]. These issues include the mission requirements, the known support capacities of all the various components, the Division and Corps equipment complement, and the known minimum component requirements (such as, any network must have at least one System Control Center). In addition, issues such as the combined total number of components, and connectivity doctrine are included to help distinguish competing individual solutions. In the final analysis, the shopping list with the fewest components meeting the exact mission requirements, with the perfect component mix is the one we seek. The fitness of such a configuration would dominate configurations with less than perfect component mix, or those that did not exactly match the mission requirements (not too much support yet not too little support).

Our simple GA ran using the above representation and fitness function. We also used the age-proportional mutation operator, due to early trials showing the use of naïve mutation to perform more poorly. Trials indicated that crossover rate had only a minor impact on results (in this case), so trials were run on population size (300, 400, 500) and mutation rate (5%, 10%, 15%), with 1,000 runs for each unique setting. The stopping criterion was ten generations with no improvement in the best fitness. The GA ran for an average of 34 generations and slightly less than 14,000 total fitness evaluations. In every case, increasing population size and mutation rate increased result quality, so only results using a mutation rate of 15% are discussed here. The most accurate trial was run with a population size of 500, giving near perfect results (the best solution was found 996 times in 1,000 trials, 99.6% optimal), with the average number of fitness evaluations before finding the optimum at 12,000. When decreasing the population size to 400, the result's quality was barely affected, dipping to 99.1% optimal, and the fitness evaluations for the best solution decreased to 9,200. When run with a population size of 300, result quality decreased to 98.1% optimal, with the GA finding the optimal solution after 6,200 fitness evaluations.

3.1.3 GA-FP Setup

We used an integer sequence representation for the Forest Planning problem to represent the harvest schedule. Within the western management unit we have 73 stands and each stand may be harvested during one of three time periods (or not harvested). Consequently, we used a 73-element (73-gene) representation where each position corresponds to a stand. The value of each element is an integer in the range of one to four indicating harvest time period.

The integer representation provides us with a natural harvest schedule interpretation. It also allows us to use the standard GA operators such as tournament selection and two-point crossover. Mutation is slightly modified in that when an element value is mutated, it is changed to another value within the proper range (similar to the mutation scheme used with the MSE problem).

We use the same fitness function defined in [4] where a target time period harvest (34,467 MBF or thousand board feet) is specified for each time period, and we attempt to minimize the sum of the squared time period errors across the entire schedule. In addition, we experimented with introducing a penalty into the fitness function whenever an adjacency constraint was encountered. Recall, green-up and adjacency constraints may cause a harvest schedule to be invalid such as when two adjacent stands are scheduled to be harvested during the same time period. For the GA, we want the fitness to reflect the constraint violations but still provide some genetic material in order to allow the GA to progress through the evolutionary process.

We experimented with a variety of GA parameter settings but avoided using any "special" GA features such as seeding or embedded local searching. We used typical GA parameter settings except for population size that was set somewhat high at 20,000 individuals. The crossover probability was set to 0.7 and the mutation probability was set to 0.17. Due to the complexity of the Forest Planning problem, the simple GA was only able to find a reasonably good harvest schedule but not as good as the target schedule reported by Bettinger. Their best harvest schedule produced a value of 5,500,391 whereas the GA derived schedule had a poorer value of 6,505,676.

3.1.4 GA-SIB Setup

The bit string representation, used by Diaz-Gomez, illustrates the hypercube through 2^d spaces, with each space representing a node in the hypercube [8, 9]. A one-bit means that the associated node is turned on (included in the path) while a zero-bit means that the node is turned off (not included in the path). Even if the node is turned on, however, it does not mean that the node is in a valid snake path. In order to determine which nodes are included, the snake is evaluated with a length fitness function. Starting at node zero, the function uses the bit string representation to grow the snake outward by connecting valid nodes that are turned on. In order to confirm the validity of the snake, we used Diaz-Gomez's "vector of neighbors" matrix. The neighbors matrix is created by multiplying the individual snake bit string vector with the hypercube adjacency matrix. The results indicate how many of each node's adjacent nodes are on the snake path. The resulting matrix clearly represents the nodes that have more than two neighbors; thus violating one of the constraints of a valid snake.

The individuals created for the genetic algorithm are the simple bit strings containing 256 bits. Due to the extreme complexity of the SIB problem, for most of the trials we pre-set seeding the beginning portion of the individuals with the longest snakes found by [18] for dimension 7 (using various prefix subsequence lengths). In order to preserve the seed, we made sure that the seed was not touched by crossover or mutation. Even though the search space is exponentially large in dimension 8, we kept our population sizes below 1,000 individuals at all times, with the best results occurring at population sizes of 100 and 350 individuals. Initially, roulette wheel selection was used to select parent individuals for reproduction. Later experiments utilized the more typical tournament selection. Four individuals were selected randomly for the tournament, and the individual with the highest fitness was chosen as a parent for the next generation. We used two-point crossover to exchange genetic material from two parents. Crossover did not occur all the time, however, but at different rates. Successful crossover rates varied from 40% to 80%. The new individuals, either with or without the occurrence of crossover, were then sent to a mutation function.

An adaptive mutation method was tried in order to maximize search space exploration. One version of the method started the mutation rate at 15%, with it decreasing with each generation, eventually approaching near zero after 250 generations. Failure to find a different longest length after four successive generations resulted in the mutation rate temporarily changing to 10%. Some experiments used a traditional mutation scheme, with the mutation rate fixed at 1% for the entirety of the program.

The most basic fitness function focused only on the length of the snake subsequences within the representation. Using the vector of neighbors, another fitness function incorporated lazy and isolated nodes (isolated nodes are nodes that are turned on, but are not included in the snake path; and lazy nodes are nodes that are turned off and not adjacent to the snake path at all). Since the ultimate goal is to find the longest snake, we also incorporated skin nodes of a given snake (nodes adjacent to snake nodes but not on the snake). This relates to the tightness characteristic used by [29]. Tightness reflects the idea of winding the snake through the hypercube in the tightest possible fashion. A characteristic of tightness is that individual skin nodes may be adjacent to many snake nodes; more means tighter. The function still focused on the length of the snake, but was also aided by how loose or tight the snake was wound through the hypercube.

With the bit string representation, we were able to find a snake of length 95. This came from experiments using 1,000 individuals in a population, 1% traditional mutation scheme, 70% crossover rate, tournament selection, and pre-set seeding. We were routinely able to find mid-90s length snakes within 250 generations (a low number of generations), and on those occasions when the best lengths were lower, they never fell below length 85 snakes. Based on the computational resources used to process 100 individual populations for 250 generations, the mid-90s length result was outstanding considering that the longest known snake for dimension 8 has length 97!

3.2 *Particle Swarm Optimization*

Particle Swarm Optimization models the natural behavior exhibited by various large groups of animals such as flocks of birds or schools of fish [10, 16, 17]. For birds, the movement of the flock is orchestrated by a combination of influences based on the speed and direction of the lead bird, and the speed and direction of an individual bird's immediate neighbors. Translating the metaphor to solutions within a search space of solutions, a small population of solutions can be "guided" towards the globally optimal solution based on the influence of: (1) the best member of the search space ever encountered, and (2) the "path" through the search space of each member of the population or swarm. The population "moves" in discrete time steps through the search space based on the following speed and direction characteristics for each member [26]:

1. The difference between the location of the best member and each member's location, ΔGB;
2. The difference between a member's best location and its current location, ΔPB;
3. A social influence: $C_1 \times Rnd \times \Delta PB$;
4. A cognitive influence: $C_2 \times Rnd \times \Delta GB$;
5. The new velocity, V_{new}: $inertia \times V_{old} +$ $social\ influence + cognitive\ influence$, limited by V_{min} and V_{max}; and
6. The new location: $old\ location + int(V_{new})$.

Note that *Rnd* is a random number and that we make this standard PSO scheme a discrete scheme to match the problem representations we deal with (none of them lend themselves to the typical real-valued PSO parameter setup). Based on the problem representation used, we either have a binary discrete PSO approach or an integer-based discrete PSO approach. In addition, the inertia or inertia weight is used to help balance the exploration and exploitation of the search.

3.2.1 DPSO-MFD Setup

Our discrete PSO algorithm was implemented using Python and Microsoft's Visual Basic 6, and experiments were run using the same setup as the GA experiments (i.e., same bit string representation scheme for diagnoses and same modified relative likelihood fitness function). The representation of a diagnosis lends itself directly to the application of a binary discrete PSO approach (or simply a binary DPSO approach). Note that, to map the velocity values to binary values, we use the common sigmoid approach that compares the velocity to a random value, and assigns either zero or one based on the comparison result. Initial trials revealed that the performance of the algorithm was highly sensitive to inertia values. Inertia values below 0.7 and above 1.4 reduced the reliability significantly, and only inertia values of between 0.9 and 1.2 seemed to have any promise. Recall that with MFD, reliability is the ratio of optimal diagnosis sessions to total diagnosis sessions (for all possible symptom sets). Settings for the social and cognitive parameters did not have such a drastic effect. Values for these variables between one and four were tested during our initial trials.

Setting the values between 1.9 and 2.2 gave the most promising results. From these initial trials, additional experiments were done to determine our best parameter settings for our binary DPSO and to compare these results to other search schemes. We experimented with a number of different parameters such as swarm sizes in the range of 100 to 500, and total generations in the range of 50 to 100 with stable generations set to 10.

Not surprisingly, our binary DPSO results were very good and very quick. We routinely achieved reliabilities in the upper-90s with modest DPSO parameter setting. For example, with a swarm size of 200, inertia set to 1.1, social value set to 2.2, cognitive value set to 1.9, limited to 60 generations, and with a stopping criterion of 10 stable generations, we were able to achieve 98% reliability (of the 1023 total symptom sets, we found the optimal diagnosis for 1002 of the symptom sets).

3.2.2 DPSO-MSE Setup

The integer-based DPSO-MSE setup followed the GA-MSE setup where we used an integer sequence to represent the network configuration as well as the fitness function based on connectivity, mission support, and fewer network components. After several trial runs to narrow down good values for the social term (C_1) and cognitive term (C_2), the DPSO algorithm was run a thousand times for each of 12 different parameter settings. The parameters included swarm size (100, 300); C_1 (1.5) and C_2 (4.0); inertia (0.4 to 0.9 with 0.1 steps); maximum allowable generations set to 300; a stopping criterion of no improvement for 20 consecutive generations; and V_{min} set to (-6), and V_{max} set to (6). With a swarm size of 300 individuals, C_1 set to 1.5, C_2 set to 4.0, and inertia varying from 0.4 to 0.9 (0.1 step size), we achieved 99.85% optimal results. That is, for six different inertia settings per run and each run repeated 1,000 times, DPSO found the optimal solution 5,991 out of 6,000 times (note that the second best solution was found three times and the third best solution was found twice). In the trials using a swarm size of 300, the average number of fitness evaluations was 31,500. Reducing the swarm size to 100 gave us a 98.63% optimal result. It was expected that the percent optimal would decrease, but surprisingly it only dipped slightly. This was achieved with C_1 set to 1.5, C_2 set to 4.0, and inertia varying from 0.4 to 0.9 for a total of 6,000 runs, of which we found the best solution in 5,918 runs (the second best solution was found 13 times and the third best solution was found only once). For these experiments, we found that DPSO required an average of 11,800 fitness evaluations per run to find the optimal solution.

3.2.3 DPSO-FP Setup

Initially, the integer-based DPSO-FP setup seemed very straightforward and almost identical to the DPSO-MSE setup. We started with an integer representation with 73 elements corresponding to the 73 stands in the western management unit. Each element took on a value in the range of one to four, where four represented the situation when we did not harvest the corresponding stand. Many experiments were run in search of good DPSO parameter settings with social and cognitive values ranging from two to four, inertia ranging from 0.6 to 1.2 in 0.1 steps, and swarm sizes ranging from 1,000 to 10,000 in steps of 1,000. The DPSO was allowed to run up to 2,500 generations with a stopping criterion set at 150 stable generations. We experimented with the fitness function specified by [4] as well as trying a fitness function containing a penalty value. The target harvest value to reach was in the 5.5M range.

The integer representation provided us with a natural harvest schedule interpretation and our early experiments utilized this representation extensively. Unfortunately, the results achieved using our integer-based discrete PSO search were poor. With a target of 5.5M, it was typically the case that the integer-based DPSO fitness would get only as low as about 150M (this seemed astonishingly high to us; recall we are minimizing). We also tried a functionally equivalent but implementationally different representation. Namely, we modified the representation from an integer string to a bit string where we used two bits for each of 73 stands (two bits are all that are needed to distinguish four items). In other words, our discrete integer-based PSO was converted to a binary discrete PSO. The fitness function experiments stayed the same.

The binary DPSO experiments varied swarm sizes from 1,000 to 5,000 in steps of 1,000. The social and cognitive parameters, and inertia value varied as above from two to four and 0.6 to 1.2, respectively. Generation and stable generation limits remained the same. Using this new setup, we were able to achieve fitness values in the neighborhood of 39M (much better than the 150M seen above) within 6M fitness evaluations but these results were still a long way for having the binary DPSO compete effectively with the other search schemes presented here. We tried a variety of other variations with the DPSO scheme in hopes of improving its performance with the Forest Planning problem but without success.

3.2.4 DPSO-SIB Setup

In our discrete PSO implementation, the population is moved around the search space by adding the calculated velocity for each particle to its current value. The social and cognitive (C_1 and C_2) values that we used had C_2 set considerably lower than C_1. The reason for this was that we wanted the particle to be influenced by its own personal history more so than the global best because the global best may be a snake that is counterproductive to the individual being evaluated. There are many different, good paths through an 8 dimensional hypercube and the global best may be a completely different path from our other individuals. Furthermore if the other individuals listen too closely to the global best, they could be influenced towards snake violations (with respect to where the individual currently is) that would worsen the individual to a point where it would not be able to recover. This is not to say the global best is not important, just that if an individual gives more weight to its own personal best, it is more likely to give good results because its personal best is attainable from where our individual's sequence of transitions places it in the search space.

Representation of individuals in the swarm is a key concern. In the DPSO-SIB case, the representation is an integer sequence where each position in the sequence contains a value representing the binary digit that changes when moving from one node of the hypercube to another. This is called a transition sequence. For example, the transitions followed when moving from node 0 to node 1 to node 3 and finally to node 7 are 1, 2, and 3, respectively (when identifying the node number using binary and counting from right to left starting from one). For these experiments, each individual had a length of 150. To check snake constraints, the transition sequences were converted into node sequences beginning with a common node (in this case, 0). Since every transition sequence is automatically a valid hypercube path, there was no need to check path validity for any potential solutions.

For the SIB problem, we also limited our velocity to the range -4 through 4 and tested inertia values of 0.3 to 1.2. We found that higher inertia values reduced the effectiveness of the DPSO for this problem. With no special DPSO-SIB setup features, we ran tests with the inertia set to 0.9 and the longest snake found was in the mid 60s. However, in most tests, the inertia was set much lower and the results were much better. Experiments were tried with swarm sizes from 1,000 individuals to 10,000 individuals. Typically, larger swarm sizes resulted in improved snake lengths.

To test for improved performance, we also used a slightly modified set of features for the DPSO-SIB, namely initial swarm pre-set seeding and a local growth scheme. This was necessary to help explore the very large search space (of which only a tiny portion contains valid snakes). The first technique consisted of seeding the initial swarm with fairly long snake subsequences. Since all transition sequences are valid paths, this was not necessary to the degree that it was with a node sequence representation, and only 5% of the swarm was seeded for these experiments. Using more seeds was stifling and inhibited the power of the search, yet using fewer seeds failed to produce good results. The initial seed came from dimension 7 and was of length 50. The other special feature introduced into our DPSO-SIB setup was a grow operator. This operator evaluated each individual in the swarm every generation, determined the location of its longest sub-snake, and then checked to see if any change could be made to the transitions adjacent to the sub-snake's ends in order to lengthen it. If so, the change was made to the individual. Overall, the longest snake the DPSO was able find was a snake of length 86 using relatively standard parameter settings and swarm seeding.

With the SIB problem, the goal is to find the longest snake. Although not reported here, we did run numerous experiments using the bit-string representation version of our DPSO. The results were disappointing. For example, using favorable parameter settings and allowing ample search time, the lengths of snakes found typically did not exceed the upper 30s. Achieving mid-80s with the transition sequence representation indicates that the problem representation is a very crucial aspect to consider for any knowledge engineer working on a difficult real-world problem.

4 Single Element Search Schemes

4.1 Raindrop Optimization

The raindrop optimization search scheme or process is a recently developed local search technique [4, 30] that utilizes stochastic and deterministic methods to improve the quality of solutions. The method performs updates on a single solution yet may use a number of alterations to that solution prior to performing an update on the subsequent solution.

The raindrop optimization heuristic is performed on a single solution candidate S as shown below. It is set to run for N iterations:

1. Generate an initial solution S, and set S_{best} to S.
2. Randomly select one of the components of S, S_c, and change its value randomly.
3. Assess whether any domain constraint violations have occurred.
4. If no domain constraint violations occur, go to step 7.
5. Otherwise, create a list of components of S that contribute to the constraint violations.
6. Repeat the following until the list is empty:
   a. Select and remove from the list the component S'_c that is physically closest to S_c,
   b. Deterministically change the value of S'_c so that the next best choice for it is used, and that the choice does not result in a constraint violation with components physically closer to S_c than S'_c.
   c. If this value causes further constraint violations, add the affected components to the list.
7. If S is better than S_{best}, then set S_{best} to S.
8. Otherwise, if X iterations have passed with no improvement, reset S to S_{best}.
9. If N iterations have passed, then stop, otherwise return to step 2.

The process is different from other heuristics in that it includes both stochastic and deterministic changes to a solution. Unlike many heuristics, the stochastic change is forced into the solution regardless of whether it results in constraint violations. In addition, this search process involves the use of a limited number of user-specified parameters: the number of iterations of the model, and the number of iterations X that pass before S reverts to S_{best} [30].

4.1.1 RO-MFD Setup

We use a bit string representation and the relative likelihood fitness function for the RO-MFD setup (there is no need for a modified version since symptom non-covers are considered invalid diagnoses). We build an initial diagnosis that explains all the symptoms in the patient's symptom set (a super-cover) but not necessarily a maximal super-cover (all diseases associated with each patient's symptom). When making a change to the diagnosis, we delete a disease and then, following the RO scheme, we check for constraint violations or, equivalently, a non-cover situation. Diseases are added back into the diagnosis based on an attempt to remain faithful to the notion of spatial closeness with RO as applied to Forest Planning. That is, we try to add a "close" disease back into the diagnosis that exposes the diagnosis to the fewest number of extra symptoms not in the patient's symptom set.

Experiments varied iterations and reversions, settling on 10,000 iterations and four reversions. Using more reversions intuitively seems to be attractive yet it allows the RO scheme to possibly drift away from quality solutions. Unfortunately, the RO scheme appears to be ill-suited for the MFD problem since we rarely achieved greater than 10% reliability (recall, reliability is a measure of goodness based on all possible symptom sets) and at the expense of over 10 million fitness evaluations. One observation during these experiments was the fact that the RO scheme works solely with valid diagnoses, ones that completely explain the patient's symptoms whereas the GA scheme may have actually produced diagnoses that were very close to but not covers. Our hypothesis was that using the modified relative likelihood might allow non-cover diagnoses to have higher fitness values than super-cover diagnoses. Fortunately, this turned out not to be the case but it leaves us still searching for the reason behind the poor performance.

4.1.2 RO-MSE Setup

Network configurations are integer strings where each position corresponds to a specific MSE component. The fitness is based on connectivity, component requirements, and mission requirements. No real notion of spatial closeness exists within the MSE problem but other RO facets exist. The initial configuration was randomly determined and forced to meet the mission requirements. A configuration that did not support both the radio subscriber and wire subscriber requirements was identified as violating the mission constraints (similar to the adjacency and green-up Forest Planning constraints). Components were incremented and decremented during the iterations and reversions as long as the configuration remained valid.

Iterations varied from 5,000 to 20,000 with the optimal solution being found most of the time using 20,000 iterations. Reversions tended to follow the typical RO pattern where the best solutions were found using reversions set to either 4 or 5. The best single search resulted in the optimal configuration being found after just under 14,000 fitness evaluations. As for the frequency of finding the optimal configuration for the specified mission, our results were typically in the neighborhood of 72% optimal.

4.1.3 RO-FP Setup

For the forest planning problem, the Raindrop optimization algorithm uses the fitness function and other data provided by [4]. It begins with a feasible harvest schedule, and in each iteration of the algorithm, the schedule is modified in the following fashion: A stand u is randomly chosen for modification and its treatment (i.e., whether it is cut at all, and if so in which period) is randomly modified. This modification might violate adjacency constraints. These are repaired by altering the treatments of each stand involved in the violation to restore consistency. Importantly, the stands are chosen for repair in order of increasing distance from u, and each repair must be consistent with all adjacent stands already repaired or else closer to u. The repairs thus radiate outward from u and continue until the schedule satisfies all constraints.

Our test essentially replicates the one found in [4]. The representation scheme is the same as that used by the GA. Specifically, harvest schedules were represented as a sequence of integers $a_1, a_2, \ldots a_{73}$, where each $a_i \in \{1,2,3,4\}$. The value for a_i denotes the period in which stand i is to be harvested (4 indicates that it is not to be harvested at all). For the test, data describing the forest — including adjacencies, x, y coordinates of centroids, and the timber volume obtained for each stand when harvested in a particular period — were stored in lookup tables.

In order to ensure that the algorithm yields increasingly improved schedules, the best schedule encountered thus far is stored. If after several iterations no improvement is found, the algorithm reverts to this best schedule. Following [4], the reversion rate is set at four iterations. This is based on their experiments with reversion rates ranging from 0 to 20. They conclude that a high reversion rate, which allows the algorithm more time to explore the search space, does not lead to better results. This is a conclusion we too arrived at when applying the RO search to our other problem areas.

The algorithm, implemented in Java, was run 50 times, with 100,000 iterations allowed per run. In each run, the initial schedule is simply a sequence of 4s (indicating no harvesting at all takes place). As shown in Table 1 below, the best result obtained was a schedule with fitness 5,500,391 (lower values are better). This matches the solution used as a baseline in [4] and surpasses their raindrop result. We note that 100,000 iterations appears very low to us — it was chosen solely to replicate the experimental setup of [4]. As may be expected, increasing the iterations to 1,000,000 had the effect of producing the same harvest result very often.

Table 1 Target Values

Algorithm	Objective Function
Baseline Target	5,500,391
Raindrop (Bettinger and Zhu 2006)	5,556,343
Raindrop (replicated test)	5,500,391

4.1.4 RO-SIB Setup

The raindrop optimization technique applied to the snake in the box problem produced the weakest result of all the search schemes. In an attempt to get better results, we tried using both a bit string representation and a transition sequence representation, but to no avail. Note that the transition sequence representation is the more conventional approach. For each representation, a slightly modified approach was taken for the application of raindrop optimization. Specifically the differences were with the correction of constraint violations due to the vastly different nature of the two representations. The bit string representation uses a binary array to represent the nodes of the hypercube that are "on" (in the path) and "off" (not in the path) rather than representing an actual path. This representation can make finding the longest path in the bit string difficult. For the transition sequence, each direction a path can travel from any node in the hypercube is assigned a numerical value based on the bit which changes, assuming each node is numbered in binary. The representation is then an array of values showing the direction to the next node of the path. For example, in a 3 dimensional cube a path may be 1, 3, 2, and 3 (but this would not be a length 4 snake).

In the bit string representation, we can count the number of neighbors of a node that are "on" and store this in an array of neighbors following Diaz-Gomez. Then looking at the array, we can quickly see where the "heads" and "tails" of snakes are by those entries with only one neighbor. Likewise, nodes with two neighbors are "body" nodes and any with three or more neighbors represent constraint "violation" nodes. By removing nodes that violate the longest snake in the current bit string solution, we expect the snake to grow longer. This process is the basis of the raindrop optimization applied to bit string snakes.

Starting with a random bit string with one bit for each node, a fitness function is used to locate the longest valid snake and its path is preserved. Next, a change is made by randomly turning on or off one or more nodes in the bit string, regardless of any violations to the best snake. Then we count and store neighbors, and begin correcting violations by turning off neighbors of nodes with three or more neighbors that are currently on. We make our corrections starting with nodes nearest to the changes we made and being sure to leave on nodes of the longest found snake. This continues until there are no nodes with more than two neighbor nodes turned on, radiating the changes outward from the change.

Following the radiating changes, the bit string contains either our original bit sequence snake, or numerous disjoint subsequences of the original snake. Thanks to the raindrop optimization approach, there should be no "violation" nodes. Running the fitness check once more, if the original snake has been improved and a longer snake was found, this longer snake is stored and the process repeats. If no better snake was found, we repeat the whole process for a set number of changes after which we revert back to the bit string with the best snake. Finally, we continue to repeat the process in the hope that a better snake may be found, as described in the raindrop description.

The best settings for the RO bit string representation were around 200 iterations with up to 128 changes and 51 repairs. With such a large number of changes before reversion, and the complexity of the fitness function, only rarely was a good snake found after the first 200 iterations. With these settings, we typically found snakes of length less than 55, and infrequently found snakes with lengths greater than 60, the best being 64.

For the transition sequence representation we first generate a random transition sequence and find the longest snake within it. This snake is then our best snake, and is stored as our valid transition sequence. Then we make one random change to the sequence by randomizing one of the directions taken, thereby twisting part of the snake within the hypercube. This twisting will almost certainly cause at least one constraint violation, so we must make additional changes to the snake. Starting with violations closest to the original transition, we change the direction taken at the node before the constraint violation occurs, continuing to change the transition sequence until no violations exist. Finally, we try to add one random direction to the end of the snake to improve its length (like a one-step grow operation). The rest of the methodology is the same as with the bit string representation, reverting to and storing better sequences as they are found.

Our transition sequence RO-SIB setup would not find snakes much longer than 40 and in effect shortened snakes more often than making them longer. We believe this to be due to the linear nature of the SIB problem, so we tried a minor modification to the RO scheme. Rather than correct closest constraint violations first, we only corrected violations that came after the change, thereby helping to preserve the length of the snake found earlier in the transition sequence. Finally, to further aid preserving the snake, we implemented a weighting factor to determine where to make random changes to the transition sequence, favoring changes to transitions outside of the valid snake; this was based on an adjustable probability. With settings of 10,000 iterations with 51 changes and 12 repairs and a preservation probability of 10%, we found a snake of length 65 and on average produced snakes of length 59.

4.2 Extremal Optimization

EO is a recently developed local search heuristic [5] based on the Bak-Sneppen (BS) model of biological evolution [1]. The BS model demonstrates self-organized criticality, a tendency for systems in statistical physics to organize themselves into non-equilibrium states. Unlike Darwinian evolution, the BS model addresses the evolution of entire ecosystems, called "neighborhoods", instead of individual species. These neighborhoods are made up of several species, each demonstrating relationships with its neighboring species. Thus, when the fitness of a species changes, the fitness of each one of its neighbors has the potential to change as well. Mutation is simulated by replacing the fitness of the least fit species in the neighborhood with a new random fitness. This affects the fitness of its neighbors, and causes a chain reaction through the entire neighborhood [1]. EO individuals are based on these neighborhoods, and each component of the individual represents one of these species. Hence, unlike most optimization techniques, EO focuses on removing poor components of solutions instead of favoring the good ones.

The EO heuristic is performed on a single solution candidate S as follows:

1. Generate an initial value for each component of S, and set S_{best} equal to S,
2. Determine a fitness associated with each component of S,
3. Rank the components by fitness and select the worst to be updated,
4. Assign the selected component a new value with a random fitness to create S',
5. If the fitness of S' is higher than the previous best, S_{best}, replace S_{best} with S',
6. Set S equal to S',
7. Repeat steps 2 through 6 a predetermined number of times.

Unfortunately, in some domains, step 3 can cause the heuristic to become stuck in local optima. To deal with this, Boettcher and Percus introduced a power-law distribution approach for selecting the component to be modified [5, 6]. To select a component using this approach, an integer $1 \leq k \leq N$ is determined from the probability function $P(k) \propto k^{-\tau}$, where N is the number of components in the configuration, and τ is a constant that determines how stochastic or deterministic the selection should be [6]. With higher values of τ, EO is more likely to get stuck in local optima for the same reasons it does without a selection function at all. For lower values of τ, EO sometimes replaces better components of a solution in order to explore a larger part of the search space. If τ is set to zero, EO produces similar results to random search. Because of this stochastic nature, EO can jump in and out of near-optimal solutions at any point in time.

Another issue related to using EO is the method of calculating a fitness contribution for an individual component of a solution. Sometimes this is not even possible given certain criteria in some domains. To overcome this, de Sousa et al. introduced Generalized Extremal Optimization, or GEO [28]. GEO was designed for use on bit-strings. For each component bit in an individual, GEO flips the bit and calculates the new fitness. Each bit is then ranked by the resulting fitness, where higher fitness values are lower ranked. Though GEO causes the heuristic to execute the objective function several more times (once for each bit of an individual), it allows EO to be used in many more domains.

Because of the stochasticity associated with lower values of τ, EO is likely to find an optimal (or near optimal) solution to a problem given adequate time. Unfortunately, problems with extraordinarily large search spaces may require infeasible amounts of time to find these quality solutions. Taking this into account, we created an efficiency measure to calculate how well the heuristic performs with different parameter settings. After performing several runs, an efficiency measure can be calculated by dividing the percentage of optimal solutions found by the number of iterations used as the stop criterion. For example, consider an experiment that found the optimal solution 75 percent of the time after 50 iterations and found the optimal solution 100 percent of the time after 500 iterations. The efficiency measures would be 1.5 and 0.2 respectively, thus demonstrating that even though we can find the optimal solution 100 percent of the time, we might be better off using the parameter settings that only found it 75 percent of the time and running the heuristic multiple times.

4.2.1 EO-MFD Setup

For the MFD problem, Generalized Extremal Optimization (GEO) was used. Two experiments were performed: one to find the parameter settings leading to the highest efficiency measure (for comparison within the EO section) and one to find the parameter settings leading to the highest reliability (recall, reliability is the ratio of optimal diagnoses found to the total number of diagnostic trials — 1023). To find the highest efficiency measure, GEO was used with $0 \leq \tau \leq 6$ in increments of 0.01, and the number of iterations from 10 to 40 in increments of 1. To find the parameters leading to the highest reliability, GEO was used with $0 \leq \tau \leq 3$ (in increments of 0.01), and the range of total GEO iterations went from 100 to 500 (in increments of 5). Representing a diagnosis followed the other setup schemes presented, namely, a bit string representation. The modified relative likelihood fitness function served as the basis for evaluating individual solutions as well.

Our GEO algorithm had the highest efficiency measure using $\tau = 4.66$ and after 16 iterations, producing a reliability of 77.1%. We were able to achieve 100% reliability with $\tau = 1.39$ and after 490 iterations. For comparison, the highest efficiency measure found was 0.048 however the run with the highest reliability had an efficiency measure of 0.002.

4.2.2 EO-MSE Setup

For the MSE problem, initial trials using an evaluation function that assigned a normalized weight to each component in the configuration gave poor results. To fix this, a "direction-based" scheme [20] was included. For each component in the configuration, the direction-based scheme evaluates the fitness twice, once with a decreased component value and once with a higher component value, keeping each component within Corps bounds. The components are then assigned directions of -1, 0, or 1. This corresponds to the component value that is less than, equal to, or greater than the current value and that results in a higher fitness. The EO search is carried out normally, but when a component is selected, it is modified only in the direction provided. In cases where a direction of zero is assigned, selected components are assigned a new random value. This helps to preserve the stochastic variability of the EO algorithm.

As with other EO setups, tests were performed to find the highest efficiency measure and the highest number of optimal solutions over repeated runs. For both tests, direction-based EO was used with τ in the range of $4 \leq \tau \leq 6$ (in increments of 0.01), and the range of total EO iterations was 100 to 20,000 (in increments of 100). At its most efficient point, our EO algorithm found only 27.8% optimal solutions, using 1400 iterations and $\tau = 5$. We were able to achieve perfect results over repeated runs, 100% optimal solutions, using 19,400 iterations and $\tau = 5$. For comparison, the highest efficiency measure found was 0.0199, but for the most optimal solutions, the efficiency was 0.0051. An interesting note is that even limiting the algorithm to 10,000 iterations still led to finding optimal solutions 93.3% of the time.

4.2.3 EO-FP Setup

During our experiments, the issue of whether or not to include the "no-harvest" time period came into question due to the nature of EO. Consequently, we tried two similar approaches on the Forest Planning problem. The experiments were identical, but the second experiment allowed for stands to be unscheduled (i.e., not harvested) and the first did not. Allowing for stands to be left without being harvested allows the EO algorithm to move around in the search space, since there are more valid solutions. EO was used with an integer representation, using the values 1 to 3 for the first experiment (each corresponding to which harvest time period to choose), and 0 to 3 for the second experiment, where a zero for any stand meant it was chosen to not be harvested. In both experiments, τ was tested on the range of $1 \leq \tau \leq 3$ (in increments of 0.05), and the number of iterations was set at 50,000. No tests were done on the efficiency measure because the optimal solutions are not known, only target best known solutions (with the best known produced by our RO-FP setup experiments) are available. Instead, the tests were to find the best value of τ to use in further research on the problem. For the first experiment, no solutions below 20M were found, and there was no obvious improvement for any values of τ. For the second experiment, the best configuration found had a solution of 10,597,074 using $\tau = 1.5$.

4.2.4 EO-SIB Setup

As with the EO-FP setup, the nature of EO led us to try two different representations on the SIB problem in our efforts to find a better known solution. The first was a bit string representation similar to our GA-SIB setup. EO was used with τ in the range of $1 \leq \tau \leq 3$ (in increments of 0.05). Again, because there are no known optimal solutions, the number of iterations was fixed at 10,000. The second representation used was a transition sequence representation similar to some of our other setups.

Since an integer sequence was used, EO was modified to test more possible configurations. A transition sequence of length 100 was used, and for each transition, the fitness function was run 7 times, once for each different value the transition could be (not including its current value). Thus for each iteration, the fitness function was run 700 times. The new fitness values were ranked and their corresponding transitions tracked, and τ-selection was used on all 700 possible changes. This scheme was then modified to only keep track of the transitions that were part of the current snake, since no changes after that would affect the fitness of the new configuration.

With both representations, snake length was the primary indicator of snake strength. But, as with other search schemes where we found disappointing SIB results, we also considered tightness (nodes in the snake that share more neighbors with others are more fit than those that do not), and the number of free nodes (nodes that are not part of the snake path or adjacent to the snake path). The longest snake found using the transition sequence was of length 70 using $\tau = 1.4$, and the only parameter used in the fitness function was the length of the snake. The longest snake found using bit-string representation was of length 74, using $\tau = 1.6$.

5 Results and Observations

In this section, we bring all of the best results together to show the effectiveness of the search schemes used in the experiments (see Table 2 below). Some expected results turned up but some unexpected results turned up as well. Some of the search schemes are in some ways similar to each other — the GA and DPSO search schemes both use populations of individuals to explore the search space, while the RO and EO schemes use single individuals. Similarities among the problems also existed but only with respect to their corresponding search scheme implementation. The MSE, FP, and SIB problems naturally lent themselves to integer sequence representations, while at the same time the MFD, FP, and SIB problems had equally plausible bit string representations. Fitness functions were consistent across problems except for the RO-MFD setup that used the relative likelihood measure instead of the modified relative likelihood (an issue related to diagnoses that did not explain or over explained the patient's symptoms) because of the nature of the raindrop method.

Table 2 Results

	GA Search	DPSO Search	RO Search	EO Search
MFD	87%	98%	12%	100%
MSE	99.6%	99.85%	72%	100%
FP	6,505,676	35M	5,500,391	10M
SIB	95	86	65	74

After good parameter settings were identified, all of the search techniques produced fairly good results, but each technique also had an Achilles' heel. The GA and DPSO, the population based schemes, both exhibited strong MSE and SIB results, mixed MFD results, but surprising FP results (the performance of the particle swarm algorithm is especially surprising). Both the GA and DPSO schemes achieved remarkable MSE results — achieving greater than 99% optimality, indicating that these techniques, when applied to the MSE problem, can consistently produce excellent results. We were pleased to see that the binary DPSO search also performed very well with the MFD problem (98% of the 1023 symptom sets resulted in optional diagnoses). This is the type of reliability that gives us confidence in the appropriateness of this method. Although the GA reliability reached only 87%, this is actually quite good, especially since we restricted the GA-MFD to only simple features (e.g., no embedded local search was performed, and no fancy adaptive operators were used).

With the SIB problem, both the GA and DPSO techniques again achieved excellent results. The SIB problem has by far the largest search space of any of the problems addressed here. The DPSO technique's result of a snake of length 86 is quite good, particularly for a search technique that samples such a relatively small portion of the space. We were especially pleased with the GA's performance on the

SIB problem since the known best length is currently 97. Much of our previous research into the SIB problem has been with the node sequence representation or the transition sequence representation. The use of the bit string representation—which led to such good results—has been very scarcely researched. This is very encouraging and has given us a good deal of motivation to pursue this line of research more extensively.

As for the FP problem, the GA worked fairly well, producing the second best result found, but the DPSO scheme performed well below what we expected it to. This was very discouraging, and we are still uncertain as to the reason for such disastrous results. Much more analysis of the FP problem and tailoring of the DPSO to it is in order; this is high on our list of further research. Particularly, further research into more specialized operators and strategies, such as age-proportional mutation and embedded local search, could allow the population based schemes to perform even better.

The single element techniques also performed well in some cases, while in others they exhibited sub par or otherwise interesting behavior. Clearly, the RO search was expected to perform well on the FP problem, since its origins stem from the forest planning domain and it was tailor-made for problems in that domain. Consequently, it was no surprise that our implementation of RO found the best known solution in short order. What is surprising is that RO had some difficulty with the other problem areas we tested. For example, finding the optimal diagnosis in only approximately 1-in-10 diagnosis sessions is unacceptable MFD performance. The cause is still under investigation but surely it is related to the notion of "spatial closeness", a concept well defined in the forest planning problem but which does not carry over perfectly well to multiple fault diagnosis. The notion of spatial closeness surely hindered the RO technique when applied to the SIB problem as well. We consider a snake of length 65 to be an average length for a search scheme to locate. Note that EO was the best performer when applied to the MFD problem.

The results achieved by the RO algorithm for the MSE problem were satisfactory but not nearly as good as the results from the GA and DPSO schemes. However, the main issue with the RO-MSE results deals with the fact that when RO did not find the optimal solution, no clear pattern was displayed. That is, typically when the GA or DPSO schemes did not find the optimal MSE solution, they did find either the second best or third best solution (a common pattern with these population based schemes). However, non-optimal RO-MSE results exhibited no such similarity, but rather showed quite a bit of diversity; the results ranged from extremely poor to near optimal. Again, EO was the best performer when applied to the MSE problem.

The performance of EO is in stark contrast to that of RO. For the MFD and MSE problems, EO had a perfect record of performance, with perfect reliability for MFD and a perfect results for MSE. It also performed acceptably well with the SIB problem, even though it found the second shortest snake among all of the search techniques. A snake of length 74 is nevertheless a quality result for a single element search. Clearly, however, the GA was the best "snake hunter". The EO technique did have a little trouble with the forest planning problem. We continue to investigate methods to adapt it, just as we do to improve the results obtained in the DPSO-FP and RO-MFD problems.

In summary, all the techniques worked very well with the MSE problem although the RO scheme was not really in the same league with the GA, DPSO, and EO. The population-based schemes (GA and DPSO) did much better on the SIB problem than the single element schemes (RO and EO). The GA was competitive with the RO technique when applied to the forest planning problem, a problem for which RO was custom made. EO did reasonably well with FP but DPSO performed surprisingly poorly. Interesting results came from the MFD problem area where we expected all of the techniques to do very well. Only DPSO and EO gave excellent results while the GA gave good results. The RO technique did not appear to mesh well with the MFD problem. The "silver lining" around our unexpected and unusual results is that these now represent opportunities for further research into these techniques and problems.

References

1. Bak, P., Sneppen, K.: Punctuated equilibrium and criticality in a simple model of evolution. Phys. Rev. Lett. 71(24), 4083–4086 (1993)
2. Bettinger, P., Sessions, J.: Spatial forest planning: to adopt, or not to adopt? J. For. 101(2), 24–29 (2003)
3. Bettinger, P., Chung, W.: The key literature of, and trends in, forest-level management planning in North America, 1950-2001. Int. For. Rev. 6, 40–50 (2004)
4. Bettinger, P., Zhu, J.: A new heuristic for solving spatially constrained forest planning problems based on mitigation of infeasibilities radiating outward from a forced choice. Silva Fennica 40(2), 315–333 (2006)
5. Boettcher, S., Percus, A.G.: Extremal optimization: methods derived from co-evolution. In: GECCO 1999: Proc. Genet. and Evol. Comput. Conf., pp. 825–832. Morgan Kaufmann, San Francisco (1999)
6. Boettcher, S., Percus, A.G.: Extremal optimization for graph partitioning. Phys. Rev. E 64, 26114 (2001)
7. Chang, F.L., Potter, W.D.: A genetic algorithm approach to solving the battlefield communication network configuration problem. In: Yfantis, E.A. (ed.) Intell. Sys. Third Golden West Intern. Conf. Theory and Decision Library D, vol. 15. Kluwer, Dordrecht (1995)
8. Diaz-Gomez, P., Hougen, D.: Genetic algorithms for hunting snakes in hypercubes: fitness function analysis and open questions. In: Seventh ACIS Intern. Conf. on Softw. Eng., Artif. Intell., Netw., and Parallel/Distrib. Comput, SNPD 2006, pp. 389–394. IEEE Computer Society, Los Alamitos (2006)
9. Diaz-Gomez, P., Hougen, D.: The snake in the box problem: mathematical conjecture and a genetic algorithm approach. In: Cattolico, M. (ed.) Proc. 8th annu. conf. on Genet. and evol. comput., pp. 1409–1410. ACM Press, New York (2006)
10. Engelbrecht, A.P.: Fundamentals of Computational Swarm Intelligence. Wiley and Sons, New York (2005)
11. Goldberg, D.E.: Genetic algorithms in search, optimization, and machine learning. Addison-Wesley, Boston (1989)

12. Harary, F., Hayes, J.P., Wu, H.J.: A survey of the theory of hyper-cube graphs. Comput. Math. Appl. 15, 277–289 (1988)
13. Holland, J.H.: Adaptation in natural and artificial systems. University of Michigan Press, Ann Arbor (1975)
14. Kautz, W.H.: Unit-distance error-checking codes. IRE Trans. Electron. Comp. 7, 179–180 (1958)
15. Klee, V.: What is the maximum length of a d-dimensional snake? Am. Math. Mon. 77, 63–65 (1970)
16. Kennedy, J., Eberhart, R.: Swarm intelligence. Morgan Kaufmann, San Francisco (2001)
17. Kennedy, J., Eberhart, R.: Particle swarm optimization. In: Proc. IEEE Intern. Conf. on Neural Netw., pp. 1942–1948. IEEE Service Center, Piscataway (1995)
18. Kochut, K.J.: Snake-in-the-box codes for dimension 7. J. Comb. Math. Comb. Comput. 20, 175–185 (1996)
19. Liepins, G.E., Potter, W.D.: A Genetic Algorithm Approach to Multiple Fault Diagnosis. In: Davis, L. (ed.) Handbook of Genetic Algorithms. Van Nostrand Reinhold, New York (1991)
20. Martin, M., Drucker, E., Potter, W.D.: GA, EO, and DPSO applied to the discrete network configuration problem. In: Proc. Intern. Conf. Genet. and Evol. Methods, GEM 2008, pp. 129–134 (2008)
21. MSE, Mobile Subscriber Equipment System: Reference Guide for the US Army. GTE Tactical Systems, Taunton, MA (1990)
22. Peng, Y., Reggia, J.A.: A probabilistic causal model for diagnostic problem solving, part I: integrating symbolic causal inference with numeric probabilistic inference. IEEE Trans. Syst., Man, Cybern. 17(2), 146–162 (1987)
23. Peng, Y., Reggia, J.A.: A probabilistic causal model for diagnostic problem solving, part II: diagnostic strategy. IEEE Trans. Syst., Man, Cybern. 17(3), 395–406 (1987)
24. Potter, W.D., Pitts, R., Gillis, P., et al.: IDA-NET: an intelligent decision aid for battlefield communications network configuration. In: Proc. 8th IEEE Conf. on Artif. Intell. Appl (CAIA 1992), pp. 247–253. IEEE Computer Society Press, Los Alamitos (1992)
25. Potter, W.D., Miller, J.A., Tonn, B.E., et al.: Improving the reliability of heuristic multiple fault diagnosis via the environmental conditioning operator. Appl. Intell. 2, 5–23 (1992)
26. Pugh, J., Martinoli, A.: Discrete multi-valued particle swarm optimization. In: Proc. 2006 IEEE Swarm Intell. Symp., pp. 103–110 (2006)
27. Reggia, J.A., Nau, D., Wang, P.: Diagnostic expert systems based on a set covering model. Int. J. Man-Mach Stud. 19(5), 437–460 (1983)
28. de Sousa, F.L., Ramos, F.M., Paglione, P., et al.: New stochastic algorithm for design optimization. AIAA J. 41(9), 1808–1818 (2003)
29. Tuohy, D.R., Potter, W.D., Casella, D.A.: Searching for snake-in-the-box codes with evolved pruning models. In: Arabnia, H.R., Yang, J.Y., Yang, M.Q. (eds.) Proc. Int. Conf. Genet. and Evol. Methods (GEM 2007), pp. 3–9. CSREA Press (2007)
30. Zhu, J., Bettinger, P., Li, R.: Additional insight into the performance of a new heuristic for solving spatially constrained forest planning problems. Silva Fennica 41(4), 687–698 (2007)

A Hybrid Intelligent System for Distributed Dynamic Scheduling

Ana Madureira, Joaquim Santos, and Ivo Pereira

Abstract. This chapter addresses the resolution of dynamic scheduling by means of meta-heuristic and multi-agent systems. Scheduling is an important aspect of automation in manufacturing systems. Several contributions have been proposed, but the problem is far from being solved satisfactorily, especially if the scheduling concerns real world applications. The proposed multi-agent scheduling system assumes the existence of several resource agents (which are decision-making entities based on meta-heuristics) distributed inside the manufacturing system that interact with each other in order to obtain optimal or near-optimal global performances.

1 Introduction

A manufacturing system has a natural dynamic nature observed through several kinds of random occurrences and perturbations on working conditions and requirements over time. For this kind of environment it is important to provide the ability to efficiently and effectively adapt, on a continuous basis, existing schedules according to the referred perturbations.

E-business trends are making it crucial for manufacturers to integrate sales-order entry and manufacturing scheduling. Customers want to place their orders via the Internet and have their products built to their specific needs as quickly as possible. This engineer-to-order approach requires that manufacturers build orders in small quantities to deal with an ever-increasing number of market segments. It might even require that manufacturers build orders in lot sizes of one.

New organisational and technological paradigms are needed to respond to the modern manufacturing systems' challenges. The traditional structure of manufacturing industries is constructed upon the three pillars of land, labour and capital.

Ana Madureira, Joaquim Santos, and Ivo Pereira
Computer Science Department, Institute of Engineering - Polytechnic of Porto
GECAD - Knowledge Engineering and Decision Support Research Group, Porto, Portugal
e-mail: amd@isep.ipp.pt, jpe@isep.ipp.pt, i020541@dei.isep.ipp.pt

R. Chiong, S. Dhakal (Eds.): Nat. Intel. for Sched., Plan. and Pack. Prob., SCI 250, pp. 295–324.
springerlink.com

The challenge is to move towards a new structure, which can be described as 'innovative production', founded on knowledge and capital. Future manufacturing solutions must identify multiple perspectives and linkages between novel approaches to customisation, customer response, logistics and maintenance. The current typically linear approach to research, development, design, construction and assembly will be replaced by simultaneous activities in all areas to satisfy global demand and shorten time-to-market [29].

The multi-agent paradigm is emerging as the source of development of solutions to very hard distributed computational problems. This paradigm is based either on the activity of "intelligent" agents which perform complex functionalities or on the exploitation of a large number of simple agents that can produce an overall intelligent behaviour leading to the solution of allegedly almost intractable problems. The multi-agent paradigm is often inspired by biological systems.

The planning of manufacturing systems frequently involves the resolution of a huge amount and variety of combinatorial optimisation problems with an important impact on the performance of manufacturing organisations. Meta-heuristics form a class of powerful and practical solution techniques for tackling complex, large-scale combinatorial problems producing efficient and high-quality solutions. From the literature we can conclude that they are adequate for static problems. However, real scheduling problems are quite dynamic, considering the arrival of new orders, orders being cancelled, machine delays or faults, etc. The scheduling problem in dynamic environments have been investigated by a number of authors, see for example [2, 5, 9, 27].

The main motivation of this work arises from the necessity to define simple and robust approaches to dynamic scheduling resolution taking advantages of the potentialities of meta-heuristics and multi-agent systems (MAS) paradigm.

The proposed approach is based on the decomposition of Extended Job-Shop Scheduling Problem (EJSSP) into Single Machine Scheduling Problems (SMSP). We consider that modeling a scheduling problem by means of its SMSP components is important by itself and also, among other reasons, because it can provide help and insight into the resolution, understanding, managing and modeling of more complex multi-processor problems. In fact, quite often, the single-machine problem appears as a component in larger scheduling problems [3]. Sometimes SMSP are independently solved and the results incorporated into larger and more complex problems. For example, in multistage multiple machine problems there are often critical machines, i.e. bottlenecks, whose processing capacity are lower than necessary. The analysis and treatment of bottlenecks as single-machines may determine the properties of the entire schedules for complex systems [13, 1, 6]. This treatment can be applied to manufacturing systems comprising a network of distributed manufacturing units, or virtual enterprises [7]. Here too, a critical unit may be identified in which the work schedule can restrict the properties of the work schedule in the entire network of units.

In this chapter we will model a manufacturing system by means of multi-agent systems and meta-heuristics technologies, where each agent may represent a processing entity (resource). The objective of the system is to deal with the complex problem of dynamic scheduling in manufacturing systems. Our approach shows that a good global solution for a scheduling problem may emerge from a community of machine agents locally solving their schedules while cooperating with other machine agents that share some relations between the operations/jobs.

The remaining sections are organised as follows: Section 2 summarises some literature review on meta-heuristics and multi-agent systems. In section 3 the scheduling problem under consideration is described. Section 4 presents the MASD-ScheGATS system and describes the implemented mechanisms. Section 5 presents the computational study and a discussion of the obtained results. Finally, the chapter presents some conclusions and puts forward some ideas for future research and development work.

2 Literature Review

Scheduling is an important aspect of automation in manufacturing systems. Scheduling problems arise in a diverse set of domains, ranging from manufacturing to hospitals' settings, transports, computer and space environments, amongst others. Most of these domains are characterised by a great amount of uncertainty that leads to significant system dynamics. Such dynamic scheduling is receiving increasing attention amongst both researchers and practitioners.

Supply chains are evolving to more coupled organisations like virtual enterprises, through maintaining the single entity's autonomy, adaptability and dynamism properties. Such organisations imply organisational and technological developments through agility, distribution, decentralisation, reactivity and flexibility. New organisational and technological paradigms are needed in order to respond to the modern manufacturing systems' challenges [29]. Some recent trends in manufacturing in particular and business in general lead to new approaches regarding the organisation and software architecture, mainly adopting distributed solutions. Multi-agent systems are a promising approach for developing applications in complex domains.

In spite of all previous contributions, no efficient solutions could be found for the scheduling problem. Indeed, the scheduling problem is known to be NP-complete [35]. This fact incites researchers to explore new directions and multi-agent technology and meta-heuristics have been considered important approaches for developing industrial distributed systems. Related work on MAS applications for scheduling can be reviewed in [8]. For additional contributions in scheduling through heuristic techniques, see for example [4, 16, 39, 41, 46].

2.1 Meta-heuristics for Scheduling

Frequently, classical optimisation methods are not efficient enough for the resolution of Job-Shop Scheduling problems [4, 35]. In most cases they are good for solving only some specific and small sized problems. The interesting aspect of new approaches, namely meta-heuristics such as Tabu Search, Simulated Annealing, and

Genetic Algorithms, based on local search, is that they lead, in general, to good solutions in an efficient way, i.e. short computing time and small implementation effort.

Meta-heuristics are a set of computing techniques inspired by biological systems. The distinction between nature-inspired techniques and meta-heuristics is largely counterproductive. Although there are surface-level dissimilarities, the central themes underlying these two classes of heuristics are nearly identical, e.g., intensification versus diversification, mechanisms for escaping local optimum, intelligent design of selection/mutation/crossover operators, and the structure of the fitness landscape. The family of meta-heuristics includes, but it is not limited, to Tabu Search, Simulated Annealing, Adaptive Memory procedures, Scatter Search, Soft Computing, Evolutionary Methods, Ant Systems, Particle Swarm Optimisation and their hybrids. For the literature on this subject, see [14, 46].

In the last few decades, there has been a significant level of research interest in meta-heuristics approaches for solving large real world scheduling problems, which are often complex, constrained and dynamic. Scheduling algorithms that achieve good or near optimal solutions and can efficiently adapt to perturbations are, in most cases, preferable to those that achieve optimal ones but cannot implement such an adaptation. This is the case with most algorithms for solving the so-called static scheduling problem for different settings of both single and multi-machine arrangements. This fact motivated us to concentrate on tools which could deal with such dynamic, distributed scheduling problems, even though, due to the complexity of these problems, it may be impossible to find optimal solutions.

Hybridisation of intelligent systems is a promising research field of computational intelligence focusing on combinations of multiple approaches to develop the next generation of intelligent systems. An important stimulus to the investigations on the hybrid intelligent systems area is the awareness that combined approaches will be necessary if the remaining tough problems in artificial intelligence are to be solved. Meta-heuristics, nature-inspired algorithms, neural computing, machine learning, fuzzy logic systems, evolutionary algorithms, agent-based methods, among others, are well established and their strength and drawbacks are well documented. Recently, hybrid intelligent systems are getting popular due to their capabilities in handling several real world complexities involving imprecision, uncertainty and vagueness [18, 19, 28].

2.1.1 Tabu Search

The basic concept of Tabu Search as described by Glover is "a meta-heuristic superimposed on another heuristic" [36]. The overall approach is to avoid entrapment in cycles by forbidding or penalising moves, which take the solution, in the next iteration, to previously visited points in the solution space previously visited (hence "tabu"). The method is still actively researched, and is continuing to evolve and improve.

As Glover points out, the resulting tendency to deviate from a charted course might be regretted as a source of error but can also prove to be a source of gain.

The Tabu method operates in this way with the exception that new courses are not chosen randomly. Instead the Tabu search proceeds according to the supposition that there is no point in accepting a new (poor) solution unless it is to avoid a path already investigated. This ensures that new regions of a problem solution space will be investigated with the goal of avoiding local minima and ultimately finding the desired solution. To avoid retracing the already used steps the method records recent moves in one or more Tabu lists. The original intent of the list was not to prevent a previous move from being repeated, but rather to ensure it was not reversed. The Tabu lists are historical in nature and form the Tabu search memory. The role of the memory can change as the algorithm proceeds. At initialisation the goal is to make a coarse examination of the solution space, known as diversification, but as candidate locations are identified the search is more focused to produce local optimal solutions in a process of intensification.

In many cases the differences between the various implementations of the Tabu method have to do with the size, variability, and adaptability of the Tabu memory to a particular problem domain. The process in which the Tabu Search seeks to transcend local optimality is based on an aggressive evaluation that chooses the best possible solution at each iteration even when this solution may result in a degradation of objective function value. The recent moves are penalised [36].

Related work with Tabu Search applications for scheduling resolution can be reviewed in [10, 24, 33, 37].

2.1.2 Genetic Algorithms

Genetic Algorithms (GA) were originally proposed by Holland in 1975 and are an attempt to mimic the biological evolution process for discovering good solutions to difficult problems [36]. They are based on a direct analogy to Darwinian natural selection and mutations in biological reproduction. A standard GA maintains a population of solutions throughout the search. It initialises the population with a pool of potential solutions to the problem and seeks to produce better solutions by combining the better of the existing ones through the use of genetic operators, namely selection, crossover and mutation [9].

As a population-based search algorithm, GA creates the initial population via some mechanism. Then, it generates other solutions, which tend to be better, by combining chromosomes, i.e. solutions, using genetic operators for selection, crossover and mutation. First, an initial population of N individuals, which evolves at each generation, is created. Generally, we can say that a generation of solutions is obtained from the previous generation through the following procedure: first, solutions are randomly selected from the current population. Pairs of selected individuals are then submitted to the crossover operation with a given crossover probability P_c. Each descendant is then submitted to a mutation operation with a mutation probability P_m which is usually very small. The ability of the chromosome to solve the problem is determined by its fitness function. The final step in the generation process is the substitution of "bad" individuals of the current population by the new descendants. The algorithm stops after a predefined number, $Nger$, of generations have been created. An alternative stopping mechanism is a limit on computing time.

The interesting aspect of GA is that although guarantees of optimal solutions to problems cannot be given, good solutions are likely to be obtained, within the available time.

In developing a GA, we must have in mind that its performance depends largely on the careful design and setup of the algorithm components, mechanisms and parameters. This includes genetic encoding of solutions, initial population of solutions, evaluation of the fitness of solutions, genetic operators for the generation of new solutions and parameters such as population size, probabilities of crossover and mutation, replacement scheme and number of generations. Some considerations on this topic were evaluated in [9]. Related work concerning GA applications for scheduling resolution can be reviewed in [5, 20, 11, 12, 25].

2.2 *MAS for Scheduling*

Considering the complexity inherent to the manufacturing systems, dynamic scheduling is considered an excellent candidate for the application of agent-based technology. In many implementations of MAS systems for manufacturing scheduling, the agents model the resources of the system and the tasks are scheduled in a distributed way by means of cooperation and coordination amongst agents [28, 31, 22]. When responding to disturbances, the distributed nature of multi-agent systems can also be a benefit to the rescheduling algorithm by involving only the agents directly affected, without disturbing the rest of the community which can continue with their work.

This fact incites researchers to explore new directions and multi-agent technology has been considered an important approach for developing industrial distributed systems [17, 30]. For further works developed on MAS for dynamic scheduling, see [33, 34, 43, 21].

2.2.1 Terms and Definitions

The concept of an agent has found application in a diverse range of subdisciplines of information technology, including software engineering, artificial intelligence, human-computer interaction, distributed and concurrency systems, computer-supported cooperative work, control systems, decision support and electronic commerce.

The main term of multi-agent based computing is an agent. However the definition of the term agent has no common consent. In the last few years most authors have agreed that this definition depends on the domain where the agents are used. However there is a general consensus about its two main abstractions [15, 23]:

- An agent is a computational system that is situated in a dynamic environment and is capable of exhibiting autonomous and intelligent behaviour;
- An agent may be located in an environment that includes other agents. The community of interacting agents, as a whole, operates as a multi-agent system.

The most important common properties of computational agents are as follows [20, 40, 42, 44]:

- act on behalf of their designer or the user they represent in order to meet a particular purpose;
- are autonomous in the sense that they control both their internal state and behaviour in the environment;
- exhibit some kind of intelligence, such as applying fixed rules to reasoning, planning and learning capabilities;
- interact with their environment, and in a community, with other agents;
- are ideally adaptive, i.e., capable of tailoring their behaviour to the changes of the environment without the intervention of their designer.

Additional agent properties, characteristic of particular domains and applications are mobility (when an agent can transport itself to another environment to access remote resources or to meet other agents), genuineness (when it does not forge its identity), credibility or trustworthiness (when it does not communicate false information wilfully) and sociality (when agents work in open operational environments hosting the execution of a multiplicity of agents, possibly belonging to different stakeholders (think, e.g., of agent-mediated marketplaces)).

2.2.2 MAS Technologies, Platforms and Methodologies

Multi-agent architectures are fundamental for MAS development, as they establish the system performance. As far as software technology is concerned, agents and MAS provide a wide array of models, techniques, formal modeling approaches and development methodologies that all together shape the general-purpose paradigm of agent-oriented software engineering (AOSE) [23]. Several programming languages and software development environments have been developed which not only support MAS programming, but also implement key concepts of MAS in a unified framework. Figure 1 provides a sampling of important developments from the last decade.

2.2.3 Coordination Aspects

In this section we will focus on the coordination, cooperation, and negotiation terms, which are extensively mentioned in scientific research but whose definitions are often unclear and overlapping.

Most recent publications agree that the definition of MAS depends on the domain where agents are used. We believe that a solid and generic definition is offered by [40] that defines MAS as a system capable of "*solving complex problems in a distributed fashion without the need for each agent to know about the whole problem being solved*". This definition and many others imply the existence of an organisational model, or management mechanism, that allows for interaction, communication and achievement of objectives within the group of agents, regardless of whether the system is comprised of self-interested, group-oriented, or both types of agents.

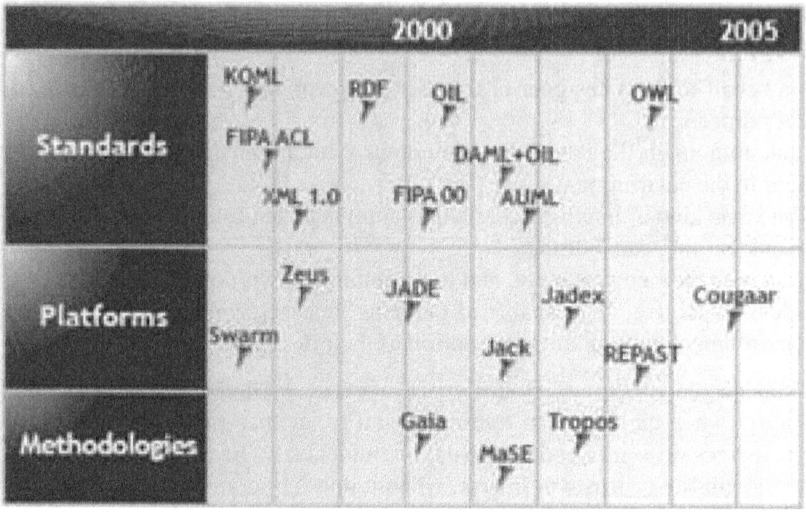

Fig. 1 Current standards, platforms and methodologies of agent-oriented software engineering [23]

A proper definition of such mechanism is given by the definition of the word coordination, particularly if one considers coordination applied to MAS. We can find several reasons why agents need to be coordinated:

- Chaos and anarchy prevention - coordination is required when the software/ algorithm design requires a well-defined separation of concerns and a correspondingly defined reassembly of partial solutions;
- Global restrictions - agents must adhere to a set of restrictions in order to perform successfully;
- Knowledge, resources and information are distributed - agents can have different capacities and specialisations. Alternatively, agents can have different information sources, resources, responsibilities and limitations;
- Dependency between agents' actions - agents' objectives are normally interdependent, requiring that their activities be coordinated;
- Efficiency - autonomous agents can work in an independent way, but at the same time they need to coordinate their actions with others. The information held by an agent can be sufficient for a group of agents to solve a problem quickly.

Cooperation and negotiation are the two main types of coordination [31] developed in the literature and implemented systems. Cooperation is generally defined in the literature as the act of combining efforts, in order to accomplish a common objective that one autonomous agent alone cannot reach by itself. In other words, "cooperation refers to a coordination among non-disputant agents". Such a process is suited when agents need to share tasks or results as a way to reach the system's objective. On the other hand, "negotiation is the coordination among competitive or simply self-interested agents". Negotiation is then the process in which at least two agents negotiate through a protocol in order to accomplish an agreement. This protocol is suitable for systems made of agents that must pursuit their own individual goals but need interaction with other agents to achieve a satisfactory outcome. The buyer-seller situation is the best analogy to demonstrate this protocol.

3 Extended Job-Shop Scheduling Problem

Real world scheduling problems have received a lot of attention in recent years. In this work we consider the resolution of realistic problems. Most real world multi-operation scheduling problems can be described as dynamic and extended versions of the classic Job-Shop scheduling combinatorial optimisation problem.

In practice, many scheduling problems include further restrictions and relaxation of others [38]. Thus, for example, precedence constraints among operations of the different jobs are common because, often, mainly in discrete manufacturing, products are made of several components that can be seen as different jobs whose manufacture must be coordinated. Additionally, since a job can be the result of manufacturing and assembly of parts at several stages, different parts of the same job may be processed simultaneously on different machines (concurrent or simultaneous processing).

Moreover, in practice, scheduling environments tend to be dynamic, i.e. new jobs arrive at unpredictable intervals, machines break down, jobs can be cancelled and due dates and processing times can change frequently.

The main elements of the scheduling problem treated in this work could be modelled as shown in the following subsections.

3.1 Jobs

A set of multi-operation jobs $J_1,...,J_n$ has to be scheduled. d_j is the due date of job J_j. t_j is the initial processing time of job J_j. r_j is the release time of job J_j. Additionally we consider:

- the existence of operations on the same job, on different parts and components, processed simultaneously on different machines, followed by components' assembly operations (multi-level jobs);
- the existence of different job release dates r_j and due dates d_j;
- the possibility of job priorities definition, reflecting the importance of satisfying their due dates, being similar to the weight assigned to jobs in scheduling theory;
- precedence constraints among operations of the different jobs;
- the existence of operations on the same job, with different parts and components, processed simultaneously on different machines;
- new jobs can arrive at unpredictable intervals;
- jobs can be cancelled;
- changes in task attributes can occur: processing times, date of delivery and priorities.

3.2 Operations

Each operation O_{ijkl} is characterised by the index (i, j, k, l), where i defines the machine where the operation k of job j is processed and l the graph precedence operation level (level 1 corresponds to initial operations, without precedents). Additionally we consider:

- precedence constraints among operations of the different jobs;
- each job J_j consists of one or more operations O_{ijkl}, where:
  - IO_{ijkl} is the time interval for starting operation O_{ijkl};
  - r_{ijkl} is the release time of operation O_{ijkl};
  - t_{ijkl} is the earliest time at which O_{ijkl} can start;
  - T_{ijkl} is the latest time at which O_{ijkl} can start;
  - p_{ijkl} is the processing time of the operation O_{ijkl};
  - C_{ijkl} is the k operation completion time from job j, level l on the machine i.
- each operation O_{ijkl} must be processed on one machine of the set M_i, where p_{ijkl} is the processing time of operation O_{ijkl} on machine M_i;
- the existence of operations on the same job, on different parts and components, processed simultaneously on different machines, followed by components assembly operations (multi-level jobs).

3.3 Machines

The shop consists of a set of machines $M_1, ..., M_n$. Additionally we consider:

- a machine can process more than one operation of the same job (recirculation);
- the existence of alternative machines, identical or not.

In this work, we define a job as a manufacturing order for a final product that could be Simple or Complex (Fig. 2). It may be Simple, like a part, requiring a set of operations to be processed. We define it as Simple Product. Complex Products, requiring processing of several operations on a number of parts followed by assembly operations at several stages, are also dealt with.

We consider the existence of two different types of tasks:

- jobs with linear structure - where operations are sequentially processed, considering that an operation can be processed when its precedent has already been finished. Job-Shop benchmark tests referred in the literature are of this type [32];
- jobs with concurrent operations - where operations of same task can be processed simultaneously. An operation can have more than one precedent operation and more than one succeeding operation (Fig. 2). This category is common in Complex Products.

Moreover, in practice, scheduling environments tend to be dynamic, i.e. new jobs arrive at unpredictable intervals, machines break down, jobs are cancelled and due

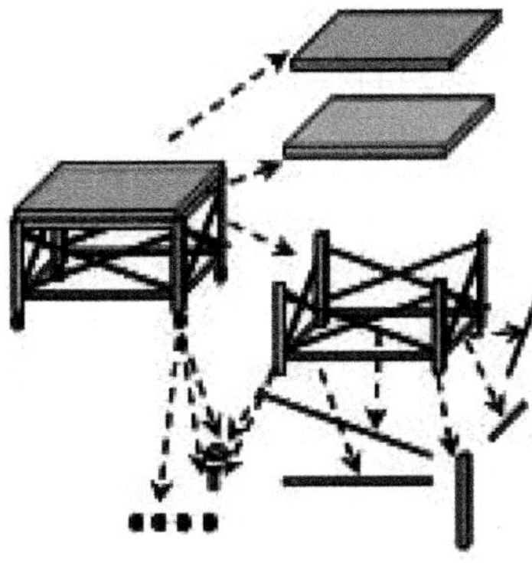

Fig. 2 Bill of materials for a Complex Job with concurrent operations

dates and processing times change frequently. This non-basic Job-Shop Scheduling Problem (JSSP) [38], which we call the Extended Job-Shop Scheduling Problem (EJSSP), has major extensions and differences in relation to the classic or basic JSSP [38]. The existence of operations on the same job, on different parts and components, processed simultaneously on different machines, followed by components' assembly operations, which characterises EJSSP [24], is not typical of scheduling problems addressed in the literature. However, such is common in practice. This approach to job definition, emphasising the importance of considering complex jobs, which mimic customer orders of products, is in accordance with real world scheduling in manufacturing.

4 Multi-agent System for Distributed Manufacturing Scheduling with Tabu Search and Genetic Algorithms

This section describes our Multi-Agent System for Distributed Manufacturing Scheduling with Genetic Algorithms and Tabu Search (MASDScheGATS) where a community of distributed, autonomous, cooperating and asynchronously communicating machines tries to solve scheduling problems.

Distributed environment approaches are important in order to improve scheduling systems' flexibility and capacity to react to unpredictable events. It is accepted that new generations of manufacturing facilities, with increasing specialisation and integration, add more problematic challenges to scheduling systems. For that reason, issues like robustness, regeneration capacities and efficiency are currently critical elements in the design of manufacturing scheduling system and encourage the development of new architectures and solutions, leveraging the MAS research results.

Many real world optimisation problems take place in constantly changing environments. Because of this, optimisation algorithms, instead of looking for single optimal solutions, should continuously track the optimum through time. Real world scheduling requirements are related with complex systems operated in dynamic environments. This means that they are frequently subject to several kinds of random occurrences and perturbations, such as new job arrivals, machine breakdowns, employees sickness, jobs cancellation and due date and time processing changes, resulting in prepared schedules becoming easily outdated and unsuitable.

The work reported in this chapter is concerned with the resolution of realistic scheduling problems, called Extended Job-Shop Scheduling Problems here. Moreover, it is concerned with the integrated scheduling of jobs which are products composed of several parts or components that may be submitted to a number of manufacturing and multi-level assembly operations.

The scheduling problem defined in section 3 is decomposed into a series of Single SMSP which are independently solved, by a meta-heuristic, and results incorporated into the main EJSSP problem. MASDScheGATS system, which is a Team-Work based MAS, assumes the existence of several resource agents that interact and cooperate with other agents in order to obtain optimal or near-optimal global performances. The idea is that from local, autonomous and often conflicting agents' behaviour, a global solution emerges from a community of resource agents locally solving their schedules through cooperative negotiation with other resource agents in order to increase schedule quality.

Resource agents (which have a meta-heuristic associated) represent a SMSP and are responsible for scheduling this resource. We can call resource agents as meta-heuristic agents, considering that each resource agent performs a meta-heuristic algorithm. As the resource agents find their respective local optima, they must negotiate with other agents in order to overcome inter-agent constraints and achieve a global schedule.

The idea is that each agent adopts the meta-heuristics (TS or GA) in accordance with the problem being solved: the method and/or parameters can change at runtime, the agents can use a different meta-heuristic technique according to problem characteristics (namely the problem size: for smaller sizes, the agents use GA; for problems with more jobs they use TS, considering efficiency constraints for example).

In the computational study section we consider two different scenario simulations, where the first is based on TS and the second is based on GA, in order to compare their performance.

4.1 MASDScheGATS Scheduling System

We start by focusing on the solution of the dynamic deterministic EJSSP problems. For solving these we developed a framework, leading to a dynamic scheduling system (Fig. 3) having a hybrid scheduling system as a fundamental scheduling tool, with two main pieces of intelligence.

One such piece is a hybrid scheduling module that could be a combination of Tabu Search and Genetic Algorithm based method and a mechanism for inter-machine activity coordination. The objective of this mechanism is to coordinate the operation of machines, taking into account the technological constraints of jobs, i.e. job operations precedence relationships, towards obtaining good schedules. The other piece is a dynamic adaptation module that includes mechanisms for

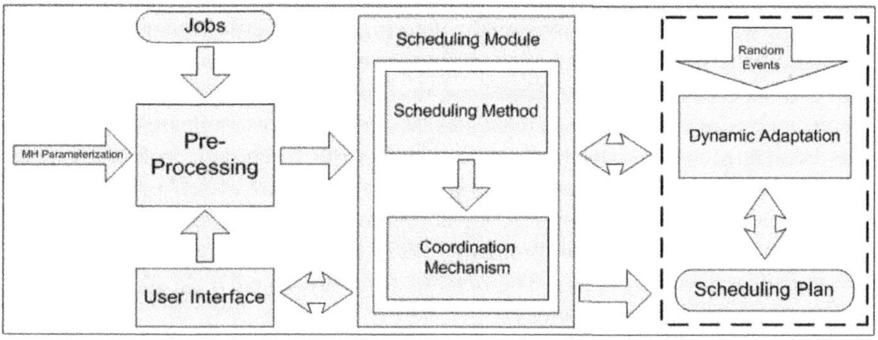

Fig. 3 MASDSCHEGATS System

neighbourhood/population regeneration under dynamic environments, increasing or decreasing it according to new job arrivals or cancellations.

4.1.1 Hybrid Scheduling Module

In this work solutions are encoded by the direct representation, where the schedule is described as a sequence of operations, i.e. each position represents an operation index with initial and final processing times. Each operation is characterised by the index (i, j, k, l), where i defines the machine where the operation k is processed, j the job it belongs to, and l the graph precedence operation level (level 1 (one) corresponds to initial operations, without precedents) as described in section 3.2.

Table 1 Notation

m - Number of machines
n - Number of jobs
l - Operation level defined on precedence graph
O_{ijkl} - Operation k from job j, to be processed on machine i with level l (defined on precedence graph)
IO_{ijkl} - Time interval for starting operation O_{ijkl}
d_j - Due date of job j
t_j - Initial processing time of job j
r_j - Release time of job j
r_{ijkl} - Release time of operation O_{ijkl}
t_{ijkl} - The earliest time at which O_{ijkl} can start
T_{ijkl} - The latest time at which O_{ijkl} can start
p_{ijkl} - Processing time of the operation O_{ijkl}
C_{ijkl} - Operation completion time from job j, level l on the machine i
L_j - Lateness ($L_j = C_j - d_j$)
T_j - Tardiness ($T_j = max\{L_j, 0\}$)

Initially, we start by decomposing the deterministic EJSSP problem into a series of deterministic Single Machine Scheduling Problems (SMSP). We assume the existence of different and known job release times r_j, prior to which no processing of the job can be done, and also job due dates d_j. Based on these, release dates and due dates are determined for each SMSP and, subsequently, each such problem is solved independently by a TS or a GA, where we need to take care of self-parametrisation as we explain in 4.1.5. Afterwards, the solutions obtained for each SMSP are integrated to obtain a solution to the main EJSSP problem instance. The scheduling method is described in Table 2. The objective is to minimise makespan, $1|r_j|C_{max}$, for SMSP subject to release dates r_j.

Table 2 Scheduling method algorithm

1st PHASE	**Finding a 1st job shop schedule based on single machine scheduling problems**
Step 1	Define completion time estimates (due dates) for each operation of each job.
Step 2	Define the interval between starting time estimates (release times) for all operations of each job.
Step 3	Define all SMSP $1\|r_j\|C_{max}$ based on information defined on Step 1 and Step 2.
Step 4	Solve all SMSP $1\|r_j\|C_{max}$ with those release times and due dates using TS or GA.
Step 5	Integrate all the obtained near-optimal solutions into the main problem.
2nd PHASE	**Check feasibility of the schedule and, if necessary apply the Inter-Machine Activity Coordination Mechanism (IMACM) coordination mechanism**
Step 6	Verify if they constitute a feasible solution and terminate with a local optimum; If not, apply a repairing mechanism.

The completion due times for each operation of a job are derived from job due dates and processing times by subtracting the processing time from the completion due time of the immediately succeeding job operation (equation 1).

$$C_{ijkl-1} = C_{ijkl} - p_{ijkl} \quad (1)$$

This procedure begins with the last job operation and ends with the first. When an operation is precedent to more than one operation, i.e. there exists a multilevel structure, the completion due time is the lower value as defined in equation 2.

$$C_{ijkl-1} = min\{C_{ijkl} - p_{ijkl}\}\forall l > l-1 \quad (2)$$

The operations' starting due time intervals $[t_{ijk}, T_{ijk}]$ are also defined considering the job release times and the operation processing times. The earliest starting time t_{ijk} corresponds to the time instant from which the operation processing can be started. The latest starting time T_{ijk} corresponds to the time at which the processing of the operation must be started in order to meet its completion due time (due date). This means that no further delay is allowed. When an operation has more than one precedent operation, i.e. there exists a multilevel structure, the interval $[t_{ijk}, T_{ijk}]$ is the interval intersection from precedent operations correlated by the respective processing times. At this stage, only technological precedence constraints of operations and job due dates will be considered for defining completion and starting times.

The starting time interval (STI) for operations without precedents is defined as follows:

$$STI_{ijkl} = [r_{ijkl}, C_{ijkl} - p_{ijkl}] \quad (3)$$

The starting time interval of an operation with one precedent operation is defined by equation 4.

$$STI_{ijkl} = [t_{ijkl-1}, T_{ijkl-1}] + p_{ijkl-1} \quad (4)$$

The starting time interval of an operation with more than one precedent operation is the intersection interval of the starting time intervals from all precedent operations correlated by the respective processing times (equation 5).

$$STI_{ijkl} = [t_{ijK1L} + p_{ijK1L}, T_{ijK1L} + p_{ijK1L}] \bigcap [t_{ijK2L} + p_{ijK2L}, T_{ijK2L} + p_{ijK2L}] \bigcap \cdots \bigcap [t_{ijKnL} + p_{ijKnL}, T_{ijKnL} + p_{ijKnL}] \text{ with } Kn < k \text{ and } L < l \quad (5)$$

Following this procedure it is easy to deal with situations where an operation has more than one precedent operation, i.e. there exists a multilevel task structure involving assembly operations. In this case the previous operations may be processed simultaneously, and therefore a relaxation of an underlining characteristic of the EJSSP problems is assumed. This situation is typical of real world manufacturing requirements. This means that a more generalised and realistic problem is dealt with the scheduling approach adopted in this work.

At this stage, only technological precedence constraints of operations and job due dates will be considered for defining completion and starting times (Fig. 4). The release date r_j correspond to the earliest starting times of each operation. The due date d_j correspond to the operation completion times. Note that at this point, the notations r_j and d_j refer to single machine problems.

4.1.2 Coordination Mechanism

The integration of the SMSP solutions may give an unfeasible schedule to the EJSSP. This happens because we only consider the precedence relationships and job due dates in each single machine. The interrelated activity of the several machines is not taken into consideration. This means that inter-machine activity coordination is not done at this stage.

This is why schedule coordination may be necessary to obtain a feasible solution. The coordination mechanism named Inter-Machine Activity Coordination Mechanism (IMACM) carries this out. The repairing is carried out through coordination of the machines' activity, taking into account job operation precedence and other problem constraints. This is done by keeping job allocation order unchanged in each machine. The IMACM mechanism establishes the starting and the completion times

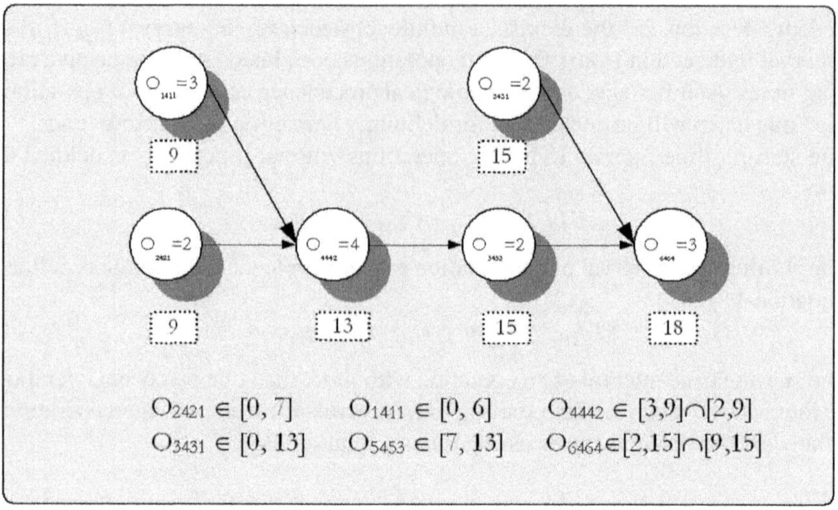

Fig. 4 Precedence Graph with operation completion due times and starting times

for each operation. It ensures that the starting time for each operation is the higher of the two following values (Table 3):

- the completion time of the immediately precedent operation in the job, if there is only one, or the highest of all if there are more;
- the completion time of the immediately precedent operation on the machine.

The method developed and just described is in line with reality and away from the approaches that deal solely with static and classic or basic job-shop scheduling problems. Thus, the method is likely to perform worse than the best available algorithms found for such problems. However, comparison is not our purpose; it would not be reasonable to rate our method against such good performing algorithms for academic JSSP. Our aim is to provide an efficient tool, which we think we have achieved with our method, for obtaining good solutions, for a variety of criteria, for many real world scheduling problems, which we named Extended JSSP. For these problems, the referred best performing algorithms are unable to solve the EJSSP as

Table 3 Inter-Machine Activity Coordination Mechanism

Step 1	Start from initial operations, without precedents, which correspond to level 1 on the sequence for processing on the machine. At this level, the starting and the completion times are the same, i. e., they are equal to those defined by the scheduling algorithm at the previous phase.
Step 2	At level 2, we will have all the operations which are immediately precedents (defined on the precedence graph), on the machine sequence already scheduled.
Step 3	The process will be repeated until all the operations have been scheduled.

described here. Furthermore, our literature survey did not reveal any similar method to solve the EJSSP as we describe here.

4.1.3 Illustration Example

Let us consider the following example with 6 machines and 4 jobs. A precedence graph representing the ordered allocation of machines for each job operation, i.e. the machine sequence, as well the processing times, release times, and due dates, is shown in Table 4.

Table 4 Illustration Example

Jobs	Machine sequence	Processing times	Release Dates	Due Dates
1	$1 \rightarrow 2 \rightarrow 3$	$p_{1111} = 10$ $p_{2122} = 8$ $p_{3133} = 4$	0	27
2	$2 \rightarrow 1 \rightarrow 4 \rightarrow 3$	$p_{2211} = 8$ $p_{1222} = 3$ $p_{4233} = 5$ $p_{3244} = 6$	0	33
3	$1 \rightarrow 2 \rightarrow 4$	$p_{1311} = 4$ $p_{2322} = 7$ $p_{4333} = 3$	0	18
4	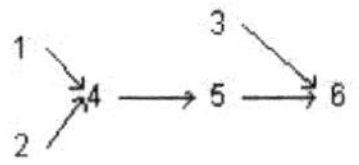	$p_{1411} = 2$ $p_{2411} = 3$ $p_{3411} = 2$ $p_{4422} = 4$ $p_{5433} = 2$ $p_{6444} = 3$	0	18

The application of the scheduling method to the example is described below.

Step 1: Define completion time estimates (due dates) for each operation of each job

For example, for job 1, the final operation completion due time is $C_{3133} = 27$ and corresponds to job due date, as can be seen in the processing precedence graph in Figure 5. The operation completion due time C_{ijkl} is calculated as shown above, and for the second operation that is processed in machine 2 the completion due time is:

$$C_{2122} = C_{3133} - p_{3133} \Rightarrow C_{2122} = 23 \tag{6}$$

Using the same procedure for the remaining operations all other completion due times can be derived (Fig. 5).

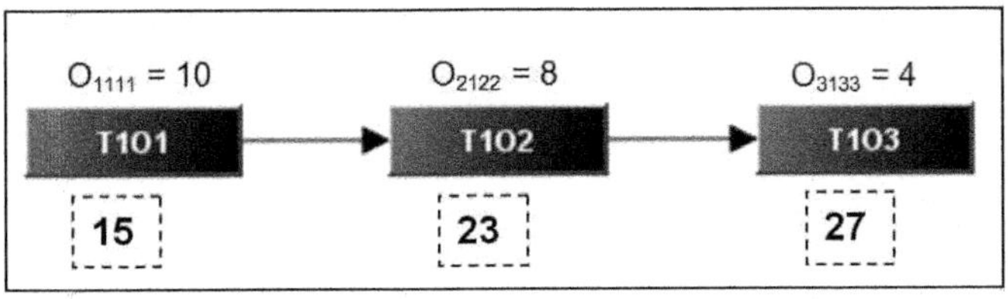

Fig. 5 Processing Precedence Graph with operation completion due times

Step 2: Define the interval between starting time estimates (release times) for all operations of each job

The starting time interval corresponds to the time interval when the operation processing must be initiated, i.e. job release must take place, in order to guarantee that job due date is achieved. The completion due time C_{1111} is 15, the release time r_{1111} is 0 and the processing time p_{1111} is 10. Thus, the processing of operation O_{1111} must be initiated within the time interval [0, 5]. Operation $O_{2122} \in [10,15]$ and $O_{3133} \in [18,23]$.

The starting time interval of an operation with more than one precedent operation is the intersection interval of the starting time intervals from all precedent operations correlated by the respective processing times.

Table 5 Operation attributes of the $1|r_j|C_{max}$

	M1				M2				M3			M4			M5	M6
Jobs	**1111**	**1222**	**1311**	**1411**	**2122**	**2211**	**2322**	**2411**	**3133**	**3244**	**3411**	**4233**	**4333**	**4422**	**5433**	**6444**
p_j	10	3	4	3	8	8	7	2	4	6	2	5	3	4	2	3
r_j	0	8	0	0	10	0	4	0	18	16	0	11	11	3	7	9
d_j	15	22	8	9	23	19	15	9	27	33	15	27	18	13	15	18

Step 3: Define all SMSP $1|r_j|C_{max}$ based on information defined in Step 1 and Step 2

The data for the $1|r_j|C_{max}$ problem corresponding to the 6 machines and shown in Table 5, is determined as previously described in the previous steps. The release times r_j are the earliest starting due times of each operation. The due dates d_j correspond to the operation completion due times. Note that using the notation r_j and d_j at this point, we consider single machine cases.

Step 4: Solve all SMSP $1|r_j|C_{max}$ with those release times and due dates using the TS or GA

Applying the Tabu Search previously described to each $1|r_j|C_{max}$ problem, we obtain the following solutions for the machines 1, 2, 3 and 4, respectively: (1311, 1411, 1111, 1222), (2411, 2322, 2211, 2122), (3411, 3133, 3244), (4422, 4333, 4233). Machines 5 and 6 do not need an optimisation process (one single solution).

Step 5: Integrate all the obtained near-optimal solutions into the main problem

The solution (O_{4422}, O_{4333}, O_{4233}) means that the processing order of operations on machine 4 is: first, one of job 4, O_{4422}, then, one of job 3, O_{4333} and finally one of job 2, O_{4323}.

At this point, it is necessary to apply a decoder to each solution, in such a way that starting times are defined for all operations.

The obtained schedules are characterised by the following starting times (Ti) and completion times (Tc) as it is shown in following Table 6.

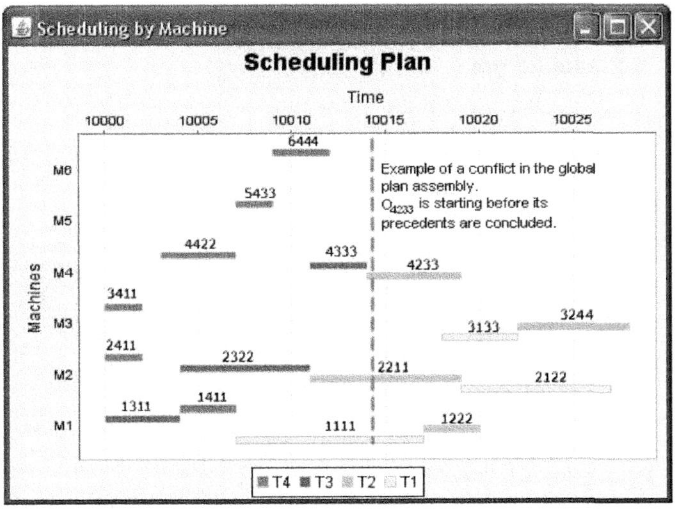

Fig. 6 Obtained solution (unfeasible)

Table 6 Obtained results by the decoder

	M1				M2				M3			M4			M5	M6
Jobs	**1111**	**1222**	**1311**	**1411**	**2122**	**2211**	**2322**	**2411**	**3133**	**3244**	**3411**	**4233**	**4333**	**4422**	**5433**	**6444**
$\mathbf{p_j}$	7	17	0	4	19	11	4	0	18	22	0	14	11	3	7	9
$\mathbf{r_j}$	17	20	4	7	27	19	11	2	22	28	2	19	14	7	9	12

In this example, the union of the several local optimum solutions, one for each $1|r_j|C_{max}$ problem, could not produce a feasible solution due to processing overlap of operations belonging to the same job, as we can see in Figure 6.

For example, O_{4233} processing is scheduled to start even before O_{1222} and O_{2211} have been finished, the precedent operations, violating the precedence constraints. An identical situation occurs with jobs 1 and 4. This happens because we only consider the precedence relationships and job due dates in each single machine. The interrelated activity of the several machines is not taken into consideration. This means that inter-machine activity coordination is not done at this stage, i.e., no mechanism for repairing the schedule has been applied.

Table 7 Obtained results

	M1				M2				M3			M4			M5	M6
Jobs	**1111**	**1222**	**1311**	**1411**	**2122**	**2211**	**2322**	**2411**	**3133**	**3244**	**3411**	**4233**	**4333**	**4422**	**5433**	**6444**
$\mathbf{p_j}$	7	19	0	4	19	11	4	0	27	31	0	22	11	7	11	13
$\mathbf{r_j}$	17	22	4	7	27	19	11	2	31	37	2	27	14	11	13	16
Level	2	3	0	1	3	2	1	0	4	5	0	4	3	2	0	0

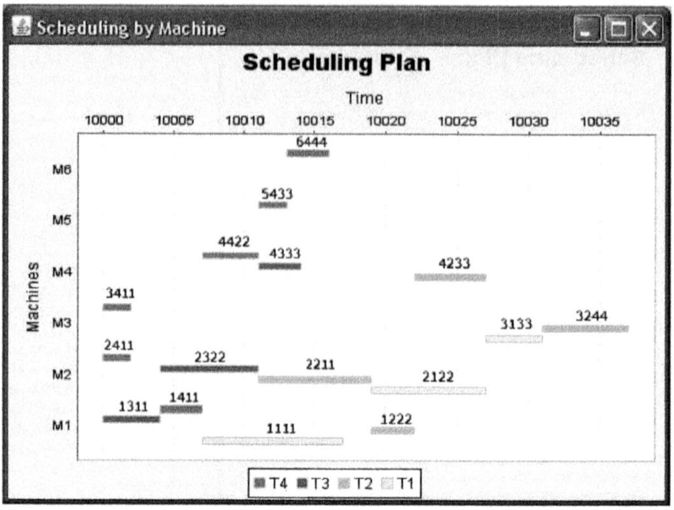

Fig. 7 Final feasible solution

Step 6: Verify if they constitute a feasible solution and terminate with a local optimum; If not, apply a repairing mechanism

We apply a repairing mechanism to implement step 6 of the algorithm, named Inter-Machine Activity Coordination Mechanism (IMACM).

Applying the coordination mechanism we obtain the values presented in Table 7, where *Ti'* is the new starting time, *Tc'* the completion time and level the processing order established by the IMACM mechanism.

As a result of the joint application of the Tabu Search and the IMACM mechanism, feasible solutions can be obtained meeting all job due dates. The feasible solution for the given example is shown in Figure 7.

4.1.4 Dynamic Adaptation Module

For non-deterministic problems some or all parameters are uncertain, i.e. are not fixed as we assumed in the deterministic problem. Non-determinism of variables has to be taken into account in real world problems. For generating acceptable solutions in such circumstances our approach starts by generating a predictive schedule, using the available information and then, if perturbations occur in the system during execution, the schedule may have to be modified or revised accordingly, i.e. rescheduling is performed.

In the scheduling system for EJSSP, rescheduling is necessary due to two classes of events [25]:

- partial events which imply variability in jobs or operations' attributes such as processing times, due dates and release times;
- total events which imply variability in neighbourhood structure, resulting from either new job arrivals or job cancellations.

Considering the processing times involved and the high frequency of perturbations, rescheduling all jobs from the beginning should be avoided. However, if work has not yet started and time is available, then an obvious and simple approach to rescheduling would be to restart the scheduling from scratch with a new modified solution which takes into account the perturbation, for example a new job arrival. When there is not enough time to reschedule from scratch or job processing has already started, a strategy must be used which adapts the current schedule taking into consideration the kind of perturbation occurred.

The occurrence of a partial event requires redefining job attributes and a re-evaluation of the schedule objective function. A change in job due date requires the re-calculation of the operation starting and completion due times of all respective operations. However, changes in the operation processing times only requires re-calculation of the operation starting and completion due times of the succeeding operations. A new job arrival requires definition of the corresponding operation's starting and completion times and a regenerating mechanism to integrate all operations on the respective single machine problems. In the case of a job cancellation, the application of a regenerating mechanism eliminates the job operations from the SMSP where they appear.

After the insertion or deletion of genes, population regeneration is done by updating the size of the population and ensuring a structure identical to the existing one. Then the scheduling module can apply the search process for better solutions with the newly modified solution.

a) Job arrival integration mechanism

When a new job arrives to be processed, an integration mechanism is needed. It analyses the job precedence graph that represents the ordered allocation of machines to each job operation, and integrates each operation into the respective single machine problem. Two alternative procedures could be used for each operation: either randomly select one position to insert the new operation into the current solution/chromosome or use some intelligent mechanism to insert this operation in the schedules, based on job priority, for example.

b) Job elimination mechanism

When a job is cancelled, an eliminating mechanism must be implemented so the correspondent position/gene will be deleted from the solutions.

c) Regeneration mechanisms

After integration/elimination of operations is carried out, by inserting/deleting positions/genes in the current solution/chromosome, population regeneration is done by updating its size. The population size for SMSP is proportional to the number of operations.

After the dynamic adaptation process, the scheduling method could be applied to search for better solutions with the modified solution as illustrated in Figure 2.

4.1.5 Meta-heuristics Self-configuration Properties

In this chapter we consider that meta-heuristics' self-parameterisation could permit a better adaptation to the dynamic situation being considered.

Meta-heuristics can be adapted to deal with dynamic problems, reusing and changing solutions/populations in accordance with the dynamism. We will use the Dynamic Adaptation Mechanisms defined in [25] for SMSP, that includes a method for neighbourhood regeneration under dynamic environments, increasing or decreasing the neighbourhood's size according to new job arrivals or cancellations.

4.2 Hybrid Multi-agent Architecture

In this chapter we describe a system where a community of distributed, autonomous, cooperating and asynchronously communicating machines tries to solve scheduling problems.

The proposed Team-Work based approach is rather different from the ones found in the literature, as we try to implement a system where each agent (Resource Agent) is responsible for optimising the scheduling of operations for one machine through TS or GA. This considers a specific kind of social interaction – cooperative problem solving (CPS), where the group of agents work together to achieve a good solution for the problem.

Each resource agent must be able to find an optimal or near optimal local solution through Tabu Search meta-heuristics (or Genetic Algorithms), to deal with system dynamism (new jobs arriving, cancelled jobs, changing attributes, etc), to change/adapt the parameters of the basic algorithm according to the current situation, to switch from one meta-heuristic algorithm to another (using a TS or GA) and to cooperate with other agents.

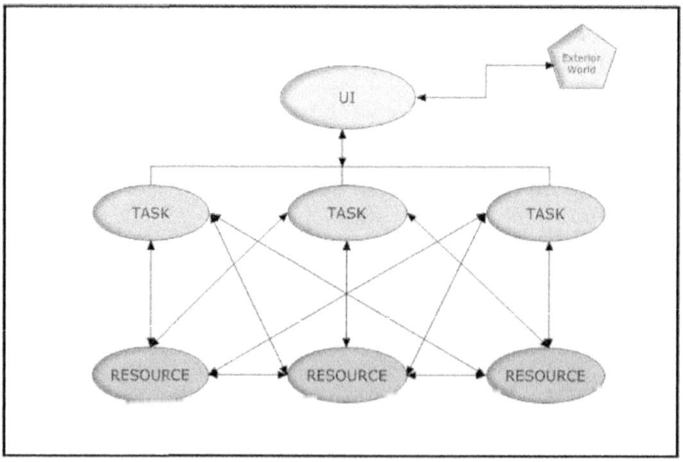

Fig. 8 MASDScheGATS System Architecture

The original scheduling problem defined in section 3 is decomposed into a series of Single Machine Scheduling Problems (SMSP) [26]. The resource agents (which have a meta-heuristic associated) obtain local solutions and later cooperate in order to overcome inter-agent constraints and achieve a global schedule.

The proposed Team-Work architecture is based on three different types of agents (Fig. 8). In order to allow a seamless communication with the user, a User Interface (UI) Agent is implemented. This agent, apart from being responsible for the user interface, will generate the necessary Task Agents dynamically according to the number of tasks that comprise the scheduling problem and assign each task to the respective Task Agent (Fig. 9).

The Task Agent will process the necessary information about the job. That is, this agent will be responsible for the generation of the earliest and latest processing times, the verification of feasible schedules and identification of constraint conflicts on each job and the decision on which resource agent is responsible for solving a specific conflict.

Finally, the resource agent is responsible for the scheduling of the operations that require processing in the machine supervised by the agent. This agent will implement meta-heuristic and local search procedures in order to find best possible

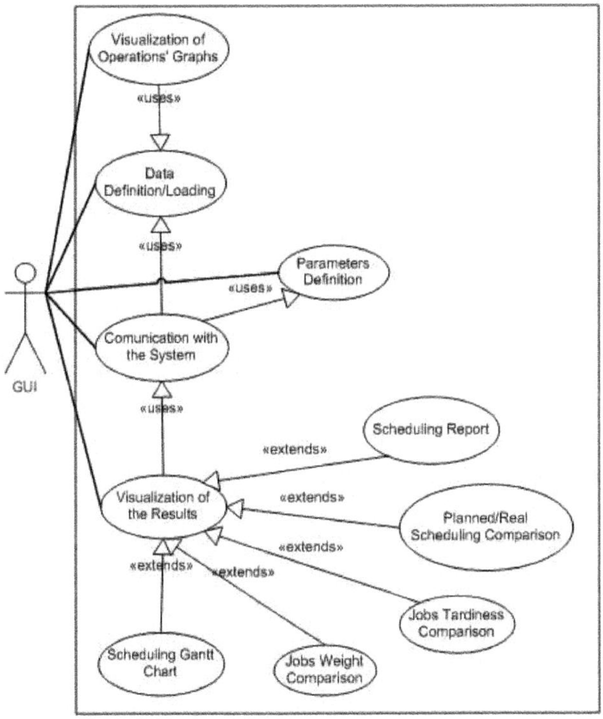

Fig. 9 User Interface Agent Functionalities

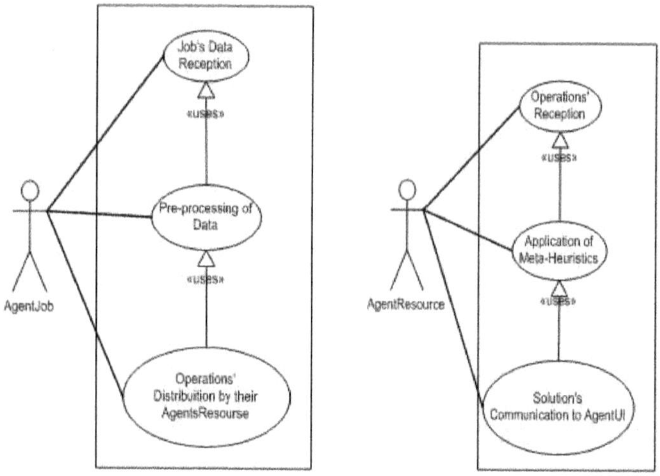

Fig. 10 Agent Job and Agent Resource functionalities

operation schedules and will communicate those solutions to the Task Agent for a later feasibility check (Fig. 10).

5 Computational Study

The proposed architecture was implemented using the Java Agent Development framework (JADE). Some computational tests were carried out to evaluate the performance of the referred scheduling systems under different manufacturing scenarios.

This section presents the results obtained by MASDScheGATS with TS and GA on the resolution of a set of academic instances of the Job-Shop problem [32], considering the difficulties in finding test problems and computational results for EJSSP. The MASDScheGATS performance will be compared with MAPS - the MultiAgent Production Planning System [45].

5.1 Tabu Search Parameterisation

In developing a Tabu Search algorithm we must keep in mind that its performance depends largely on the careful design and set-up of the algorithm components, mechanisms and parameters. This includes the representation of solutions, initial generation of solutions, and evaluation of the solutions, such as neighbourhood size, tabu list length, tabu list attributes, and stopping criteria.

Details of the algorithm parameterisation are briefly described as follows:

- **Solution Representation -** The solutions are encoded by the natural representation, where the schedule is described as a sequence of operations, i.e., each position represents an operation index;

- **Initial Neighbourhood Generation -** An initial solution is generated by a procedure where the operations are sequenced in order of non-decreasing processing level (defined on precedence graph), giving priority to operations that are processed earlier. Thus, we expect to generate a good initial solution from which an initial neighbourhood will be obtained;
- **Tabu list attributes and length -** It uses a tabu list that stores the pairs of jobs involved in exchanging of positions (on the neighbourhood generation process), with length 4;
- As **stopping criteria** in the Tabu Search algorithm, we used a maximum of 100 iterations.

5.2 Genetic Algorithms Parameterisation

In developing a genetic algorithm, we must keep in mind that its performance depends largely on the careful design and set-up of the algorithm components, mechanisms and parameters. This includes genetic encoding of solutions, initial population of solutions, evaluation of the fitness of solutions, genetic operators for the generation of new solutions and parameters such as population size, probabilities of crossover and mutation, replacement scheme, and number of generations.

Details of the algorithm parameterisation are briefly described as follows:

- **Solution Encoding -** In this work, solutions are encoded by the natural representation [9]. In this representation each gene represents an operation index. The gene position in a chromosome represents the operation position in a sequence, therefore defining, the operation processing order or priority. The number of genes in the chromosome represents the number of operations in a solution.
- **Genetic Operators -** Individuals, i.e. solutions, are randomly selected from the population and combined to produce descendants in the next generation. Depending on the problems to be solved and their encoding, several crossover operators may be used: one point, two points, uniform and order crossover [9]. Here, we use the single point crossover operator with probability $P_c = 0.8$. The single point crossover operator will be applied to M pairs of chromosomes randomly chosen, with $M = N/2$, where N is the size of the population. The mutation operator is applied with probability $P_m = 0.001$, to prevent the lost of diversity. Thus, a single point in a chromosome is randomly selected, and the currently selected resource for the task is replaced by another in the set of alternatives resources.
- **Replacement Scheme -** When creating a new population by crossover and mutation we must avoid losing the best chromosomes or individuals. To achieve this, the replacement of the less fit individuals of the current population by offspring is based on elitism [9]. Thus, the best individuals, i.e. solutions, will survive into the next generation.
- As **stopping criteria** in the Genetic Algorithm we use a maximum of 100 generations.

5.3 Computational Results

This section presents and compares the results obtained by MASDScheGATS through TS and GA with those obtained by the MAPS system [45]. For our experiments, we consider some benchmark problems [32]. For release dates, we consider zero for all instances. Due dates are considered to be the optimal *makespan* value.

The obtained results (Table 8) with our method based on Tabu Search (TS) and Genetic Algorithms (GA) are compared with those obtained using the MAPS system, which presents the Tightly Coupled Scheduling (TCS) and the Loosely Coupled Scheduling (LCS) approaches [45]. Moreover, MASDScheGATS was carried out for *makespan* (C_{max}), maximum lateness (L_{max}) and weighted tardiness (WT) minimisation.

Table 8 Computational Results

Instance	n	m	C_{max} Optimum	MAPS - TCS			MAPS - LCS			MASDScheGATS TS				MASDScheGATS GA		
				Average	Best	Worst	Average	Best	Worst	F. Obj.	Average	Best	Worst	Average	Best	Worst
										C_{max}	61.4	59	64	78.4	65	97
FT06	6	6	55	69.6	62	82	65.8	57	76	WT	63.2	59	68	74	65	89
										L_{max}	99	99	99	74.6	66	96
										C_{max}	1519	1408	1612	1476.4	1448	1540
FT10	10	10	930	1203.4	1164	1242	1277	1172	1373	WT	1368.8	1319	1416	1627.2	1427	1816
										L_{max}	1634.8	1542	1739	1504.6	1399	1611
										C_{max}	1715	1707	1726	1761.4	1692	1952
FT20	20	5	1165	1603.4	1513	1723	1470	1417	1503	WT	1678.8	1664	1714	1696	1662	1755
										L_{max}	1784.6	1686	1889	1740.6	1714	1833
										C_{max}	812.2	764	858	919.8	810	1160
La01	10	5	666	782	725	832	861	822	891	WT	772.6	740	795	865	782	921
										L_{max}	1059	947	1122	1049.2	888	1164
										C_{max}	869.6	839	920	914.4	847	985
La02	10	5	655	886.6	840	911	821.6	744	873	WT	859.4	839	876	940.8	839	1117
										L_{max}	968.4	849	1099	979	871	1152
										C_{max}	834.6	776	916	888	804	997
La03	10	5	597	778.8	726	823	793.6	716	858	WT	794.2	749	830	849	809	894
										L_{max}	947.4	875	990	898.2	823	988
										C_{max}	865.8	798	924	904.8	746	979
La04	10	5	590	835	801	894	786.2	724	819	WT	875.2	767	989	846.6	773	950
										L_{max}	959.2	842	1085	915.4	822	1003

With a simple implementation of the TS and GA and a small parameterisation effort, it was possible to achieve good performance for most instances of the problem when compared with MAPS system, considering C_{max} minimisation.

It is important to emphasise that our scheduling framework, which uses TS and GA here, is flexible in several ways. It can accommodate other local search metaheuristics and can optimise schedules based on practically any performance measure.

Moreover, the framework is not restricted to a specific type of scheduling problem, as is the case with many methods.

Both MAPS and MASDScheGATS results are obtained after five processing runs for each problem considered. The best, worst, and average are represented for both systems and different approaches.

Comparing our MASDScheGATS results based on a simple implementation of TS and GA with a small parameterisation effort, it was possible to achieve good performance for most instances of the problem compared with the MAPS system.

It is possible to conclude, based on obtained results, that MASDScheGATS with TS is more effective than with GA on the resolution of weighted tardiness (WT) and makespan ($C_m ax$), considering best and average values. When comparing results for lateness minimisation ($L_m ax$), MASDScheGATS through GA was more effective. MASDScheGATS through TS is more efficient in all instances analysed.

One novel approach, rarely addressed in the literature but very important in practice, considered in our scheduling framework is the ability to schedule jobs with complex processing structures, i.e. with both parallel processing of product component parts followed by their assembly at several stages.

Additionally the proposed coordination mechanism shows a very simple implementation. We expect that with a more effective cooperation mechanism (whose development is ongoing) it is possible to improve MASDScheGATS' performance.

We recognise the need for further testing, particularly to evaluate the suitability of the proposed framework and mechanisms for the EJSSP under dynamic environments. We realise, however, that this is not an easy task because it is difficult to find test problems in the literature with the job structure that we selected and consider important in practice. This structure assumes a job made of several parts processed and assembled through several assembly operations and stages.

6 Conclusions and Future Work

This chapter presented the MASDScheGATS Scheduling System that relies on the combination of different meta-heuristics and multi-agent systems. To solve the scheduling problem, Machine Agents and Task Agents must interact and cooperate with other agents in order to obtain optimal or near-optimal global performances through meta-heuristics. The idea is that from local, autonomous and often conflicting objectives of agents, a global solution emerges from a community of machine agents solving their schedules locally while cooperating with other machine agents. Agents have to manage their internal behaviour and their relationships with other agents via cooperative negotiation in accordance with business policies defined by the user manager.

We consider our approach as described in this chapter to be a valuable contribution for the resolution of more realistic scheduling problems, specifically of Extended Job Shop Problems. The main characteristic of our approach is the procedure to schedule operations, as each machine will first find local optimal or near-optimal solutions, succeeded by the interaction with other machines through cooperation mechanism as a way to find an optimal or near-optimal global schedule.

In most practical environments, scheduling is an ongoing reactive process where the presence of realtime information continually forces reconsideration and revision of pre-established schedules.

Considering that natural evolution is a process of continuous adaptation, it seemed appropriate to us to consider Genetic Algorithms and Tabu Search for tackling dynamic scheduling problems. Thus, the MASDScheGATS based scheduling system adapts the resolution of the deterministic problem to the dynamic one in which changes may occur continually. A population/solution regenerating mechanism is put forward, for adapting the population/neighbourhood of solutions, according to disturbances, to a new population, which increases or decreases according to new job arrivals or cancellations.

Additionally, we envisage to develop a learning mechanism that will permit resource agents, supported by a knowledge base, to recognise scheduling patterns and therefore improve the overall efficiency of the system.

Acknowledgements

The authors would like to acknowledge FCT, FEDER, POCTI, POSI, POCI 2010 for their support to R&D Projects and the GECAD Unit.

References

1. Adams, J., Balas, E., Zawack, D.: The shifting bottleneck procedure for job shop scheduling. Management Science 34(3), 391–401 (1988)
2. Aytug, H., Lawley, M.A., Mckay, K., Mohan, S., Uzsoy, R.: Executing production schedules in the face of uncertainties: A review and some future directions. European Journal of Operational Research 161(1), 86–110 (2005)
3. Baker, K.R.: Introduction to Sequencing and Scheduling. Wiley, New York (1974)
4. Blazewicz, J., Ecker, K., Trystram, D.: Recent advances in scheduling in computer and manufacturing systems. European Journal of Operational Research 164, 573–574 (2005)
5. Branke, J.: Efficient evolutionary algorithms for searching robust solutions. In: Proceedings of the Fourth International Conference on Adaptive Computing in Design and Manufacture, pp. 299–308. Springer, Heidelberg (2000)
6. Brownie, J., Harhen, J., Shivnan, J.: Production Management Systems. Addison-Wesley Publishing Co., Reading (1988)
7. Camarinha-Matos, L.M., Afsarmanesh, H.: The virtual enterprise concept. In: PRO-VE 1999: Proceedings of the IFIP TC5 WG5.3 / PRODNET Working Conference on Infrastructures for Virtual Enterprises, pp. 3–14. Kluwer Academic Publishers, Deventer (1999)
8. Caridi, M., Cavalieri, S.: Multiagent systems in production planning and control: An overview. Production Planning and Control 15(2), 106–118 (2004)
9. Davis, L.D., Mitchell, M.: Handbook of genetic algorithms. Van Nostrand Reinhold (1991), `http://citeseerx.ist.psu.edu/viewdoc/summary?doi=0.1.1.87.3586`

10. Dell'Amico, M., Trubian, M.: Applying tabu search to the job-shop scheduling problem. Annals of Operations Research 41(1-4), 231–252 (1993)
11. Dorn, J., Kerr, R., Thalhammer, G.: Reactive scheduling: improving the robustness of schedules and restricting the effects of shop floor disturbances by fuzzy reasoning. International Journal on Human-Computer Studies 42, 687–704 (1995)
12. Fang, H.-l., Ross, P., Corne, D.: A promising genetic algorithm approach to job-shop scheduling, rescheduling, and open-shop scheduling problems. In: Proceedings of the Fifth International Conference on Genetic Algorithms, pp. 375–382. Morgan Kaufmann, San Francisco (1993)
13. Goldtratt, E., Fox, R.: The Race. North River Press (1986)
14. Gonzalez, T.F.: Handbook of Approximation Algorithms and Metaheuristics. Chapman & Hall/Crc Computer and Information Science Series. Chapman & Hall/CRC, Boca Raton (2007), `http://www.amazon.ca/exec/obidos/redirect?tag=citeulike09-20%&path=ASIN/1584885505`
15. Horling, B., Lesser, V.: A survey of multi-agent organizational paradigms. The Knowledge Engineering Review 19(4), 281–316 (2005), `http://mas.cs.umass.edu/paper/366`
16. Jain, A.S., Meeran, S.: Deterministic job-shop scheduling: Past, present and future. European Journal of Operational Research 113, 390–434 (1999)
17. Kádár, B., Monostori, L., Csáji, B.: Adaptive approaches to increase the performance of production control systems. In: Proceedings of 36th CIRP International Seminar on Manufacturing Systems, Saarbrücken, Germany, pp. 305–312 (2003)
18. Kamrul Hasan, S., Sarker, R., Cornforth, D.: Hybrid genetic algorithm for solving job-shop scheduling problem. In: 6th IEEE International Conference on Computer and Information Science (ICIS 2007), pp. 519–524. IEEE Computer Society, Melbourne (2007)
19. Kim, B.I., Heragu, S.S., Graves, R.J., Onge, A.S.: A hybrid scheduling and control system architecture for warehouse management. IEEE Transactions on Robotics and Automation 19(6), 991–1000 (2003)
20. Lee, C.-Y., Piramuthu, S., Tsai, Y.-K.: Job shop scheduling with a genetic algorithm and machine learning. International Journal of Production Research 35, 1171–1191 (1997)
21. Lee, Y.-H., Kumara, S.R.T., Chatterjee, K.: Multiagent based dynamic resource scheduling for distributed multiple projects using a market mechanism. Journal of Intelligent Manufacturing 14, 471–484 (2003)
22. Lu, T.-P., Yih, Y.: An agent-based production control framework for multiple-line collaborative manufacturing. International Journal of Production Research 39, 2155–2176 (2001)
23. Luck, M., McBurney, P., Shehory, O., Willmott, S.: Agent Technology: Computing as Interaction (A Roadmap for Agent Based Computing) (2005), `http://www.agentlink.org/roadmap/al3rm.pdf`
24. Madureira, A.: Meta-heuristics application to scheduling in dynamic environments of discrete manufacturing. PhD thesis, University of Minho, Braga, Portugal (2003) (in Portuguese)
25. Madureira, A., Ramos, C., Silva, S.C.: A genetic algorithm for the dynamic single machine scheduling problem. In: Proceedings of the IFIP TC5/WG5.3 Forth IFIP/IEEE International Conference on Information Technology for Balanced Automation Systems in Manufacture and Transportation, pp. 315–324. Kluwer, B.V, Deventer (2000)
26. Madureira, A., Ramos, C., Silva, S.: An inter-machine activity coordination based approach for dynamic job-shop scheduling. International Journal for Manufacturing Science and Production 4(2), 121–131 (2001)

27. Madureira, A., Ramos, C., Silva, S.: Toward dynamic scheduling through evolutionary computing. WSEAS Transactions on Systems 3, 1596–1604 (2004)
28. Madureira, A., Santos, J., Gomes, N., Ramos, C.: Proposal of a cooperation mechanism for team-work based multi-agent system in dynamic scheduling through meta-heuristics. In: Proceedings 2007 IEEE International Symposium on Assembly and Manufacturing (ISAM 2007), Ann Arbor, Michigan, USA, pp. 233–238 (2007)
29. Manufuture, A vision for 2020, report of high level group. European Commission (2004)
30. Monostori, L., Váncza, J., Kumara, S.R.T.: Agent based systems for manufacturing. CIRP Annals - Manufacturing Technology 55(2), 697–720 (2006)
31. Nwana, H.S., Lee, L.C., Jennings, N.R.: Coordination in software agent systems. British Telecom Technical Journal 14(4), 79–88 (1996), `http://eprints.ecs.soton.ac.uk/2109/`
32. OR-library (1990), `http://people.brunel.ac.uk/~mastjjb/jeb/info.html`
33. Ouelhadj, D., Cowling, P.I., Petrovic, S.: Utility and stability measures for agent-based dynamic scheduling of steel continuous casting. In: Proceedings of the IEEE International Conference on Robotics and Automation, Taiwan, pp. 175–180 (2003)
34. Papakostas, N., Mourtzis, D., Bechrakis, K., Chryssolouris, G., Doukas, D., Doyle, R.: A flexible agent based framework for manufacturing decision-making. In: Proceedings of the 9th Flexible Automation and Intelligent Manufacturing Conference, pp. 789–800 (1999)
35. Pinedo, M.L.: Planning and Scheduling in Manufacturing and Services. Springer Series in Operations Research and Financial Engineering. Springer, Heidelberg (2007)
36. Pirlot, M.: General local search heuristics in combinatorial optimization: a tutorial. JORBEL Belgian Journal of Operations Research 32, 7–68 (1992)
37. Ponnambalam, S.G., Aravindan, P., Rajesh, S.V.: A tabu search algorithm for job shop scheduling. The International Journal of Advanced Manufacturing Technology 6(10), 765–771 (2000)
38. Portmann, M.C.: Scheduling methodology: optimization and compu-search approaches. In: The Planning and Scheduling of Production Systems, ch. 9, pp. 271–300. Chapman and Hall, Boca Raton (1997)
39. Raman, N., Brian Talbot, F.: The job shop tardiness problem: A decomposition approach. European Journal of Operational Research 69(2), 187–199 (1993)
40. Russell, S.J., Norvig, P.: Artificial Intelligence: A Modern Approach. Pearson Education, London (2003)
41. Sabuncuoglu, I., Bayiz, M.: Analysis of reactive scheduling problem in a job shop environment. European Journal of Operational Research 126(3), 567–586 (2000)
42. Shen, W., Maturana, F., Norrie, D.H.: Learning in agent-based manufacturing systems. In: Proceedings of AI & Manufacturing Research Planning Workshop, pp. 177–183. AAAI Press, Menlo Park (1998)
43. Váncza, J., Márkus, A.: An agent model for incentive-based production scheduling. Computers in Industry 43/2, 173–187 (2000)
44. Weiss, G.: Multi Agent Systems - A Modern Approach to Distributed Artificial Intelligence. MIT Press, Cambridge (1999)
45. Wellner, J., Dilger, W.: Job shop scheduling with multiagents. In: Workshop "Planen und Konfigurieren", University of Würzburg (1999)
46. Xhafa, F., Abraham, A.: Metaheuristics for Scheduling in Industrial and Manufacturing Applications. Springer, Heidelberg (2008)

Author Index

Index

GPSR Compliance

The European Union's (EU) General Product Safety Regulation (GPSR) is a set of rules that requires consumer products to be safe and our obligations to ensure this.

If you have any concerns about our products, you can contact us on ProductSafety@springernature.com

In case Publisher is established outside the EU, the EU authorized representative is:

Springer Nature Customer Service Center GmbH
Europaplatz 3
69115 Heidelberg, Germany

Batch number: 10755284

Printed by Printforce, the Netherlands